Scientific Data Mining

Scientific Data Mining

A PRACTICAL PERSPECTIVE

Chandrika Kamath

Lawrence Livermore National Laboratory
Livermore, California

siam.

Society for Industrial and Applied Mathematics
Philadelphia

10 9 8 7 6 5 4 3 2 1

The galaxy images on the cover are from the Faint Images of the Radio Sky at Twenty centimeters (FIRST) survey, available at sundog.stsci.edu.

Library of Congress Cataloging-in-Publication Data
Kamath, Chandrika.
Scientific data mining : a practical perspective / Chandrika Kamath.
p. cm.
Includes bibliographical references and index.
ISBN 978-0-898716-75-7
1. Data mining. 2. Science--Databases. 3. Engineering--Databases. I. Title.
QA76.9.D343K356 2009
502.85'6312--dc22 2008056149

To my parents and Sisira

Contents

Preface

Advances in sensors, information technology, and high-performance computing have resulted in massive data sets becoming available in many scientific disciplines. These data sets are not only very large, being measured in terabytes and petabytes, but are also quite complex. This complexity arises as the data are collected by different sensors, at different times, at different frequencies, and at different resolutions. Further, the data are usually in the form of images or meshes, and often have both a spatial and a temporal component. These data sets arise in diverse fields such as astronomy, medical imaging, remote sensing, nondestructive testing, physics, materials science, and bioinformatics. They can be obtained from simulations, experiments, or observations.

This increasing size and complexity of data in scientific disciplines has resulted in a challenging problem. Many of the traditional techniques from visualization and statistics that were used for the analysis of these data are no longer suitable. Visualization techniques, even for moderate-sized data, are impractical due to their subjective nature and human limitations in absorbing detail, while statistical techniques do not scale up to massive data sets. As a result, much of the data collected are never even looked at, and the full potential of our advanced data collecting capabilities is only partially realized, if at all.

Data mining is the process concerned with uncovering patterns, associations, anomalies, and statistically significant structures in data. It is an iterative and interactive process involving data preprocessing, search for patterns, and visualization and validation of the results. It is a multidisciplinary field, borrowing and enhancing ideas from domains including image understanding, statistics, machine learning, mathematical optimization, high-performance computing, information retrieval, and computer vision. Data mining techniques hold the promise of assisting scientists and engineers in the analysis of massive, complex data sets, enabling them to make scientific discoveries, gain fundamental insights into the physical processes being studied, and advance their understanding of the world around us.

Why focus on scientific data?

There are several books available on data mining. Many focus on the specific task of finding patterns in the data through techniques such as decision trees or neural networks, while others focus on commercial data and are targeted at the business communities. Pattern recognition techniques are necessary in the mining of scientific data; however, they form only a part of the entire data mining process. Texts focusing on pattern recognition do not discuss how to convert raw scientific data in the form of images or meshes into a form that can be used as input to a pattern recognition algorithm. On the other hand, texts on mining business applications focus on data that have been cleaned and are in a database, a situation that is rarely true for scientific data. Further, the assumptions made for business data may not hold for scientific data. For example, in a targeted marketing application, it is possible to use historical data to generate a large training set, with equal number of positive and negative examples. In contrast, training sets in scientific data tend to be quite small as they are generated manually. They are also frequently unbalanced, with far more examples of one type than another.

The data mining problems encountered in science and engineering applications may seem very different at first glance. However, there are several common threads among them. This book focuses on the identification of these common problems and their potential solutions. It considers the end-to-end process of data mining, starting with the raw data in the form of images or meshes, preprocessing the data, and finding useful information in the data that are then shown to the scientist for validation. There is greater focus on the pre- and postprocessing steps as these are often the more critical and time-consuming parts of mining scientific data. In the process, this book brings together topics that are often spread out among books focusing on image processing, information retrieval, or mathematical optimization. This book has also been written with an emphasis on the practical aspects of mining scientific data and introduces the reader to topics not often covered in academic texts such as data mining systems.

Any book must be limited in scope if it has any hope of seeing the light of day (or the printer's ink); therefore, several topics are beyond the scope of this book. These include:

- **Database technology:** In many scientific disciplines, metadata representing aspects of the data, such as the time it was collected and the circumstances under which it was collected, may be stored in a database. However, the data itself is usually stored in flat files, with perhaps a pointer from the associated metadata in the database. As a result, database technology is often not very relevant to scientific data. In fact, when many scientists use the term "database," they are referring to their data as a whole; these data may be in flat files if they are online or they may just be a collection of tapes in a file cabinet.

- **Online Analytical Processing (OLAP):** This includes tasks such as querying the database and extracting slices or cubes from the data that satisfy certain constraints. Such techniques are used in some scientific applications for tasks such as exploratory data analysis and subsetting the data for further analysis. Tools to support this are provided either through visualization software or by domain- or problem-specific software.

- **Parallel implementations:** While these are often necessary to make the problem of mining massive data sets tractable, the subject is broad enough that several books can be devoted to it.

- **Collection of the data:** The process used for collection of the data, whether from experiments, observations, or simulations, is often an important aspect of data mining. In particular, it can dictate the choice of algorithms to use as well as the confidence in the conclusions being drawn from the data. However, it is also a topic that is very dependent on both the problem and the application domain, and is therefore beyond the scope of this book.

- **Storage and access of the data:** While these are topics germane to mining scientific data, they are more appropriate for a text on data management.

In addition, since there are several excellent texts on pattern recognition algorithms, I will discuss this topic only briefly in Chapter 11, focusing on issues more relevant to practical applications and providing pointers as appropriate.

What is in this book?

This book is focused on the practical aspects of scientific data mining. Over the last ten years, as I analyzed data from various scientific domains, I encountered a diversity of problems. Often, the solutions were not found in standard texts on data mining, but in texts on different domains, or on techniques developed in the context of a different problem in a different application area. In many cases, I had to modify existing techniques or innovate and come up with new approaches which exploited domain-specific information. In the process, I realized that there are many pitfalls in mining scientific data, and one must apply techniques with care, while being aware of the characteristics of the data and the assumptions of the different algorithms. The end goal of finding scientifically meaningful information in the data cannot be achieved by simply applying techniques blindly to the data.

This book tries to bring together, in a coherent whole, the different techniques which can be used in solving a variety of problems in the analysis of data from various scientific domains. Not all the techniques are useful in all problems and some techniques developed in the context of one application domain may find use in an entirely different application domain.

The chapters in the book are divided into five broad areas.

- **Data mining in scientific applications:** The first two chapters focus on an introduction to data mining (Chapter 1) and the role it plays in science and engineering applications (Chapter 2).

- **The data mining process:** Chapter 3 identifies the common themes in the applications discussed in Chapter 2. It also describes the data types used in scientific applications—these are used to motivate the data mining process described in Chapter 4. This chapter discusses how we start from the raw data in the form of images or meshes and the steps we take to extract useful information from them.

- **The tasks in the data mining process:** Chapters 5 through 12 are devoted to discussing the specific tasks in the data mining process. Chapter 5 discusses various ways of reducing the size of the data through techniques such as sampling and multiresolution. Chapter 6 discusses data fusion, or how we can combine complementary data from different sources. Chapter 7 describes how to remove noise from data, followed by algorithms for identifying objects in the data in Chapter 8. Once the objects have been identified, there are various representative features that can be extracted from the objects. These are discussed in Chapter 9. At the end of this extraction step, we have a list of objects that have been identified in the data, with each object described by a number of features. Often, these features number in the tens or hundreds. Ways of reducing this number are discussed in Chapter 10. The original raw data have now been reduced to a form that can be input to various pattern recognition algorithms, as described in Chapter 11. Finally, the patterns identified in the data are visualized for validation by the domain scientist, as discussed in Chapter 12. These same visualization techniques can also be used in validating the results from any task in the data mining process, as well as in the initial exploratory analysis of the data which is done to determine either the next task or an appropriate algorithm for a task.

In each of these chapters, the intent is to introduce the reader to some commonly used techniques for a particular task, as well as some of the more advanced techniques that have been proposed recently. While the pros and cons of the traditional approaches are well known, the recent techniques are still topics of active research. As appropriate, suggestions for further reading are recommended in each chapter.

The techniques which are listed for the different tasks are by no means exhaustive. These are just the techniques I have found useful based on my experiences thus far. The omission of a particular solution approach is indicative either of my ignorance, or the space and time limitations which accompany the writing of any book.

- **Data mining systems:** Chapter 13 describes several data mining systems that have been developed for science and engineering data. These illustrate the challenges faced in building an end-to-end system and the various approaches taken to address these challenges.

- **Lessons learned, challenges, and opportunities:** Chapter 14 summarizes, both from my personal perspective as well as those of other data miners, the lessons learned in mining scientific data. It also describes the challenges and opportunities that await a data miner who takes on the task of analyzing scientific data.

The chapters in this book have been written to be as self contained as possible, so that a reader can focus on a topic of interest without having to read the entire book from the beginning. However, it is my hope that the reader who perseveres through the entire book will come away with an appreciation of the technical challenges in scientific data mining, the opportunities to borrow ideas from other fields, and the potential for making exciting scientific discoveries.

In this book, I have included the URLs for any information available on the Web. While I realize that some of these links may be short-lived, I find it helpful to include the information, as a Web search can provide the new location if a link has changed. I did not want to take the alternative of excluding the links, as it meant that I could not refer to the wealth of information easily accessible to all via the Web.

Several of the diagrams appearing in this book have been inspired by similar diagrams appearing elsewhere, as follows:

- Figure 5.1 is derived from the article by Witkin [629].

- Figure 5.2 is derived from Figure 2.2 in the text by Lindeberg [382].

- Figure 5.3 is derived from Figure 2.1 in the text by Stollnitz, DeRose, and Salesin [568].

- Figure 7.2 is derived from the article by Smith and Brady [553].

- Figure 11.2 is derived from Figure 4.2 in the book by Mitchell [430].

Finally, despite my best efforts, I am sure there are some unintentional errors in the book, either typographical errors, ambiguous statements, or incorrect statements reflecting my less-than-thorough understanding of a topic. In the absence of a coauthor to whom I could attribute these errors, I assume full responsibility for them. The opinions presented in the book represent my experiences with scientific data mining thus far; these opinions may not necessarily reflect the views of my colleagues, funding sources, employers, or the domain scientists I have worked with over the years. As I continue my education in the many fields that comprise this exciting subject, it is very likely that my thoughts will evolve as well. I also acknowledge that the book is biased toward techniques and topics I have considered in addressing the problems I have worked on in the last decade. I am sure there are others I have overlooked in my ignorance. I will gratefully receive all errata as well as suggestions for improvement.

Acknowledgments

It is my great pleasure to acknowledge all those who contributed, directly or indirectly, to this book.

This book grew out of tutorials I presented at the SIAM International Conference on Data Mining in 2001 and 2003, as well as an overview tutorial at a week-long program I organized on Scientific Data Mining at the Institute for Pure and Applied Mathematics, University of California, Los Angeles, in 2002. I would like to thank Linda Thiel of SIAM for suggesting that I expand the course material presented in these tutorials. She helped me to get started with the logistics of writing the book, providing many ideas for improvement along the way. It has been my pleasure to work with her, Sara Murphy, and the many members of SIAM staff; I truly appreciate their support, encouragement, and patience over the years.

I was introduced to the subject of data mining by Shivakumar Vaithyanathan, who helped me to realize that not only was this topic very different from data bases, it also used some of the techniques I knew from numerical linear algebra. My education in the many fields which comprise scientific data mining continued through interactions with the members of my project, Sapphire, at the Lawrence Livermore National Laboratory. It was a wonderful experience to work with all of them, especially Erick Cantú-Paz, Samson Sen-Ching Cheung, and Charles R. Musick who contributed to many stimulating discussions. This book reflects many of the ideas and efforts of the Sapphire team members.

The Sapphire project was funded by several different sources from 1998 to 2008, including the ASC and SciDAC programs at the US Department of Energy, the LDRD program at Lawrence Livermore National Laboratory, and the US Department of Homeland Security. I am very grateful to these programs, and the individuals associated with them, for supporting my work and thus indirectly contributing to this book.

I gained much of my practical experience in scientific data mining by working with real data sets and solving real problems. I owe a significant set of debts to all the scientists who generously shared their data, their time, and their domain expertise with my project team. It was my good fortune (or perhaps serendipity) that the first data set we analyzed was the Faint Images of the Radio Sky at Twenty centimeters (FIRST) data. The FIRST astronomers—Robert H. Becker, Richard L. White, Michael D. Gregg, and Sally A. Laurent-Muehleisen—taught me much about scientific data analysis and provided a gentle introduction to collaborating with domain scientists. Many others provided a fascinating glimpse into the "science" aspect of scientific data mining, including Charles Alcock, Joshua Breslau, Keith H. Burrell, William H. Cabot, Andrew W. Cook, John A. Futterman, Omar A. Hurricane, Scott A. Klasky, Zhihong Lin, Ricardo J. Maqueda, Alfredo A. Marchetti, Paul L. Miller, Donald A. Monticello, Neil Pomphrey, Benjamin D. Santer, Brian K. Spears, Daren P. Stotler, Frederick H. Streitz, Michael A. Walker, Nathan G. Wimer, and Stewart J. Zweben. These scientists gave me a wonderful opportunity to apply analysis techniques to a diverse set of challenging problems.

I owe a profound intellectual debt to those who contributed toward my education—my parents, my brother Jayant, my late brother Hemant, my husband Sisira, and my teachers at the Lady Irwin School in New Delhi, India; the Indian Institute of Technology, Bombay, India; and the University of Illinois at Urbana-Champaign, USA.

Finally, I would like to thank the three people who, perhaps unknowingly, contributed immensely to this book—my mother, Sharada, for sharing with me her love of mathematics; my late father, Taranath, for teaching me English grammar and writing skills (I hope I have not let him down!); and my husband, Sisira Weeratunga, for his incredible support over the years, his patience as this book took far longer to complete than anticipated, and the countless technical discussions, which were both stimulating and educational. I dedicate this book to them, in gratitude.

November 2008 *Chandrika Kamath*

Chapter 1

Introduction

The purpose of computing is insight, not numbers.

—Richard Hamming [247, p. v]

Advances in sensor technology have enabled us to collect large amounts of data in fields such as remote sensing, astronomy, medical imaging, and high-energy physics. Complementing these data sets from experiments and observations are data sets from numerical simulations of complex physical phenomena such as the flow around an airplane, global climate modeling, and evolution of galaxies in astrophysics. With computers having faster processing units, more memory, and many thousands of processors, it is now possible to simulate phenomena of a size and a complexity which were unthinkable a decade or two ago. The data generated by these simulations are very large, being measured in terabytes, and are also very complex, often with time-varying structures at different scales.

Complementing these advances in sensor technology and high-performance computing are advances in storage technology which have allowed more information to be stored on disks as well as in file systems which allow rapid access to the data. To realize the full potential of these advanced data collection and storage technologies, we need fast and accurate data analysis techniques to find the useful information that had originally motivated the collection of the data.

1.1 Defining "data mining"

The term "data mining" has had a varied history [186, 555]. As with many emerging fields that evolve over time, borrowing and enhancing ideas from other fields, data mining too has had its share of terminology confusion. Part of the confusion arises from the multidisciplinary nature of data mining. Many who work in areas that comprise data mining, such as machine learning, exploratory data analysis, or pattern recognition, view data mining as an extension of what they have been doing for many years in their chosen field. In some disciplines such as statistics, the terms "data mining" and "data dredging" have a negative connotation. This dates back to the time when these terms were used to describe extensive searches through data, where if you searched long and hard enough, you could find patterns where none existed (see [252, Section 1.7]). Such negative connotations have had the unfortunate consequence that statisticians have tended to ignore the developments in data mining, to the detriment of both communities.

Data miners too have contributed to the confusion surrounding the term "data mining." The term originally referred to a single step in the multistep process of Knowledge Discovery in Databases (KDD) [185] . KDD is defined as the "nontrivial process of identifying valid, novel, potentially useful, and ultimately understandable patterns in data." In this viewpoint, the data mining component of KDD is the means by which patterns are extracted and enumerated from the data. The other steps in KDD include data preparation, data selection, and interpretation of the results. As the acronym KDD suggests, this viewpoint assumes that the data reside in a database, which is rarely true for scientific data sets.

Other authors [548] refer to data mining as the "process of extracting valid, previously unknown, comprehensible and actionable information from large databases and using it to make crucial business decisions." The data mining process is applied to a data warehouse and is composed of the four basic steps of "data selection, data transformation, data mining, and result interpretation." This somewhat circular definition of the term "data mining" considers the data mining process, not KDD, to be the overall process of extracting useful information from data. Further, by focusing on data warehouses, it restricts data mining to commercial and business applications, which typically use databases and data warehouses.

In what may be a wise choice, some data miners have distanced themselves from the terminology debate, choosing instead to focus on how the ideas from different fields can be combined and enhanced to solve the problems of interest in data analysis. For example, the authors of a 1999 report of a NASA Workshop on Issues in the Application of Data Mining to Scientific Data [30] observed that "it was difficult to arrive at a consensus for the definition of data mining, apart from the clear importance of scalability as an underlying theme." They, like many other practitioners and proponents of data mining, agreed that data mining is a multidisciplinary field, borrowing ideas from diverse technologies including machine learning, statistics, visualization, image processing, high-performance computing, databases, and so on. It is difficult to clearly distinguish data mining from any single component discipline. What is new is the confluence of the mature offshoots of these technologies at a time when we can exploit them for the analysis of massive complex data sets. In addition, as data mining techniques are applied to new forms of data, such as text and electronic commerce, newer disciplines such as natural language processing and issues such as handling of privacy are added to the fields that comprise data mining.

In this book, I consider data mining to be the process concerned with uncovering patterns, associations, anomalies, and statistically significant structures in data. I do not differentiate among terms such as "knowledge discovery," "KDD," and "data mining," nor do I attempt to address the subtle differences between data mining and any of its component disciplines. Instead, I focus on how these disciplines can contribute toward finding useful information in data. As the data analysis problems in science and engineering are often rather difficult, we need to borrow and build upon ideas from several different disciplines to address these challenging problems.

1.2 Mining science and engineering data

The mining of data from science and engineering applications is driven by the end goals of the scientists, in particular their reasons for analyzing the data. Often, the scientists and engineers who collect or generate the data start the process with the goal of addressing a particular question or a set of questions. Sometimes, the question is well formulated, with a clearly identified solution approach. In other cases, the question may be well formulated, but it may not be clear at all how one would proceed to answer the question. Usually, the process of answering one question gives rise to others, and the data may often be analyzed to answer questions completely unrelated to the one which motivated the collection of the data in the first place. In other situations, the scientists may not have a specific question, but may be interested in knowing what is in the data. This is often the case when the science of what is being observed or simulated is not completely understood.

A poorly formulated question or data, which may not be the most appropriate for answering the question at hand, are not the only challenges in mining science and engineering data sets. The characteristics of the data themselves can be a major impediment, whether they are data from experiments, simulations, or observations. A key issue is that the data are usually available in the "raw" form. This can be as images, or mesh data from simulations, where several variables are associated with each point in a computational grid. There is often both a spatial and a temporal component to the data. Sometimes, data collected from more than one sensor may need to be mined together to exploit the complementary information obtained by the sensors. As a result, a substantial amount of preprocessing is required to identify the objects in the data, and find appropriate representations for them, before we can identify patterns in the data.

Another characteristic of the data is that they may be very large in size, being measured in terabytes or petabytes. Such data are usually stored on tapes, possibly as part of a high-performance storage system, making access to them nontrivial. The data may also be distributed, either in separate files or even in different geographic locations. Mining these data accurately and efficiently may need the development of new algorithms or the implementation of parallel or distributed versions of existing algorithms.

As a result of their size and complexity, much of the data from science and engineering applications are never analyzed, or even looked at, once they have been collected. This is, of course, disconcerting to scientists who wonder about the science still left undiscovered in the data. More often than not, scientists are very interested in the analysis of their data and realize that the current techniques being used are inadequate for the information they want to extract from the data. But, they may not know what expertise is necessary to accomplish this analysis, or where they can find this expertise.

This provides a unique opportunity for data miners (and I use the term in a very generic sense) to collaborate with scientists and engineers not only to address their analysis problems and aid in scientific discovery but also, in the process, to identify challenging problems whose solution will advance the state of the art in analysis techniques.

1.3 Summary

Advances in sensor technology, high-performance computing, and high-density storage have all resulted in massive complex data sets in various domains in science and engineering. The multidisciplinary field of data mining holds the promise of allowing scientists and engineers to analyze these complex data sets and to find useful information in them. Scientific data mining borrows ideas from a multitude of disciplines including image and video processing, machine learning, pattern recognition, information retrieval, and statistics. By combining techniques from these diverse disciplines, our goal is to help scientists and engineers fully exploit the benefits of the increased computational and data gathering capabilities now available.

In the next chapter, I describe the many ways in which data mining techniques are being used in various science and engineering applications. I then use these examples to motivate the end-to-end process of scientific data mining.

Regarding terminology, for ease of description I use the term “scientific data” to mean data from both science and engineering applications.

Chapter 2

Data Mining in Science and Engineering

Science was originally empirical, like Leonardo making wonderful drawings of nature. Next came the theorists who tried to write down the equations that explained the observed behaviors, like Kepler or Einstein. Then, when we got to complex enough systems like the clustering of a million galaxies, there came the computer simulations, the computational branch of science. Now we are getting into the data exploration part of science, which is kind of a little bit of them all.

—Alex Szalay [573]

The analysis of data has formed a cornerstone of scientific discovery in many domains. While these domains, and the problems of analysis therein, may appear very different at first glance, there are several common themes among them. This chapter is an overview of the many ways in which data mining techniques are being used in science and engineering applications. The intent of this survey is severalfold. First, it illustrates the diversity of problems and the richness of the challenges encountered in scientific data mining. Second, it highlights the characteristics of the data in various scientific domains, characteristics which are discussed further in Chapter 3. Third, it serves as a prelude to identifying the common themes and similarities in scientific data analysis, and thus motivates the tasks described in the data mining process in Chapter 4. Finally, by describing how scientists use data analysis techniques in several very different domains, this survey shows that techniques developed in the context of one domain can be applied or extended to other domains easily. I discuss these techniques further in later chapters, with the goal of enabling crosscutting exchange of ideas across several disciplines.

The survey in this chapter is by no means an exhaustive one. It is not intended to cover all possible applications of data mining in scientific domains or all scientific domains with large, complex data sets. It is, however, intended to show how the more recent data mining techniques complement existing techniques from statistics, exploratory data analysis, and domain-specific approaches to find useful information in data.

The sections in this chapter are organized as follows. Sections 2.1 through 2.7 use illustrative examples to describe the various ways in which data mining techniques are being used in applications ranging from astronomy to remote sensing, biology, and physics.

These examples use techniques which are described later in the book. For those unfamiliar with these techniques, it may be helpful to read this chapter again to better understand the application of the solution approaches. Each section also contains a subsection summarizing the characteristics of the data in that application domain. Section 2.8 briefly mentions some of the newer application areas where data mining techniques are beginning to make inroads. Finally, Section 2.10 includes several pointers for additional information.

2.1 Astronomy and astrophysics

Ever since humans first looked at the skies and observed obvious patterns such as the waxing and waning of the moon and the equinoxes, there has been an interest in understanding the universe through observations. The tools used for these observations have ranged from the simple, but surprisingly accurate, instruments of the ancient cultures to sophisticated telescopes, both earth-based, such as the Very Large Array (VLA) [606], and space-based, such as Hubble [281] and Chandra [96], to name a few. As the tools for collecting the data have become more complex, so have the techniques for analyzing the data.

Astronomers have been using techniques from the component disciplines of data mining long before the term itself became popular. In the 1970s, they realized that computers could be used for the automated construction of catalogs, which are systematic collections of astronomical object descriptions satisfying specified selection criteria. The benefits of automated catalog construction are many, including the ability to process large numbers of objects, the consistent application of selection criteria, and the ability to make exhaustive searches for all objects satisfying these selection criteria. In particular, the completeness of such catalogs formed an important factor in the analysis of the data using statistical techniques.

In addition to catalog generation, astronomers also realized that automated techniques could be used for the detection and analysis of objects in images on astronomical plates. As Jarvis and Tyson observed in their seminal paper [304], the use of an automated detection and classification system can avoid many potentially devastating and subtle systematic errors that occur in statistical astronomical studies which investigate properties of images over many magnitudes. These errors arise from the effects of different sample selections as well as variations in reduction and analysis techniques. It is interesting to note that in the 1970s and early 1980s, even though the raw data was in hardcopy in the form of astronomical image plates, the use of automated techniques for catalog construction and image analysis was gaining acceptance in the astronomy community.

The paper by Jarvis and Tyson [304] on the Faint Object Classification and Analysis System (FOCAS), though written in 1981, provides an interesting insight into the early beginnings of data mining in astronomy. The goals of FOCAS were many, including reliable classification of all objects as stars, galaxies, or noise; an accurate position for each object; and a characterization of object shape.

First, the raw data in the form of image plates were digitized into 6000×6000 pixel images which were then processed for object detection, classification, and catalog manipulation. The detection of objects was done using image segmentation. A low-pass filter (see Section 7.2.1) was first applied to the image. Then, any pixel whose value exceeded the local sky density value at that pixel by a specified threshold, was kept as an object pixel. The size of the low-pass filter was chosen to meet the constraints on computational time and memory requirements of the systems being used (initially a Digital Equipment Corporation

PDP 11/45, followed by a VAX 11/780). Once a pixel was identified as belonging to an object, groups of neighboring pixels were combined to form an object whose description was passed to an object evaluation module.

This module extracted various parameters or features that characterized each of the objects. The average sky density at the object location was first calculated using domain-specific techniques and converted into an image intensity value. The densities of the object pixels were also converted into intensities which were then used in the calculation of the features. The main features used for each object were the Hu moments [279], which are discussed in more detail in Section 9.2. FOCAS used only three of the seven Hu moments as they were scale-, rotation-, and translation-invariant. Additional features included the peak intensity of the object which was defined as the intensity difference between the local sky intensity and the average intensity in a 3×3 pixel window centered on the object centroid; an effective radius of the object; a measure of how closely the object matched an ideal star image; and the object magnitude.

During the evaluation of these shape features, several consistency checks were made to ensure that an object was a valid image of a star or galaxy. While these checks did not remove all noise effects, they did reduce the size of the data that were sent to the classifier. Additional checks were also made to separate closely spaced objects that might have been treated as a single object in the object detection phase. It is interesting to note that FOCAS maintained separate files to store the output of the image segmentation step. It was a computationally expensive step and the astronomers did not want to repeat it if any changes were required in subsequent steps of the analysis. Also, in addition to the shape features, each object in the object list contained information on its position in the image plate, area in pixels, magnitude, and flags indicating exceptional conditions such as multiple objects in an area. These data formed the basic catalog generated from the image data.

Once the object list was generated for each object on an image plate, it was input to a nonparametric statistical classifier [25]. A two-stage classification was used. The noise objects were first identified using a single feature—the effective radius for bright objects and the peak intensity for dim objects. However, as this set of decision surfaces could not adequately separate the stars and galaxies, a second level of classification was used that empirically combined classification and clustering techniques (see Chapter 11). First, a training set of approximately 800 stars and galaxies was created. The features for the objects in this set were scaled and normalized to reduce the variation in the features. A clustering algorithm, in this case, the ISODATA algorithm [17, 18], was repeatedly applied until uniformly homogeneous clusters (to within a threshold) were obtained. Next, decision surfaces in the form of hyperellipsoids were generated to separate the star regions from the galaxy regions in the seven-dimensional feature space. The parameters of the hyperellipsoids were obtained using an iterative method based on minimizing the misclassification on the training set. Once the hyperellipsoids were generated, they were used to classify all the objects. Hyperellipsoids were chosen, as they performed better than simpler surfaces using the metric that the relative proportion of the two classes did not change significantly if another iteration of the classifier was made.

The approach used by the astronomers in FOCAS illustrates several key ideas and issues in scientific data mining. First, there is the focus on converting the low-level image data to higher-level information in the form of features through the process of segmenting the images to identify the objects, followed by the extraction of features which are scale-, rotation-, and translation-invariant. Second, the astronomers took care to ensure that the input to the classification algorithm was of high quality by scaling and normalizing the features prior to classification, separating closely spaced objects, using flags to indicate exceptional conditions in the catalog, incorporating consistency checks to reduce the data input to the classifier, and carefully choosing the examples in the training set. Third, they paid attention to the results from the classifier and selected their options to obtain scientifically meaningful results. As we shall see in later chapters, these issues play an important role in the practical implementation of scientific data mining techniques.

Following the early work of Jarvis and Tyson on FOCAS, several astronomers used data mining techniques for the analysis of their data. For example, Odewahn et al. [456, 455] used neural networks (see Section 11.2.4) to discriminate between stars and galaxies in the Palomar Sky Survey. They observed that the most time-consuming part of constructing a neural network classifier was the manual classification of the training set; this should be as free of misclassifications as possible and span the full range of possibilities. Other potential factors that were identified as contributing to lower accuracies of classification were inadequate image parameters and problems in neural network construction. Either the image parameters used in classification did not represent enough pertinent image features or the original image did not contain enough information to make an accurate classification. Further, the neural network may have had too few or too many nodes and may have been under- or overtrained. In addition, the astronomers observed that neural networks did not provide direct information on why an object was assigned a particular classification. This indicates another characteristic of scientific data mining, namely, the need of the scientist to understand how a decision was made. This is referred to as the interpretability of the models built by the classifier.

Other examples on the use of neural nets in classification problems in astronomy include the morphological classification of galaxies [571, 2, 444], as well as spectral classification [570, 556], where principal components analysis is used to reduce the number of features used as input to the neural networks.

More recently, data miners have found astronomy to be a rich source of analysis problems. The Sky Imaging Cataloging and Analysis Tool (SKICAT) used classification techniques, specifically decision trees, to automate the reduction and analysis of the Digital Palomar Observatory Sky Survey (DPOSS-II) data. Using a modified version of the FOCAS, they identified objects in the images using image segmentation techniques followed by the extraction of 40 base-level attributes or features representing each object. However, these base-level features by themselves did not exhibit the required variance between different regions of an image plate or across image plates. As Fayyad, Djorgovski, and Weir [184] observed, in classification learning, the choice of attributes used to characterize the objects was by far the single most determining factor of the success or failure of the learning algorithm. So, in addition, they used four new normalized attributes that were derived from the base-level attributes and possessed the necessary invariance between and across image plates. They also calculated two additional attributes representing a point-spread-function (PSF) template. This allowed the astronomers to compensate for the blurred appearance of point sources (stars) due to the turbulence of the earth's atmosphere. These derived attributes, along with the base-level attributes, were used in classification. The training set was generated manually, using higher-resolution images from a separate telescope with a higher signal-to-noise ratio for the fainter objects. A decision tree classifier, trained on this higher-resolution data, was able to correctly classify objects obtained from the lower-resolution DPOSS-II images.

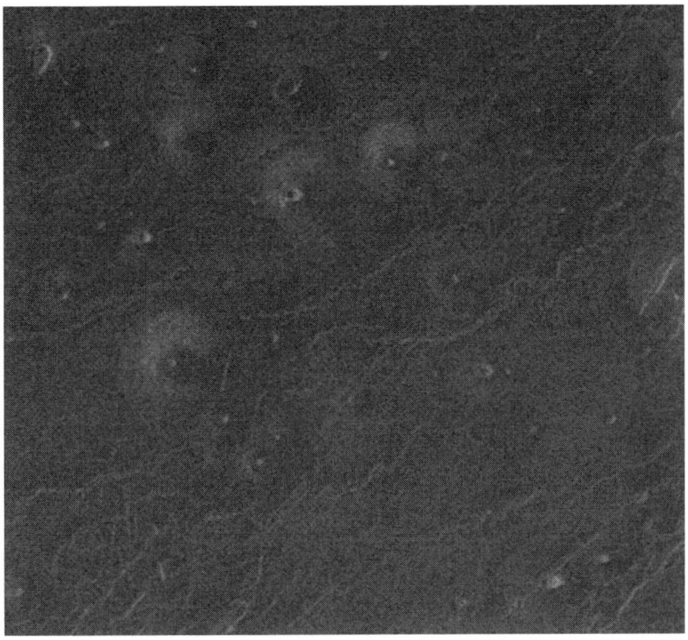

Figure 2.1. *Subset from img3 from the "Volcanoes on Venus" data set in the UCI KDD archive* [272]. *The data set is the one used in the JARtool project* [71] *which developed an automatic system for cataloging small volcanoes in the large set of images of the surface of Venus returned by the Magellan spacecraft.*

Several important conclusions can be drawn from the experiences of the SKICAT scientists. First, it was possible to create an automated approach that was accurate enough for use by the astronomers. Second, through the use of robust features and a training set from higher-resolution imagery, it was possible to obtain classifiers whose accuracy exceeded that of humans for faint objects. Third, the conversion of the data from high-dimensional pixel space of images to the lower-dimensional feature space transformed the problem into one solvable by learning algorithms. The use of derived features in addition to the base-level features resulted in high accuracy within and across plates. Finally, SKICAT showed the potential of data mining techniques in the semiautomated analysis of very large data sets containing millions of objects.

Other recent efforts in mining astronomy data include the JPL Adaptive Recognition Tool (JARTool) [71] for cataloging volcanoes on Venus (see Figure 2.1) and its follow-on, the second generation tool Diamond Eye [73], which provided an application independent infrastructure for mining image collections. A similar application-independent approach was also used in the more recent Sapphire project which built a set of tools for various tasks in scientific data mining. These tools were then used in different applications, such as finding bent-double galaxies in astronomical surveys [315]. The system architectures for some of these efforts are discussed in further detail in Chapter 13.

Much of the focus in astronomy is on observation data from surveys, though it is also possible to mine data from simulations of astrophysical phenomena such as the evolution of a star [579]. This topic will be covered in Section 2.5 which discusses the analysis of data from computer simulations.

2.1.1 Characteristics of astronomy data

There are several characteristics which are common across data sets from observational astronomy, and the analyses conducted on these data sets, regardless of the instruments used to collect the data or the frequency at which the sky is being observed. These characteristics include:

- **Massive size of the data:** Astronomical surveys, whether conducted through ground-based telescopes such as the VLA [606], or space-based telescopes such as the Hubble Space Telescope (HST) [281], all result in very large amounts of data. For example, the Massive Compact Halo Objects (MACHO) survey [405], which concluded in 2000, resulted in 8 terabytes of data, and the Sloan Digital Sky Survey (SDSS) [528] will have nearly 15 terabytes of data when complete. Telescopes currently being planned, such as the Large Synoptic Survey Telescope (LSST) [396], are expected to collect more than 8 terabytes of data per night, producing several petabytes (a petabyte is 10^{15} bytes) of data per year, once it is operational around 2010.

- **Data collected at different frequencies:** Astronomers survey the sky at different wavelengths including optical, infrared, X-ray, and radio frequencies. Since celestial objects radiate energy over an extremely wide range of wavelengths, important information about the nature of the objects and the physical processes inside them can be obtained by combining observations at several different wavelengths. For example, the Faint Images of the Radio Sky at Twenty centimeters (FIRST) [174] surveyed the sky at radio frequency, while LSST is an optical telescope, and the Chandra telescope [96] surveys the sky at X-ray frequency.

- **Real-time analysis:** While much of the astronomical data are collected once and analyzed many times, usually off-line, there is an increasing need for real-time analysis in telescopes under development. For example, it is the goal of the LSST to make quality assurance data available to the telescope control system in real time to ensure correct operation of the telescope. In addition, as the image data are collected, they will be moved through data analysis pipelines that will compare the new image with previous images of the same region, generate prioritized lists of transient phenomena, and make the information available to the public in near real time for further observation.

- **Noisy data:** Astronomical images are often noisy due to factors such as noise from the sensors and distortions due to atmospheric turbulence. For example, Figure 2.2 shows the "noise" patterns in the FIRST data that arise due to the Y-shaped arrangement of the telescopes in the VLA [606]. In addition, astronomical images may have missing data due to a sensor malfunction or invalid pixel values caused when an extremely bright object makes the pixels around it artificially bright.

- **Lack of ground truth:** In astronomy, a key challenge to validating the results of analysis can be the lack of ground truth. In other words, how do we verify that a

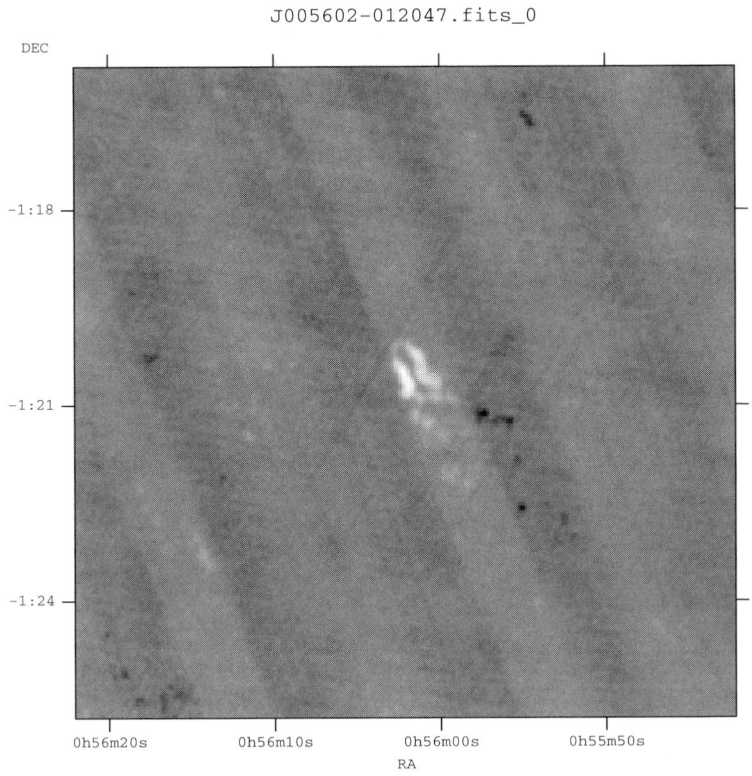

Figure 2.2. *Image from the FIRST Survey* [174], *extracted from the FIRST cutout server (sundog.stsci.edu) at RA* = 00 56 2.084 *and DEC* = −01 20 47.41, *using an image size of* 10 *arc minutes and maximum intensity of scaling of* 10 *mJy. Image is displayed using a log scale for clarity. Note the "Y" pattern of the noise in the image. The brighter wavy object in the center is a bent-double galaxy.*

region in an image of the surface of Venus is really a volcano, or that a galaxy is really a bent-double galaxy? This issue can be even more challenging for the negative examples; for example, is a galaxy non-bent-double because it is truly one, or is it just labeled as one because in the image, which is a two-dimensional projection of the three-dimensional universe, it appears to be one?

- **Temporal data analysis:** In many astronomical surveys, the goal is to study transient phenomena. For example, the goal of the MACHO survey [405] was to detect microlensing, which was indicated by changes in the intensity of an object over time. The LSST survey, on the other hand, is interested in detecting near-earth objects that move fast and must be detected in multiple exposures. Such temporal analyses require that images of a region taken at different times must first be aligned before any comparison with previous images can be made.

Several other characteristics of astronomy data are also worth mentioning. These data are often available publicly, usually as soon as the initial processing of the raw data from the telescopes is completed. Often, data once collected are analyzed many times, sometimes for reasons other than the one that motivated the initial collection of the data. Careful attention is paid to metadata, which is information describing the data. Items such as the time when the data were collected, the telescope settings, and the techniques used to process the data, are meticulously described and associated with any derived data products. Further, as more advanced algorithms for processing the data become available, the data are reanalyzed and new versions of derived data products created.

Since many astronomical surveys cover the same region of the sky using different instruments and wavelengths, the astronomical community is developing "Virtual Observatories" [173, 597]. The goal of these virtual observatories is to enable astronomers to mine data which are distributed across different surveys, where the data are stored in geographically distributed locations. Both the European and the American efforts are currently addressing the major challenge of having the data in all these different archives correspond to the same format. Of greater interest will be the mining of the data across surveys, which would include statistical analyses such as correlations across data taken at different resolutions, different wavelengths, and different noise characteristics.

2.2 Remote sensing

In contrast to astronomy, where the data are "look-up" data, remote sensing data can be considered as "look-down" data. Some of the earliest images of the earth were taken by cameras strapped on pigeons. More sophisticated images taken from balloons and aircraft soon followed, and current technology now provides us access to high-resolution satellite imagery, taken in several spectral bands, at several resolutions [527, 381].

Remote sensing systems are invaluable in monitoring the earth. Their uses include global climate change detection through the identification of deforestation and global warming; yield prediction in agriculture; land use and mapping for urban growth; resource exploration for minerals and natural gas; as well as military surveillance and reconnaissance for the purpose of tactical assessment. The era of earth remote sensing began with the Landsat multispectral scanning system in 1972 which provided the first consistent high-resolution images of the earth. The data were collected in four spectral bands, at a spatial resolution of 80 meters, with coverage being repeated every 18 days. Systems deployed since then have improved the coverage, both in terms of the number of spectral bands as well as the spatial resolution. For example, data from the IKONOS satellite [212], which was launched in September 1999, are available in the red, green, blue, and near-infrared bands at 4 meters resolution and in grey-scale or panchromatic at 1 meter resolution. Data from the Quickbird satellite [146] are available at an even higher resolution of 61 centimeters in panchromatic and in 2.4 meters in four bands (red, green, blue, and near-infrared). Note that, at such resolution, it is possible to clearly identify everyday objects such as cars, highways, roads, and houses in the satellite imagery. This opens up potential new applications such as the detection of human settlements. It also presents an important challenge in storing, managing, retrieving, and analyzing these large data sets.

A key task in the mining of remote sensing imagery is the identification of man-made structures such as buildings, roads, bridges, airports, etc. Some of the early work in this area using aerial imagery focused on exploiting user-specified information about the properties of these structures and their relationships with other more easily extracted objects. For example, in their 1982 paper [450], Nevatia and Price consider the task of finding airports in an aerial image of the San Francisco Bay area in California. They exploit the knowledge that some of the airports have runways near or projecting into the water. They first use a combination of two simple image segmentation techniques (see Chapter 8)—a line detector to find roads and a region segmentor to identify land-water boundaries. Each region or line is considered as an identifiable object, and features, such as average color and intensity values of a region, the height to width ratio of a region, and the location of the center, are extracted. User-specified information in the form of a rough sketch of the region identifying the major structures and their location relative to each other is used to map the regions in the aerial imagery with the corresponding objects in the sketch. The sketch is represented as a graph with the nodes identifying the objects (for example, San Francisco Bay North and Angel Island) and the edges indicating the relationship between the objects (for example, is-a-neighbor-of). The sketch is general and can be applied to any view of the area. Finally, Nevatia and Price used graph-matching techniques to create an annotated aerial image of the San Francisco Bay area.

With the availability of high-resolution satellite imagery from different sensors, the analysis of such imagery for man-made structures has also become more sophisticated. For example, Mandal, Murthy, and Pal [413] analyze data from the Indian Remote Sensing (IRS) satellite, with a resolution of 36.25 meters. They focus on the green and infrared bands as they are more sensitive in discriminating various land-cover types. First, they classify the image pixels into 6 classes (such as concrete structures, open space, etc.), followed by further processing using heuristics to identify man-made structures. For example, a heuristic may be that an airport should have a runway, which is a discrete narrow linear concrete structure region at least 30 pixels in length. In [270], Henderson and Xia provide a status report on the use of Synthetic Aperture Radar (SAR) imagery in human settlement detection and land-use applications. They observe that radar imaging can provide information which is complementary to the information provided by systems operating in the visible and near-infrared regions. In Lee et al. [367], the authors consider the problem of extracting roads from the 1-meter IKONOS satellite imagery [212]. They first segment the imagery using a variant of the watershed algorithm (see Section 8.2), identify road candidates using information about the regions such as elongatedness, and then expand the list of identified candidates by connecting close-by roads. IKONOS imagery has also been used in a multilevel process for the identification of human settlements using decision trees and statistical techniques [323].

A recent work describing the practical issues in applying machine learning techniques to remote sensing imagery focuses on the detection of rooftops in aerial imagery [412]. The authors discuss issues such as the importance of choosing features relevant to the task, the problems with labeling the rooftop candidates including consistency in labeling and disagreements among experts, and the effect of having a severely skewed training set with many more negative examples than positive ones accompanied by a higher cost of errors made in misclassifying the minority class.

In addition to land-use applications, remote sensing imagery has also been used for meteorological data mining. For example, Kitamoto [342] describes the use of techniques from the informatics communities, such as pattern recognition, computer vision, and information retrieval, to analyze and predict typhoons using satellite images that capture the cloud patterns of the typhoon. In contrast to mathematical models used in numerical weather prediction, where the model is deduced from partial differential equations describing the dynamics of fluids in the gravity field, Kitamoto considers pattern recognition and the past experience of experts as an indispensable part of the approach. The data used are from the Japanese GMS-5 geostationary satellite in one visible and three infrared bands, the latter enabling observation of clouds even during night-time. Cloud patterns are identified by dividing a 512×512 pixel image into 8×8 pixel blocks and assigning a "cloud amount" value to the data in each block. This gives a 64×64 or 4096-dimensional cloud-amount vector which is then used in subsequent analysis using techniques such as principal component analysis, clustering with the k-means algorithm, and self-organizing maps. While early results with the use of data mining techniques appear promising, Kitamoto cautions that one must not forget the fundamental difficulty in predicting atmospheric events, namely the chaotic nature of the atmosphere.

A task related to classifying and predicting typhoons is that of tracking them over time. For example, Zhou et al. [648] use motion analysis to track hurricanes in multispectral data from the Geostationary Operational Environmental Satellite (GOES). Since the data are collected in a super-rapid scan sequence at 1 minute intervals, it is possible to observe hurricane dynamics in the imagery. Using a nonrigid motion model and the Levenberg–Marquardt nonlinear least-squares method, Zhou et al. fit the model to each small cloud region in the image and then estimate the structure and motion correspondences between the frames of the data.

Another related work using data from GOES is the detection of cumulus cloud fields by Nair et al. [445], who compare the accuracy of two techniques. They found that a structural thresholding method based on the morphology of cloud fields outperformed an approach based on classification algorithms applied to edge and spectral features extracted from the imagery. Classifiers such as neural networks have also been used in cloud classification using texture features extracted from Landsat imagery [368].

A different example of mining remote sensing data is the Quakefinder system described by Stolorz and Dean [569]. This analyzed panchromatic data collected by the French SPOT satellite at 10 meter resolution to detect and measure tectonic activity in the earth's crust. Using the 1992 Landers earthquake in Southern California, Stolorz and Dean found that the ground displacements varied in magnitude up to 7 meters, which was smaller than the pixel resolution, making naïve pixel-to-pixel change detection methods inapplicable. They had to use sophisticated methods to infer the motion of a single pixel, a process that had to be repeated for each pixel in the image. As even a modest-sized image contained 2050×2050 pixels, this required the use of a massively parallel processor. In addition, the authors found that they had to incorporate corrections for slight differences in spacecraft trajectories caused by differences in height above the target region as well as the yaw, pitch, and roll of the spacecraft. Spurious differences between scenes due to differences in the sun angle and the view angle were found to be negligible as the satellite was in a sun-synchronous orbit.

Other interesting applications of mining remote sensing imagery include the detection of fire smoke using neural networks [380], the automatic detection of rice fields [589], and the monitoring of rice crop growth [355]. Kubat, Holte, and Matwin [352] illustrate several aspects of practical data mining in their work on the detection of oil spills. In particular, they discuss issues related to the scarcity of the data, as only a few images contained oil spills, which in turn, resulted in an unbalanced training set. They also observed that their examples showed a natural grouping in batches, where the examples in a batch were from a single image, leading to greater similarity among them. In addition, they found that they had to carefully consider which images to present to the human who can control the trade-off between false alarms and missed positives.

2.2.1 Characteristics of remote sensing data

Remotely sensed images, whether obtained from satellites or aerial photography, are a very rich source of data for analysis. The characteristics of these data include:

- **Massive size:** The increasing sophistication of remotely sensed systems has resulted in an explosive growth in the amount of data being collected. NASA's Earth Observing System (EOS) Project [171] represents an extreme of this trend. Starting in 1997, several satellites have been documenting the earth's processes and global change, acquiring 1.6 terabytes of images and other data per day. This will result in over 11,000 terabytes of data over the 15-year lifetime of the project.

- **Multiresolution data:** Over the years, as the earth has been observed by various systems ranging from Landsat to IKONOS and Quickbird, we have acquired imagery at resolutions ranging from a few hundred meters to under a meter. As a result, we have several images of an area taken over time at different resolutions. When such data, collected over time, must be analyzed to understand changes that have occurred over long periods, we will need to use analysis techniques which can exploit the different resolutions of the data.

- **Multispectral data:** While some of the very early remotely sensed data were collected in only one band (i.e., panchromatic), data collected more recently are often in several spectral bands. These range from the 4-band multispectral data from IKONOS and Quickbird to the 224-band hyperspectral data from the Advanced Visible/Infrared Imaging Spectrometer (AVIRIS) [11]. This acquisition of imagery at various spectral bands simultaneously provides multiple "snapshots" of spectral properties which can contain far more information than a single band alone. This allows a scientist to assign a spectral signature to an object when the resolution is too low for identification using shape or spatial detail. Note that the increase in the number of spectral bands also increases the size of the data collected.

- **Multisensor data:** In addition to the resolution or the number of spectral bands, remote sensing systems are also characterized by the sensors used, which in turn influences the way in which the data are processed. Several factors must be considered, including the orbit of the platform which affects the angle at which the image is taken and the number of days between imaging of nearby swathes; the platform attitude, that is, the roll, pitch, and yaw of the platform which may distort the image; the scan geometry which determines the order in which the scene is scanned and the number of detector elements used in scanning; topographic distortions; and so on [527].

- **Spatiotemporal data:** Since remote sensing platforms, especially satellites, cover the same region of the earth several times, there is an implicit temporal aspect to the data collected. As a result, change detection is an important application in remote sensing, not just using data collected over time by a single sensor, but also data collected by different sensors. Thus image registration, that is, aligning the satellite images, becomes an important task in mining remotely sensed data.

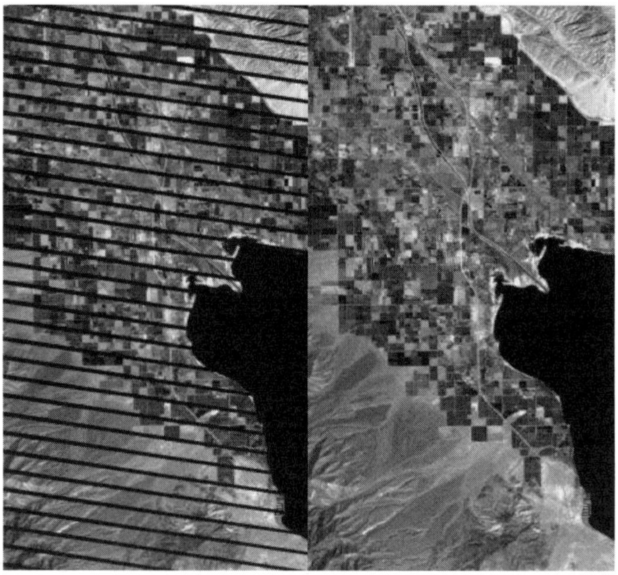

Figure 2.3. *An image from the Landsat Gallery illustrating how missing data due to sensor malfunction may be handled in remote sensing imagery. The left image is path 39 row 37, acquired over Salton Sea in southern California on 9/17/2003, and it shows the scan gaps caused by the failed Scan Line Corrector. The right image is the same data, but with the gaps filled by using data acquired on 9/14/2002. Image: Grey-scale version of Landsat_Gallery_394_1_450.jpg from http://landsat.usgs.gov/gallery/main/2/.*

- **Missing data:** As with any other observational science, remote sensing data can suffer from missing data problems due to sensor malfunction (see Figure 2.3). Filling in these gaps must be done with care not only to avoid the introduction of artifacts, but also to ensure the integrity of the image.

- **Noisy data:** Data in remotely sensed imagery can be noisy due to sensor noise or extraneous objects such as clouds in optical imagery that may obscure the scene on the earth (see Figure 2.4). However, what is considered as noise may depend on the problem being solved; clouds in a scene could be considered as the signal in an application involving the classification of clouds.

Several other characteristics of remotely sensed data are worth mentioning. Since the raw data sets are very large, they may be available after various levels of processing. For example, EOS data products are available at levels ranging from Level 0 to Level 4 [167]. Level 0 data products are raw EOS instrument data at full instrument resolution. At higher levels, the raw instrument data are converted into more usable parameters and formats of interest to the scientific community. For example, Level 3 data products are variables mapped on uniform space-time grids, usually with some completeness and consistency. At Level 4, the parameters are further refined through the use of models. Many of the data sets from remote sensing are publicly available. Some, such as the data from EOS, are free, while others are available for a fee.

Figure 2.4. *Atmospheric vortices over Guadalupe Island showing the cloud cover obscuring the land mass underneath. If the task is to obtain more information about the land mass, then this image would be inappropriate. If, however, the task is to identify the type of clouds, this image would be an ideal one. Image courtesy of NASA's Earth Observatory, http://earthobservatory.nasa.gov/Newsroom/NewImages/Images/.*

In many remote sensing applications, ground truth can be difficult or expensive to obtain. For example, soil samples may need to be collected to evaluate analysis techniques that determine mineral content from remotely sensed imagery. The areas with these samples may be inaccessible, or there may be too many of them widely distributed geographically to make the collection of samples feasible.

While much of the discussion in this section has focused on remote sensing of the earth, it is worth mentioning that the same techniques can be applied to remotely sensed data from other planets such as Mars, where such analysis can be an invaluable precursor in planning for "in situ," that is, in-place sensors.

Remote sensing systems are also a key part of surveillance and reconnaissance [365], where the purpose is to acquire military, economic, and political intelligence information. Reconnaissance refers to preliminary inspection, while surveillance maintains close observation of a group or location. The latter involves frequent or continuous coverage and involves the collection of video data, further aggravating the challenges of handling and analyzing massive data sets. This topic is further discussed in Section 2.4.

No discussion of remote sensing would be complete without a mention of Geographic Information System (GIS) [350]. In addition to being a cartographic tool to produce maps, a GIS stores geographic data, retrieves and combines this data to create new representations of geographic space, and provides tools for spatial analysis. Examples of data in a GIS include information on streets, rivers, countries, states, power lines, school districts, office buildings, and factories. The results from the analysis of remotely sensed imagery, when used together with the complementary information in a GIS, can form an invaluable tool.

2.3 Biological sciences

A very rich source of challenging data mining problems are the fields of biological sciences, including medical image analysis, clinical trials, and bioinformatics. While it is possible to treat each topic as a separate section (and each is broad enough to cover several books!),

I have chosen to include them in one section, as the lines between them are beginning to blur—advances in genomics and proteomics are leading to an improved understanding of systems-level cellular behavior with potential benefits to clinical research.

2.3.1 Bioinformatics

Bioinformatics is the science of managing, mining, and interpreting information from biological sequences and structures. It came into prominence through the Human Genome Project [603, 358]. Fueled by advances in DNA sequencing and genome mapping techniques, the Human Genome Project resulted in large databases of genetic sequences, providing a rich source of challenges for data miners. More recently, the field of proteomics promises an even more radical transformation of biological and medical research. The proteome defines the entire protein complement in a given cell, tissue, or organism. In a broad sense, proteomics can be considered to be everything "postgenomic" and includes protein activities, the interactions between proteins, and the structural description of proteins and their higher-order complexes. By studying global patterns of protein content and activity, we can identify how proteins change during development or in response to a disease. This not only can boost our understanding of systems-level cellular behavior, but also benefit clinical research through the identification of new drug targets and the development of new diagnostic markers. An excellent introduction to various aspects of this emerging field of proteomics is given in a series of articles in the special issue of Nature [448], as well as in the more recent survey of the computational techniques used in protein structure prediction [468].

Data mining techniques have found an important role in genomics where they are extensively used in the analysis of data in the genetic sequence and protein structure databases [609]. For example, the text by Stekel [566] is a good introduction to microarray bioinformatics, providing an overview of microarray technology; the use of image processing techniques to extract features from microarrays; the role of exploratory data analysis techniques in cleaning the data; and the use of statistics and machine learning techniques, such as principal component analysis, clustering, and classification, to study the relationship between genes or to identify genes or samples that behave in a similar or coordinated manner. Baldi and Brunak [16] provide a more machine learning focus to bioinformatics and discuss how techniques such as neural networks, hidden Markov models, and stochastic grammars can be used in the analysis of DNA and protein sequence data. Their text also includes a chapter on Internet resources and public databases. Other introductory texts in this area include the ones by Baxevanis and Ouellette [27] and Lesk [371].

The challenges in proteomics far surpass those in genomics and offer many opportunities for the use of data mining techniques. For example, Zaki [643] discusses how data mining techniques, such as association rules and hidden Markov models, can be used in protein structure prediction. Using a database of protein sequences and their three-dimensional structure in the form of contact maps, he shows how one can build a model to predict whether parts of amino acids are likely to be in contact and to discover common nonlocal contact patterns.

Proteomics is also having an effect on drug discovery—as most drug targets are proteins, it is inevitable that proteomics will enable drug discovery [251]. For example, Deshpande, Kuramochi, and Karypis [138] use the discovery of frequent subgraphs to classify chemical compound data sets. They observe that graph-based mining is more suitable for handling the relational nature of chemical structures. By combining methods for finding frequent subgraphs in chemical compounds with traditional machine learning techniques, such as rules and support vector machines, they found that they could improve the classification accuracy on problems such as the discovery of carcinogenic compounds and compounds that inhibit the HIV virus.

The area of computational biology is using computational systems to simulate complex structures and processes inherent in living systems. Instead of focusing on just one level, say genomics or proteomics, there is an increasing interest in mining data across different levels, ranging from various genome projects, proteomics, and protein structure determination to digitized versions of patient medical records. Just as in physics, computers are now being used to simulate biological processes from the atomic and molecular level to cells and parts of organs. For example, the Blue Brain project [420, 46] is using massively parallel computers to create a biologically accurate functional model of the brain. The goal is to increase our understanding of brain function and dysfunction and help explore solutions to various neurological diseases. The use of computer simulations in understanding biological phenomena comes with all the issues associated with the use of simulations in any science, namely, the analysis of massive amounts of complex data, the understanding of the sensitivity of the results to various input parameters, the verification of the computer models, and the validation of the simulations by comparisons with experiments or observations.

Another interesting application of data mining in bioinformatics is in automating the collection, organization, summarization, and analysis of data present in the literature [351]. This requires the use of text mining and optical character recognition to understand the text in documents, image mining to extract information from images and plots in documents, and the use of machine learning and statistical techniques to find the connections between various documents.

2.3.2 Medicine

Statistical techniques have had a long history in the field of clinical trials where they were used for drawing conclusions about the efficacy of a drug used in the treatment of a disease or for finding patterns in the occurrences of diseases [232]. With the availability of other forms of data, such as images from X-rays and magnetic resonance imaging (MRI), the field of medical imaging is also providing data that can be used in the diagnosis of diseases [558, 37, 20].

There are several examples where data mining techniques are being used in medical imagery. A compelling example is the use of neural network technology in the detection of breast cancer in mammograms [219]. Breast cancer is a serious problem with early detection having a dramatic effect on raising the survival rate. While mass radiological screening is the only reliable means of early detection, the examination of the mammograms is very tedious and inefficient. When a single radiologist has to examine a large volume of mammograms in a short time, it can lead to missed cancers. A computer-assisted approach can be invaluable in such circumstances, leading not only to reduced radiologist time per mammogram, but also to improved early detection as the computer system could draw attention to lesions that have been overlooked in the mammogram. A practical implementation of such technology is the ImageChecker [288] which uses neural networks to provide a potential improvement of over 20% detection of breast cancer.

A more detailed look at the issues encountered in computer-aided mammographic screening is provided by Kegelmeyer et al. [333]. This work built upon earlier work in the detection of spiculated lesions in mammograms [334] which showed that by the application of appropriate image processing techniques to the mammograms, it was possible to extract features (that is, measurements from the images) which, when input to a binary decision tree, improved the probability of detection of the lesions by almost 10%. This work was then extended to the detection of two other important cancer signs: microcalcifications and circumscribed masses.

In contrast to the more traditional approaches in mammography, where regions of interest are first identified in the image before measurements are made on these regions, Kegelmeyer et al. processed the entire image for the extraction of features. While this was more time consuming, it addressed the problem of a missed detection due to a lesion not being included in a region of interest. Further, they adopted a methodology of iterative refinement, wherein they would start with a variety of features, use the decision tree to identify the most promising ones, refine them further or extract new features, and keep iterating until an effective subset of features was found.

For the microcalcification detection, Kegelmeyer et al. first segmented the mammograms using a variety of image segmentation algorithms (see Chapter 8), such as adaptive thresholding and edge detection, and then extracted features, such as the number of pixels in the object, the shape of the object which provided a measure of how close the object was to a circle, the standard deviation of the pixel intensities within the object, the standard deviation of the pixel intensities of the background around the object, and the texture of the object. They found that the best results were obtained when a mix of segmentation algorithms and features extracted from the segmented images were used, as different segmentation algorithms resulted in different objects, thus affecting the features.

For the mass detection problem, Kegelmeyer and his coauthors focused extensively on the feature extraction part using techniques such as flattening the breast-boundary brightness gradient to make the images more uniform, reducing the contrast variation across the image, edge detection, and the circular Hough transform to detect circumscribed masses which can be approximated to first order by circles. This resulted in a feature image indicating the relative likelihood of a circumscribed mass at each pixel in the image.

While mammogram interpretation, and the semiautomatic detection of cancers, can be challenging, these images are relatively easy, as there are no other tissues such as bones or other irrelevant structures to complicate the task. In other problems, for example, the detection and progression of diseases of the respiratory systems [507], the images can include different types of tissues such as lungs, bones, heart, and so on. Tissues not of interest in the analysis may be considered as noise and must be removed. Further, several imaging modalities are available, including X-ray computed tomography, positron emission computed tomography (PET), single photon emission computed tomography (SPECT), and MRI. Since each modality gives a different insight into the anatomy of the patient, it is important that the complementary information available be exploited whenever possible by means of data fusion. In addition, whenever there is a need to follow events dynamically, for example, the collapse of airways as a patient breathes out, or tracking the growth of a tumor over longer time intervals such as months or years, there is a need to register or align images taken at different times. This need for registration algorithms is fundamental in brain image analysis, where brain atlases are used to study how the brain varies across age, gender, disease, imaging modalities, and over populations [582]. Sophisticated mathematical algorithms from computer vision, pattern recognition, visualization, and statistics are being used to shed light on the complex structural and functional organization of the human brain.

This proliferation of data modalities and the resulting challenges in analysis have not gone unnoticed by data miners. For example, machine learning is being used in the analysis of functional MRI data [122, 645], PET scans [237], and SPECT images [519, 567]. As the challenges in mining each type of data are better understood and addressed, there will be a need to mine across the different modalities to improve and validate the analysis results.

2.3.3 Characteristics of biological data

Biological data share many characteristics common to other large scientific data sets, such as massive size, multimodal data, and spatiotemporal data. Though such data are very diverse, ranging from medical images to protein structures, there are some unique characteristics common to many biological data sets:

- **Unstructured data in heterogeneous databases:** Some bioinformatics data, unlike most other scientific data, are stored in databases. Many of these are publicly available and encode different views of the biological domain using different data formats, different database management systems, and different data manipulation languages. As these databases are constantly evolving and being enhanced, mining across them can be challenging. Another example of unstructured heterogeneous data is the literature in bioinformatics which includes unstructured text images, plots, images, tables, as well as links to other documents. Identifying connections among different experiments conducted on the same or different data sets by connecting the dots in the scientific literature is still an open problem.

- **Information in the form of images:** While the field of medical image analysis by its very nature has to deal with all the issues associated with the extraction of useful information from images, the field of bioinformatics is also beginning to work with image data. For example, Leherte et al. [369] discuss the use of image processing and computer vision techniques in protein structure determination by formulating it as a problem in scene analysis.

- **Poor quality of data:** Data in medical image analysis is often of poor quality due to sensor noise, presence of various types of tissues that respond differently to sensors, and occlusions by tissues. In clinical trials, data can often be missing or input incorrectly. Also, in bioinformatics, the quality of data from various sources can vary in accuracy. All these factors, along with the fact that biologists have incomplete knowledge of many aspects of the field, imply that one must be careful in drawing conclusions based on the results of data analysis. This is especially true in medical data analysis, where a conclusion could be a matter of life or death.

- **Privacy protection:** Medical data, unlike data from most other scientific applications, come with the constraint of privacy. Though a detailed discussion of privacy preserving data mining is beyond the scope of this book, I have mentioned it here as it can have implications on how the data are analyzed.

2.4 Security and surveillance

Another broad area where data mining techniques are being used and gaining acceptance is security and surveillance. This includes topics as diverse as biometrics with fingerprint, iris, face, signature, and voice recognition; automated target recognition in aerial and satellite imagery; video surveillance; and network intrusion detection. Though security applications may not strictly be considered as scientific, they have much in common with scientific applications and are therefore included in the book. I next briefly discuss each of these topics.

2.4.1 Biometrics

Biometrics is the automated identification or verification of an individual by using certain physiological or behavioral traits associated with the person [469]. One of the earliest and best known biometrics technologies is fingerprint identification which has been used in forensic applications for many decades. The large number of fingerprints in the FBI database (200 million fingerprint cards at 10 megabytes each in 2002) has led to the use of sophisticated techniques using wavelets (see Section 5.2) to compress and transmit these images while preserving the key features necessary for identification [62].

More recently, fingerprint technology is being used in place of password-based security, for example, for accessing computer systems [627, 147]. Many of these biometric identification systems, whether based on fingerprints, iris, face, voice, gait, or signature, have two phases. In the enrollment phase, the biometric characteristic of a person is scanned to acquire a digital representation, which is then further processed by a feature extractor to generate a more compact representation. In the recognition phase, the same characteristic of the individual to be identified is acquired, processed by the feature extractor, and compared with the feature on record. For example, in fingerprint recognition, the features may be derived from the local orientations of ridges, with some postprocessing using the Karhunen–Loeve transform (see Section 10.2.1) to keep only the key features [78]. Or, in face recognition, each face is represented in terms of a basis, called eigenfaces, which is obtained by performing principal component analysis (see Section 10.2.1) on a database of face images and keeping only the eigenvectors corresponding to the largest 100 or so eigenvalues [341, 595].

While biometric systems are being developed and deployed commercially, there are still several issues regarding their use. A key issue is the negative perception associated with the perceived violation of the privacy rights of an individual. On the technical front, a concern with biometrics is the issue of accuracy—while a password-based system always accepts a correctly entered password, a biometric system may not guarantee correct authentication due to sensor noise, limitations of the processing methods, as well as variability in the biometric characteristic and presentation. For example, face recognition methods tend to perform poorly if the lighting is poor or the individual is wearing a disguise such as a hat or glasses, or is not presenting a frontal view. One solution to this lies in combining multiple biometric modalities [300]. In addition, it may be difficult to create a cost-effective biometrics system that is easy to use, is not too intrusive, and takes a short time to learn and operate.

2.4.2 Surveillance

Surveillance is defined as maintaining close observation of a group or locations [365]. While the classic application of surveillance is in a military setting, it is increasingly being used in commercial applications such as monitoring activity near automated teller machines or at traffic intersections, and for security in parking lots, stores, and around homes. Associated with surveillance is reconnaissance, or making a preliminary inspection. In a military application, this could include determining the lay of the land or the location of enemy troops, while in a civilian setting, reconnoitering can be used in detection of regions with oil or other minerals. The goal in both surveillance and reconnaissance is to image an area or object on the ground and extract information from the image. The imaging can be done either aerially through an unmanned aerial vehicle or via satellite, using radar, infrared, or visible systems.

While there is much that is similar between analysis problems in remote sensing (Section 2.2) and in surveillance and reconnaissance, there are some key differences. The subject being observed in surveillance may go to great lengths to avoid being seen or identified. Also, in many cases, surveillance is done in a persistent mode for long periods of time, for example, a video camera in a store for security or at a traffic intersection to identify vehicles that do not stop at a red light. The video acquired in these situations is at a frame rate such that we can identify motion in the scene. This temporal aspect of the imagery allows us to detect and track moving objects in a scene, which in turn enables us to build a model of normal behavior in the scene and flag events that deviate from the normal.

Detection and tracking of moving objects is key to video mining. Typically the detection is done by maintaining a "background" image of a scene and subtracting this from each frame to identify objects that have moved [587, 565, 111, 112]. This background image has to be constantly updated to account for changes in illumination, motion of the camera, objects that were moving but have stopped for a while, objects that were stationary but which have started moving, and so on. These background subtraction algorithms (see Section 8.4.1) can use simple techniques, such as the difference of two frames or the moving average of frames in a time window around the current frame, to more complex models using the Kalman filter or mixtures of Gaussians, or a combination of simpler techniques.

Once the moving objects have been detected, it is possible to extract features such as velocity, intensity of pixel values, and shape, and use them to track the objects from frame to frame. This can be done using techniques such as the Kalman filter, motion correspondence, or particle filters [501, 44, 6, 151]. Tracking can be difficult in the presence of occlusions, in low contrast imagery, in poor lighting conditions such as fog or rain, or when the platform from which the imagery is taken is moving. Once the tracks for an object have been obtained, they can be used to build models of the scene, identify the activity [600, 401], and indicate anomalous behavior.

2.4.3 Network intrusion detection

With the Internet becoming ubiquitous, the problem of protecting the computers on the Internet from unauthorized use or malicious attacks has become increasingly important in recent years. The networks which connect computers to each other, and to the Internet, are constantly monitored to determine if they are working correctly or if they are under attack. This monitoring of the traffic on the Internet produces vast quantities of streaming data which must be analyzed in real time, so that a system under attack can be identified and the problem can be addressed before it spreads to other computers.

There are several different approaches to detecting network intrusions. Certain types of network attacks have a known signature which can be described by experts; such intrusions are relatively easy to detect. However, if a new type of attack is used to penetrate a computer network, a system based only on signatures will not be able to detect the attack. In such cases, solutions based on data mining techniques such as anomaly detection and misuse detection [172] or statistical approaches [419] can be quite useful. These approaches either build models of normal and intrusive behavior using training data or build models of normal behavior and identify any deviations from it as anomalies or potential intrusions. The latter approach is more general as it does not depend on a training set containing examples of normal and intrusive behavior. This is especially important, as novel ways of attacking computer networks are constantly being developed, and what is considered intrusive changes over time. This aspect makes network intrusion detection an active area of data mining research.

2.4.4 Automated target recognition

Automated Target Recognition (ATR) is closely related to surveillance and remote sensing. Given a remotely sensed image, the goal of ATR is to identify the objects of interest and assign them to a class. This is a classic data analysis problem, where objects are first identified in imagery through image processing techniques such as segmentation and edge detection. Next, features, such as the size (height, width, area), shape (aspect ratio, Fourier descriptors), and radar signatures are calculated for each object in the image and then used in various statistical and machine learning–based classifiers to identify the object. Often data from more than one sensor can be used to reduce the rate of false alarms.

2.4.5 Characteristics of security and surveillance data

The data that form part of many surveillance applications have characteristics that are common in scientific applications. These characteristics include:

- **Massive size:** Surveillance data can often be very large, for example, the FBI fingerprint collection is over 100 terabytes, and large field-of-view aerial imagery can easily approach terabytes and beyond when the temporal aspect of the video is taken into account. The data from computer networks can also be very large, given the large number of computers connected to the Internet.

- **Multiresolution, multisensor, multispectral data:** In many ATR applications, more than one sensor (for example, radar and visible) is often used to minimize false alarms by exploiting the complementary information gathered by the sensors.

- **Spatiotemporal data:** The temporal aspect of data is key to several applications such as video mining, where the temporal information in the video frames is exploited to model the interactions among the objects in a scene and flag anomalous events.

- **Noisy data:** The data in security and surveillance applications are often noisy. This can be the result of poor illumination in applications such as face recognition; changing illumination in the tracking of moving objects at a traffic intersection; or problems in the collection and distribution of data, such as poor fingerprint quality due to dirt on a person's finger or compression artifacts that arise when a fingerprint is compressed for faster transmission.

- **Real-time response:** The very nature of security and surveillance applications requires a real-time response, whether it is for the purpose of authentication in biometrics, preventing a robbery at a store in video surveillance, or stopping a computer network attack from spreading to other computers.

- **Privacy protection:** The privacy of the people being observed is an important issue and one which must be resolved for biometric and civilian surveillance systems to be deployed successfully.

- **Streaming data:** Several of the applications in security and surveillance result in data sets which are streaming in nature; that is, there is a constant stream of data being received at the sensors. This is especially true in video surveillance and network intrusion detection. Such data can place additional constraints on both the sensor, which must be able to keep up with the flow of data, and the analysis techniques, which must analyze the data as they arrive.

2.5 Computer simulations

Computer simulations are increasingly gaining acceptance as the third mode of science, complementing theory and experiment. The idea behind such simulations is to understand complex phenomena by analyzing mathematical models on high-performance computers. This approach is often taken when the phenomena are too complex to be solved by analytical methods or too expensive, impractical, or dangerous to be studied using experimental means. For example, we may use computational fluid dynamics to study the flow around an airplane or the turbulent mixing of two fluids; environmental simulations to understand the effect of volcanic eruptions on global climate or to model the flow of pollutants underground in an accidental discharge from a reservoir; and structural mechanics simulations to study the effects of car crash tests or the bending of tall structures due to strong winds.

Computer simulations often generate large data sets whose sheer size and complexity make them difficult to analyze. Analysis is usually done by visual inspection, which is impractical when the output from simulations is measured in terabytes and petabytes. As a result, much of these data are never really explored by scientists tasked with finding the science in the data.

The visualization community has made great strides in addressing these concerns through the use of advanced displays, such as power walls, for large volumes of data; novel data representations that allow out-of-core progressive visualization of data; and high-accuracy volume rendering frameworks. However, analyses done through these traditional means have been mainly qualitative. Scientists are now interested in more quantitative analyses, as well as ways in which they can directly focus on the areas of interest, instead of first browsing through vast quantities of data.

There are many different ways in which data mining is playing a role in the analysis of simulation data sets [314], as discussed next.

- **Detection of coherent structures:** There are several examples of the use of data mining techniques to identify and track coherent structures. For instance, Ferre-Gine et al. [188] analyze the data from a wind tunnel using a fuzzy ARTMAP neural network to identify eddy motions in a turbulent wake flow. They considered frames of 3×3 adjacent velocity vectors and represented patterns by normalized velocity vectors. From this data, they manually identified patterns which represented clockwise and counterclockwise structures and used them as the training set for a neural network.

Other applications involve the identification and tracking of coherent structures over multiple time steps. Traditionally, tracking is done by a human or by computationally expensive heuristics which track objects through all intermediate time steps. Banerjee, Hirsch, and Ellman [19] proposed an alternative tracking method based on machine learning techniques and described an application to track vortices which change shape, split, merge, appear, and cease to exist unexpectedly. Their method begins by constructing a training data set composed of vectors which describe related and unrelated vortices after a time gap. The training set is used to create a decision tree, which is then used to identify related vortices in different time steps. As expected, the accuracy degrades as the time gap increases. The decision tree performs worse than the tracking approach for small time gaps, but over intermediate and long time gaps, the decision tree performs better than a traditional tracking program.

Another technique that has been extensively used for the analysis of coherent structures is wavelets [181]. As wavelets operate on different scales, the information they extract from the data are ideal for identifying structures at different scales in problems such as turbulent flow. For example, Farge and Schneider [182] describe a nonlinear procedure to filter wavelet coefficients to separate the coherent vortices from the background flow. Their work shows that very few of the wavelet modes contain most of the enstrophy (the variable of interest in this problem) and that selecting the top modes (2 or 3%) is sufficient to select the coherent structures. In a similar manner, Siegel and Weiss [547] use wavelet packet techniques, again in the context of fluid flow, to separate signals into coherent and noncoherent parts. They too observe that by keeping only a small subset of the wavelet coefficients they are able to extract individual coherent structures in turbulent flow. A slightly different approach is taken by Hangen et al. [253], who combine wavelets with the more traditional template-matching approach from pattern recognition to match patterns at different scales.

- **Dimension reduction:** Principal Component Analysis (PCA) is an important analysis tool used in many fields. In simulation data, it can be used to analyze the spatial or temporal variability of physical fields. The technique has been used extensively under the name of Empirical Orthogonal Functions (EOFs) in atmospheric sciences [483] and as Proper Orthogonal Decomposition (POD) in turbulence [399]. In particular, it has been used to build low-dimensional models by Lumley and Blossey in [398], who exploit the fact that in certain problems, some parts of the domain are dominated by large-scale coherent structures. These can be resolved by a low-dimensional model of the region, while parameterizing the smaller scale effects. This low-dimensional model is generated by keeping a few eigenfunctions obtained from the POD.

  Nonlinear extensions to PCA, based on autoassociative neural networks, have been studied by Sengupta and Boyle [532]. They found that the nonlinear method effected somewhat greater data reduction and that the leading nonlinear mode captured the seasonal cycle of precipitation more clearly than the leading linear mode. They concluded that nonlinear techniques should be considered especially when linear PCA fails to uncover physically meaningful patterns in climatological data analysis.

  A different application of dimension reduction techniques has been used in the Sapphire project to separate signals in climate data [199]. Depending on the simulation, the data may contain the effects of many different sources, such as El Niño and volcano signals; these must be removed in order to make meaningful comparisons between different climate models. This separation is made complicated by the fact that recent volcano eruptions coincided with El Niño events. Our experiences showed that by combining the traditional PCA with the newer Independent Component Analysis [286], we obtained better separation than by just using PCA alone [199].

As we shall see later in this section, dimension reduction techniques are also useful in building code surrogates, which are data-driven models used to predict the output of a simulation based on the input parameters. Such code surrogates are built using a training set of input and output parameter values. As the number of input parameters is often large, a very large number of simulations would have to be run to span this parameter space well. Dimension reduction techniques offer the promise of providing insights into which of the input parameters are important and therefore should be used in building the code surrogate.

- **Information retrieval in simulation data:** Finding objects of interest in simulation data is a task that is essential to analyzing the data. The traditional solution to this problem has been the use of visualization techniques for "feature" extraction. In contrast to data mining, where the term "feature" is a descriptor of an object in the data, the visualization community uses the term to refer to the object itself. For example, Machiraju et al. [404] describe the detection of swirl, which is a scalar quantity used to identify features such as vortices, which are characterized by swirling flow. The swirl is a well-defined mathematical quantity derived from the velocity and velocity gradients at each grid point. The authors illustrate how these quantities can be approximated directly from the compressed simulation output without first uncompressing the data. They borrow ideas from multigrid algorithms and finite-difference approximations to incorporate correction terms which account for the different spacing between the grid points at different levels in the wavelet representation of the compressed data.

  A more recent development in the analysis of simulation data is the use of techniques from the Content-Based Information Retrieval (CBIR) and related communities. In CBIR, the task is to find images in a database that are similar to a query image (see Section 2.7). Many of the current CBIR systems focus on photographic, remotely sensed, medical, or geological images. However, simulation data are different from such image data. First, unless the underlying grid is a simple Cartesian mesh, simulation data are not available at regular points in a rectangular domain. So, many of the image processing algorithms cannot be directly applied. Second, there are often several physical quantities associated with each grid point, unlike an image, which is just a single quantity. Third, the "objects" in a simulation rarely correspond to physical objects such as a car or a person. In addition, the "objects" in an image often change shape as the simulation evolves, and may even appear or disappear with time. Finally, much of the CBIR work focuses on extracting features for, and retrieving, the entire image, while in simulation data, we are interested in parts of an image. This implies that the image must be segmented to identify these parts prior to the extraction of features.

  Despite these differences, a couple of efforts have used CBIR techniques in the context of simulation data. The Feature Extraction and Evaluation for Massive ASCI

Data Sets (FEEMADS) project [79] extracted features using the data from a hydrodynamics simulation. The grid used is from an adaptive mesh refinement scheme (see Section 3.1). The authors consider two scenarios: one in which the adaptive mesh data are resampled on a uniform grid so image processing techniques can be directly applied, and another in which the data are directly obtained from the oct-tree data structure used to store the unstructured mesh. In the latter scenario, the authors exploit the fact that in the areas of interest, the mesh is refined, and the level of refinement can be used to limit the search. An approach more closely resembling the traditional CBIR systems is taken in the Sapphire project. This work is described in more detail in Section 2.7.

- **Code validation:** Another area where data mining techniques are being applied in simulation data is code verification and validation. Verification assesses the degree to which a code correctly implements a chosen physical model, while validation is the degree to which a code describes the real world. In other words, verification addresses the question, "Have we built the code right?" while validation addresses the question, "Have we built the right code?" Validation usually involves comparing simulations to experiments often by extracting higher-level characteristics from the data. If a scientist can identify which of two simulations (sometimes also called numerical experiments) is closer to a real experiment using an appropriately defined metric, he or she can use the information to improve the predictive capabilities of the simulation codes. Then, when these codes are used to design an experiment, there is greater confidence in the experiment behaving as expected.

  Scientific data mining techniques have just started making inroads into this important area of code verification and validation [510, 653, 192, 322, 321]. It is clear that with high-quality experiments becoming available in many domains, such quantitative comparisons will become more commonplace.

- **Understanding simulations:** Data mining methods can also be used to interpret results from simulations. For example, Mladenić et al. [431] describe the use of regression trees and a rule inducer to interpret results from discrete event simulations of a supermarket. The simulator takes user-defined parameters, such as the number of cashiers and arrival rates, as inputs and produces summary statistics, such as cashier utilization and length of customer queues, as outputs. The goal is to examine which input parameters affect the output of the simulators. The simulators are executed multiple times with different parameters and their outputs are recorded. Each execution produces a training example that is used as input to the tree or rule inducers. Both the regression tree algorithm and the rule inducer created accurate models that could be interpreted by humans.

  Mladenić et al. [432] have also applied decision trees to a simulation of a steel production plant with the goal of identifying the reasons for excessive waste. Decision trees have also been used to analyze the massive data sets from simulations in [53], while image analysis techniques have been used in understanding the behavior of coherent structures in Rayleigh–Taylor instability [318, 319]. This instability occurs when an initially perturbed interface between a heavier fluid on top of a lighter fluid, is allowed to grow under the influence of gravity. Fingers of the lighter fluid penetrate the heavier fluid in what are known as bubbles, while spikes of heavier fluid enter the lighter fluid. Understanding both the static and the dynamic behavior of these bubble and spike structures provides an insight into the mixing of the two fluids and the process of turbulence [215].

- **Code surrogates or building predictive models:** An area which has been gaining popularity in recent years is one of building code surrogates, also called code emulators, surrogate models, or metamodels. These are essentially models which, given the inputs to a code, predict the outputs of interest. While model-driven emulators such as polynomial-based regression have long been used to derive a function relating an output to the set of inputs, data-driven models, such as regression trees (see Section 11.3.2) or Gaussian processes [504], are also being used for prediction.

  A good code surrogate can be invaluable in several respects. If the code takes a long time to run, utilizing a large amount of computer resources, a code surrogate, being much simpler, can give an approximation to the output in far less time, using far fewer computer resources. It can also shed light on which input parameters are important or if the input parameter space is such that in some parts of the space, parameters which may have been relatively unimportant elsewhere suddenly become important. This may indicate that we need to use multiple models to fit the response surface accurately. The process of identifying the key input parameters, and building the predictive model, can also provide scientists insights into the phenomena being modeled. In addition, code surrogates can also be used in validation and sensitivity analysis [273, 588, 457].

  The building of code surrogates is a relatively recent use of data mining in computer simulations. Several challenges must be addressed to make these surrogates useful, including feature selection to identify key input parameters; building models which are accurate and interpretable; understanding ways of improving the data so that they span the input parameter space better; and using techniques such as active learning (see Section 11.2.7) to enhance the data in an intelligent way so that the information gained by the addition of any new data points is maximized [74].

  The subject of code surrogates is closely related to the field of design of experiments [437]. However, instead of designing physical experiments, the goal is the design and analysis of computer experiments (DACE) [520, 344, 125, 522]. Many of the methods proposed for building code surrogates tend to be statistical in nature, though the problem can be solved using data mining techniques as well. Another related area worthy of mention in this context is response surface modeling [443] which is used for process and product optimization.

2.5.1 Characteristics of simulation data

The output from computer simulations has characteristics that are somewhat different from the data we have encountered so far. These include:

- **Spatial data at mesh points:** The output from computer simulations is usually in the form of floating-point values of variables at grid points or mesh points in two or three spatial dimensions. Unlike image data, which can be considered as pixel values at regularly spaced points in the x- and y-directions, the grid points in a simulation can be irregularly spaced. Even if the points are regularly spaced in each dimension, the spacing between points may not be the same for all dimensions. This spatial distribution of points, which is discussed in more detail in Section 3.1, can cause problems in extracting objects of interest from the data.

- **Multivariate data:** The output of a simulation typically consists of several variables, such as pressure, density, and velocities, at each mesh point. This diversity of data can be exploited in the analysis; however care should be taken in the interpretation of the results. Such multivariate data are different from the multispectral data in remote sensing, where all variables can be interpreted as the response of the scene to different frequencies.

- **Distributed data:** For large simulations which are run on parallel computers, the output data are usually available in a distributed form. Each output file may have data corresponding to a subset of the points in the whole domain. We may not always have the computational resources to put these subsets back together as a single domain for analysis. In such cases, a parallel implementation of the analysis algorithm may need to be used on multiple processors. Alternatively, the analysis algorithm could be applied to each subset independently, possibly using data from neighboring subsets, and the results for each subset patched back together to generate the result for the whole domain.

- **Spatiotemporal data:** Implicit in many computer simulations is the idea of a phenomenon evolving over time. To understand the phenomenon, the values of the variables at each mesh point in the domain are output at each time step. Temporal analysis thus plays an important role in the understanding of simulation output.

- **Massive size:** As has been indicated earlier, the size of the output from a simulation can be quite large, making it a challenge to output the data and read it back again for analysis. In such cases, reducing the size of the data by preprocessing the output before it is written out may be an option for reducing the data-access time. However, care must be taken to preserve the key aspects of the data during their reduction. This can be difficult in problems where the goal of the simulation is to understand a physical phenomenon and the key aspects of the data are not known beforehand. We must also ensure that the algorithms and the parameter settings used in the preprocessing do not result in the loss of any important information in the reduced output.

- **Simple, real-time analysis tools:** Since large computer simulations often run for weeks on a parallel machine, it is important that the results be checked periodically to ensure that the simulation is running smoothly. If a check indicates that there is something amiss with the simulation, it can be stopped without wasting additional computational resources. Or, if something interesting is observed in the simulation, additional data can be output for analysis later on. Therefore, there is the need for simple analysis tools which can support real-time analysis of a simulation in progress.

2.6 Experimental physics

An area which generates a lot of data is the field of experimental physics, which includes particle colliders and plasma physics experiments. For example, the Stanford Linear Accelerator Center (SLAC) [561] has long focused in the area of high-energy physics and particle astrophysics. Their BaBar detector, built to study B mesons, has resulted in one of the largest databases in the world, containing almost 900 terabytes at the end of 2004. While this may seem massive, the Large Hadron Collider (LHC) at CERN [376], which is an accelerator that brings protons and ions into head-on collisions at higher energies than ever achieved before, will generate 10 petabytes of data per year. Managing and mining such massive data sets can be a challenge, especially as high-energy physics experiments are usually international collaborations with several thousands of scientists participating from countries all around the world. These scientists often need to access various subsets of the data for analysis. To support such access and analysis, experiments often have their own specialized software, such as the ROOT system [515], which is a widely accepted, interactive data analysis framework built using object-oriented techniques. It includes tools for graphing and plotting of different types of data, creating histograms, finding peaks, as well as relevant statistical functions for various distributions and techniques such as PCA. It also supports track finding and fitting, which is one of the key tasks in many high-energy physics experiments [40].

Another source of experimental physics data is the plasma physics devices such as the DIII-D Tokamak [148], as well as several experiments at the Princeton Plasma Physics Laboratory [485], such as the National Spherical Torus Experiment (NSTX) [447] which is designed to test physics principles of spherically shaped plasmas, leading to the development of smaller, more economical, fusion reactors.

The data analysis problems associated with these plasma physics devices can be very varied, ranging from understanding the simulation and experimental data, to using the experimental data to refine the theory and to validate the simulations. The goal of the scientists is to keep the plasma confined at a high enough temperature, for a long enough time, so that fusion can occur, resulting in the energy output from the process becoming greater than the energy input to it. There are several challenges to reaching this goal and many of the data analysis problems arise as a result of these challenges.

For example, physicists have observed fine-scale turbulence near the edge of the confined plasma, leading to the leakage of plasma from the center of the device towards the edge. This can result in significant heat loss from the plasma, loss of confinement of the particles, as well as potential vaporization of the containment wall of the reactor. To understand the process of turbulence at the edge of the plasma, and its role in the transport of heat and particles, scientists working with NSTX are using ultra-high-speed, high-resolution cameras to image the plasma. They are interested in characterizing and tracking the blobs of plasma over time in these experimental image sequences so that the results can be compared with theory and the theory can be validated or refined. One approach to addressing this problem is to use image analysis techniques to segment the images in a sequence, thus identifying the blobs [392], which are then tracked over time. A key challenge is the variation across the images in a sequence, making it difficult to select a segmentation algorithm, and a set of associated parameters, which will work well for all the images in the sequence. Using different algorithms or parameters in the analysis of the frames of a single sequence is not an option as it would not be clear if the results reflect the data, or are an artifact of the changing algorithms or parameters. Another challenge is that the plasma physicists have a poor empirical understanding of the blob structures, which are also not well understood theoretically. This makes the analysis more challenging as we cannot let our current understanding of the theory influence the results as the goal of the analysis is to validate the theory.

Another problem in plasma physics with a very different flavor from the problems usually encountered in scientific applications, is the analysis of Poincaré plots. A plot (see Figure 2.5) is composed of a number of orbits, where each orbit is described by a

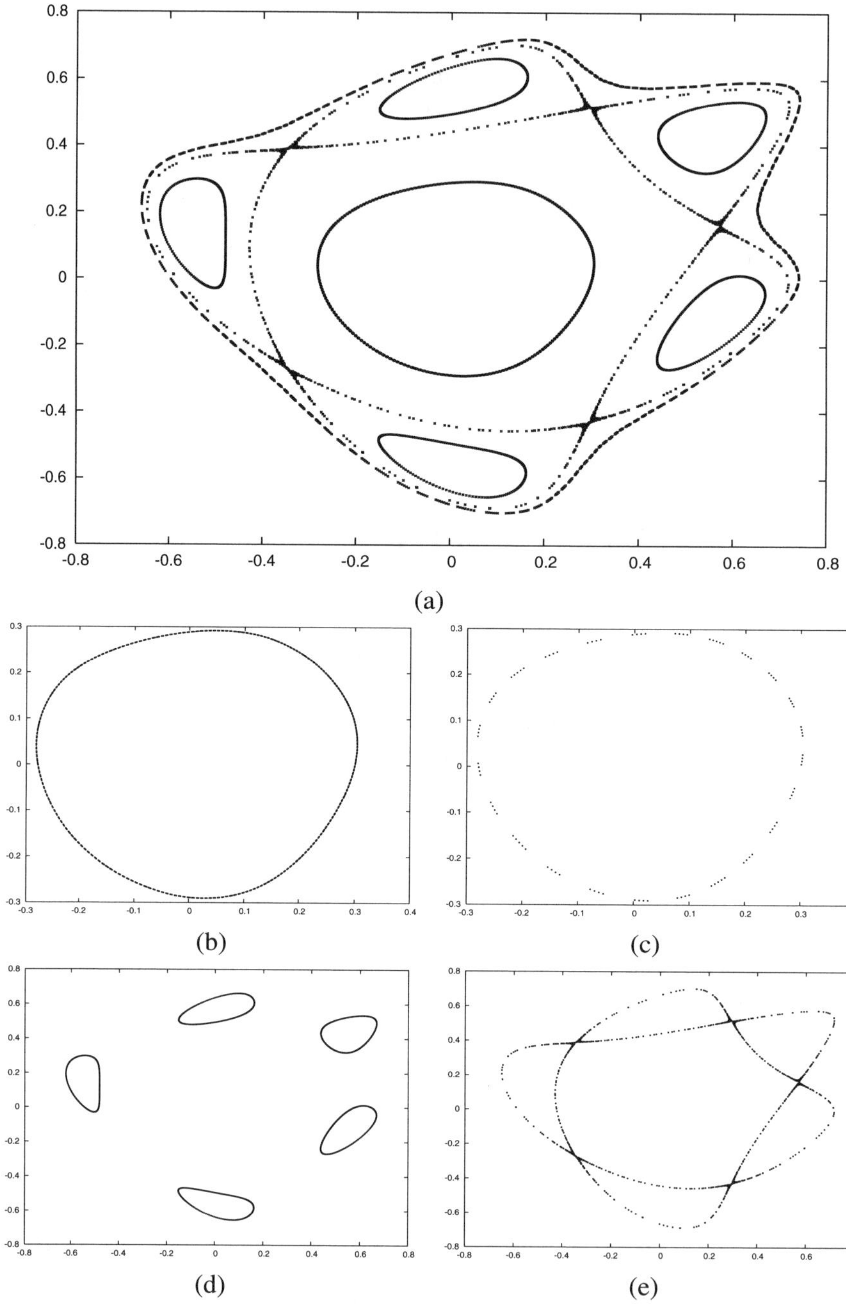

Figure 2.5. (a) *A Poincaré plot with four orbits produced using a Henon map* [638]. (b) *A quasi-periodic orbit with* 1000 *points;* (c) *the same orbit using only the first* 100 *points.* (d) *A* 1000*-point island chain orbit with five islands.* (e) *A* 1000*-point separatrix orbit, with five x-points or cross points between the lobes.*

sequence of points in two-dimensional space. These points represent the intersection of a particle with a poloidal plane, which is a plane perpendicular to the magnetic axis of a tokamak. As the particle goes around the tokamak, it intersects this plane at different points, creating orbits with different shapes, which are given labels such as quasi-periodic, island chain, and separatrix. For example, Figure 2.5(a) shows an example of a plot with four orbits, while panels (b), (d), and (e) of the figure show examples of three different types of orbits, each created using 1000 points.

The actual shape of a given class of orbits may vary widely, though all orbits of a particular class, say quasi-periodic, have characteristics which are unique to it and differentiate it from the other classes. For example, in Figure 2.5(a), the innermost quasi-periodic orbit is almost circular, while the outermost orbit, which is also quasi-periodic, has a wavy structure due to the large-lobed island chain and separatrix which are right next to it. In the case of a stellarator, where the cross section is not necessarily circular like a tokamak, all the orbits could have a shape similar to the cross section, which is more triangular than circular or oval. The island-chain orbits might have islands with a very narrow width, appearing more like a crescent moon rather than the rather wide islands in Figure 2.5(d). This is true of the separatrix orbit as well, where an orbit may have many cross points, closely spaced, with the width of the orbit between cross points being very small. In such cases, a separatrix orbit would closely resemble a quasi-periodic orbit. Further, when an orbit is described by very few points, it may appear to mimic another orbit. For example, a quasi-periodic orbit, with few points (see, for example, Figure 2.5(c), which is the same orbit as Figure 2.5(b), created using 100, instead of 1000, points), may appear as a sequence of broken curves, each curve is indistinguishable from an island with a very narrow width. The problem becomes even more challenging when we consider stochastic versions of the three types of orbits, where the location of the points appears to be perturbed by the addition of random noise. If the noise level is relatively small, it is very difficult to differentiate between orbits of a given class, one with stochasticity and one without.

There are several goals in the analysis of Poincaré plots. First, given an orbit, physicists want to associate a label, such as quasi-periodic or island chain, to it. If the orbit is an island chain, they want to identify the number of islands, and the width of each at its widest point. In the case of a separatrix, they are interested in the number of cross points and the width of the "lobes" of the separatrix (which is the part of the orbit between cross points). A plot can then be characterized by the locations and attributes of the island-chain and separatrix orbits. This enables us to create a database of plots and their characteristics and associate each plot with the experimental setup used to create it. This can be done for plots generated by both computer simulations and experiments. Such a database could then be used in tasks such as the validation of the simulations and the design of the experiments.

Poincaré plots are related to dynamical systems, as discussed by Yip in [638]. He proposes an approach to classifying the different types of orbits based on considering the minimal spanning tree of the graph formed by the points, extracting certain characteristics of the graph, and then using simple heuristics. An extension of this approach in the context of data from simulations is given in [14]. A key challenge in the classification of the orbits is the variation across the orbits of a single class, especially, when it is described using few points, or there is noise in the data, or the orbit displays incipient stochasticity. This problem is different from other problems in scientific data mining as the data are not in the form of images, mesh data, or time series data from sensors, but as points in space which form a pattern.

The sensor data from the plasma physics devices can also be analyzed to shed insight into the phenomena being observed. For example, the data from the sensors in the D-III D tokamak were analyzed to identify the quantities which were relevant to the observation of a certain phenomenon, called edge-harmonic oscillations, in the plasma [83]. This phenomenon is observed whenever the plasma is in the quiescent H-mode, which is the desired mode of plasma operation. By identifying the sensor measurements which are key to the edge-harmonic oscillations, physicists hope to understand what leads to the quiescent H-mode, so they can replicate the conditions.

2.6.1 Characteristics of experimental physics data

The data from experimental physics have characteristics which are very similar to the data we have encountered so far in other scientific applications. These characteristics include:

- **Large size:** As mentioned earlier, experimental physics has resulted in some of the largest data sets so far. In some extreme cases, the raw data is processed and only the reduced version, which is much smaller, is stored. However, not all experimental physics data are massive. Some, like the Poincaré plots, are relatively small as each orbit consists of several thousand pairs of floating-point numbers. The image data from NSTX are also moderate in size, consisting of 64×64 pixel images, with 300 images in the shorter sequences and 7000 in the longer ones.

- **Spatiotemporal data:** The data in experimental physics are usually the result of observing a phenomenon over time and thus have an inherent spatiotemporal aspect to them.

- **Missing and noisy data:** As the data are collected by sensors, similar to other experimental domains, it suffers from noise, randomly missing values, and possibly, outliers. Such data quality issues may be the result of sensors which are not working or are not working correctly.

- **Need for remote access:** Since most experiments are international collaborations, there is often a need to provide remote access to the data. This can be particularly challenging given the large size of the data.

It should be noted that the analysis of data in experimental physics is not just restricted to the data obtained from the experimental setup, but also involves a comparison with simulations, often resulting in a refinement of the simulations, as well as a change of the experimental setup, if so indicated by the simulations. The simulations are used to guide the experiments, which, in turn, are used to validate the simulations. Good quality code surrogates, which are discussed in Section 2.5, as well as models of failure modes built from simulation data, can be invaluable in quickly identifying a cause if an experiment fails to perform as expected.

2.7 Information retrieval

Content-Based Information Retrieval (CBIR) is the task of finding images in a database which are similar to a query image. While many of the early CBIR systems focused on photographs, the techniques have also been applied to remotely sensed images, medical

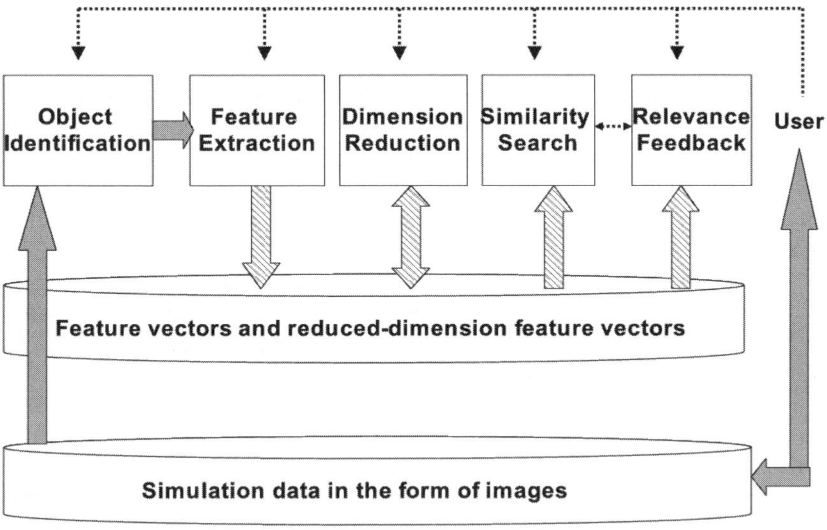

Figure 2.6. *A schematic diagram of the SBOR system used for information retrieval in the context of simulation data. Solid arrows indicate the flow of image data; hatched arrows indicate the flow of feature vectors; and dotted lines indicate the flow of control by the user.*

images, simulation data sets, and geological images. Excellent reviews of different CBIR systems can be found in [176, 477, 201, 475, 636, 92, 601, 150].

The idea behind a CBIR system is simple—given an image, or a region in an image, the goal is to find other similar images or regions in a collection of images. The images or regions are represented by higher-level features or descriptors and similarity is modeled through the use of mathematical distance functions between the feature vectors for different images or regions. Extensive research has been performed to derive compact, representative features and distance functions to model visual cues such as color, texture, and shapes, especially in the context of photographic images.

As an example of a CBIR system, I describe the Similarity-Based Object Retrieval (SBOR) system, which was developed as part of the Sapphire data mining project [314]. While the main application of this system was the exploration of data from computer simulations, it borrowed several ideas from the traditional CBIR community which focuses mainly on image data. The early work in SBOR considered simulation data from regular Cartesian meshes, which are very similar to image data in structure. Multiple floating-point values are available at each mesh point in the simulation domain, describing the various variables used in the simulation.

The SBOR system operated as follows. A scientist would browse through results from a two-dimensional simulation, find a region of interest, and highlight it by drawing a window around the region, thus identifying the query region. Next, the scientist would select the options to be used in the similarity search. When the results from the search were returned and presented to the scientist, they could either refine the search by providing feedback on the relevance of the results, or return to a previous step and select more appropriate options.

The SBOR system (Figure 2.6) consisted of five major functional modules: object identification, feature extraction, dimension reduction, similarity search, and relevance feedback.

The object identification module essentially extracted objects from the images. We supported two types of objects. In the simpler case, we considered the region of interest highlighted by the scientist to be a rectangular tile object. Such tile objects are relatively easy to extract from the images by simply overlaying an image with tiles of the same size as the query tile. These tiles could be overlapping or nonoverlapping and could be either rectangular or circular. In addition, we considered the more complex situation where the scientist wanted the query to be the object in the region of interest, not the entire region highlighted in the query window. In other words, the region of the query window outside the object was not of interest. In this case, image segmentation techniques (see Chapter 8) were first applied to isolate the object from the background.

Once the objects or tiles had been identified in the images, we extracted features representative of the objects. These features could be as simple as statistical quantities such as the mean, variance, and histogram, or more complex features such as texture features based on Gabor or wavelet filters and shape features derived from the Angular Radial Transform (see Chapter 9). After the feature extraction step, each object was represented by a feature vector containing all the features requested by the scientist.

Often the number of features extracted was quite large, and not all features were relevant in discriminating among the objects. A large number of features would cause problems in retrieval due to the "curse of dimensionality" encountered in the similarity search, where the number of features extracted was the dimension of the search space. Dimension reduction techniques, such as principal component Analysis (PCA) or feature selection techniques (see Chapter 10), were used to identify the key discriminating features which were then used in the search.

The similarity search essentially found feature vectors which were close to the feature vector representing the query. The idea was that if the features were selected carefully so that they accurately represented the query object, other objects close to it in feature space would be similar to the query object. Similarity was defined by considering a distance metric, such as the Euclidean distance, between two feature vectors. The SBOR system then returned the best matches, ordered by their distance from the query object.

Of course, given a single query object, it was unlikely that all the results returned would be good matches. The results could then be refined by relevance feedback, where the scientist assigned a score to each result returned indicating its relevance, that is, how well it matched the query. The system then used these scores to perform another iteration of the similarity search in the hope that the results returned were a better match to the query. The second iteration could be done either by changing the weights on the features or by using the scores from the user feedback to build a training set which was used to create a classifier that could then be used to identify the good matches to the query. The idea behind relevance feedback was to iteratively refine the results of the query based on input from the scientist.

The SBOR system included several options targeted toward characteristics of simulation data, including support for querying on multiple variables so all the variables in the simulation data could be used; the ability to extract tiles at different scales using multiresolution analysis which enabled the retrieval of similar objects at different scales; and the option of precomputing features using fixed-size tiles to allow searches through very large data sets.

2.7.1 Characteristics of retrieval problems

Information retrieval problems share several characteristics common to other applications. These include large sizes of the data sets, high dimensionality of the feature vectors extracted from the data, and the need to identify objects in an image. A characteristic perhaps unique to information retrieval is that it is mainly an exploratory tool and is used by the scientist to browse through the data. This makes the problem more interactive, requiring near real-time performance.

2.8 Other applications

There are several other science and engineering application areas where data mining techniques are gaining acceptance. A brief description of some of these newer application areas follows.

2.8.1 Nondestructive testing

Nondestructive testing is used to study the inside of an object without taking it apart or otherwise destroying it. Such techniques are extremely useful when the contents of an object may be dangerous, for example, a storage container with potentially harmful chemicals; or when it is more cost effective not to dismantle the object as in the periodic inspection of bridges to estimate the effect of deterioration; or when it is infeasible to study the object by any other means, for example, land mine detection or the characterization of the defects in a block of concrete to determine if it meets safety standards.

Nondestructive testing and evaluation techniques can be applied at various stages throughout the life cycle of an object. These techniques can be used at the beginning of the manufacturing of a part to find flaws and fabrication defects, during the use of a part to determine if it has deteriorated, and the end of the life of a part to determine if it can be reused or not [372, 12, 108, 64].

Nondestructive testing usually involves studying an object using X-rays, ultrasound, or optical sensors. The resulting signals are then analyzed using techniques from image and signal processing to characterize the objects. In problems where there are several parts to be examined, for example, in quality control during manufacturing, such automated techniques can also be useful in reducing the turnaround time required for the detection of defects. Further, they do not have any of the drawbacks of manual inspection such as subjectivity, inaccuracies due to fatigue on the part of the human inspector, or inconsistencies due to concept drift.

2.8.2 Earth, environmental, and atmospheric sciences

This is an area which has a strong overlap with two other application areas: remote sensing, in the context of observational data, and computer simulations to understand, for example, the spread of a pollutant in the atmosphere or in the ground [294, 291]. A key characteristic of such data is their temporal aspect; the problems of interest in earth and atmospheric sciences typically involve how a pattern or phenomenon varies over time for example, weather prediction, tracking of cyclones, and so on. In addition, the simulations may incorporate observed data to improve their accuracy, in a process known as data assimilation [375].

2.8.3 Chemistry and cheminformatics

An area where data mining techniques are increasingly playing an important role is combinatorial chemistry, which is a technology for generating molecules en masse and testing them for desirable properties. Instead of working with one molecule at a time, combinatorial chemistry provides a better way to discover new drugs and materials by essentially parallelizing the process. Needless to say, this generates a large amount of data which could benefit from data mining techniques [269, 139]. Data mining techniques are also being used to identify quantitative structure activity relationships, where a suitable description of the structure of a molecule is used to associate it with various properties [607]. This association between the structure of a molecule and its biological activity has extensive applications in the field of drug discovery and development. It is at the core of the new field called cheminformatics, which includes the organization, storage, retrieval, mining, and analysis of chemical information [364].

2.8.4 Materials science and materials informatics

Another application area which exploits the connection between the structure and properties is materials science. Traditionally, prior knowledge was used to establish common structure-property relationships across different classes of materials. However, in recent years, a more data-driven approach based on data mining techniques is providing an alternative which allows one to explore, with no prior assumptions, several different types of data to identify what influences a given property of a material [498, 499].

2.8.5 Manufacturing

Manufacturing is another area where data mining techniques have been used extensively, often to evaluate the quality of the products being manufactured [580]. For example, Goodwin et al. [228] describe the use of data mining techniques in semiconductor manufacturing where the data sets are not only large and high dimensional with millions of observations and tens of thousands of variables, but also complex with a mix of categorical and numeric data, nonrandomly missing data, noise and outliers, as well as non-Gaussian and multimodal relationships.

2.8.6 Scientific and information visualization

Finally, two areas which can benefit from data mining are scientific visualization, which is used extensively in the context of simulation data, and the more recent information visualization [598], which is used for viewing complicated data structures, such as graphs, and the relationships among any substructures.

As simulation data reaches the petabyte regime, it is clear that there will be major challenges in visualizing such a large amount of data. We will not only encounter limits to the amount of data which can be displayed at any one time, we will also run into problems just moving the data from disk to the display. Further, as scientific visualization is typically an interactive process, changing the view of a massive data set in real time will become rather difficult. In such instances, data mining techniques can be used to identify the more interesting parts of a simulation so that only these parts, rather than the entire simulation, can be displayed. This will reduce the size of the data which must be displayed and provide real-time response by focusing the visualization.

The area of information visualization [611, 107, 598] is closely tied to data mining as the data being visualized often results from data mining. For example, information visualization is being used to look at the results from clustering massive data sets and to investigate the connections in data represented as graphs, such as the Internet, various social networks, or the electric power grid.

2.9 Summary

In this chapter, I described problems in several different scientific domains which are being solved using data mining techniques. These problems range in size from several kilobytes to several terabytes. They arise from scientific simulations, observations, as well as experiments. Many, if not all, are rather complex, making it a challenge to find techniques which work well. In some problems, the domain scientists may not quite know what they are looking for, while in others, it is a challenge to extract the information from the data in a robust manner so that the results reflect the data and not the algorithms used to extract the information. Even within an application domain, the type of problems may vary widely. At the same time, there is a lot of similarity among problems in different domains.

The list of problems described in this chapter is by no means exhaustive. In fact, I have barely begun to scratch the surface. The use of data mining is increasing in scientific data analysis, and new problem domains, as well as new problems in existing domains, are benefitting from data mining techniques. Through this diversity of problems, I hope to identify some common themes in scientific data mining which I will use to outline the scientific data mining process. These will form the basis of the next two chapters.

2.10 Suggestions for further reading

There are several journals and tutorials which focus on the application areas described in this chapter. These include the IEEE Transactions on GeoScience and Remote Sensing, the IEEE Transactions on Medical Imaging, Research in Nondestructive Evaluation, the Astronomical Journal, Bioinformatics, Applied Bioinformatics, etc. Two excellent tutorials in the area of remote sensing are available from the Canada Center for Remote Sensing and NASA [77, 544].

Conferences are another source of information on how data mining techniques are being used in application areas. For example, the Astronomical Data Analysis Software and Systems (ADASS) conference focuses on the analysis of astronomy data while the SPIE Medical Imaging conference focuses on all aspects of medical imaging from the instruments to the analysis of the data. SPIE also organizes conference series in the areas of defense and security, remote sensing, and astronomical telescopes. Several data mining conferences, such as the ones organized annually by SIAM, ACM, and IEEE, often have workshops and tutorials focusing on the mining techniques used in different application areas such as bioinformatics or time series analysis.

Further, in domains where the data are collected by instruments (such as astronomy and remote sensing), the Web pages describing the instruments are a rich source of information on the instruments themselves, their characteristics, as well as the initial data processing which is done before the raw data from the instruments is made available for analysis.

Chapter 3

Common Themes in Mining Scientific Data

When a scientist doesn't know the answer to a problem, he is ignorant. When he has a hunch as to what the result is, he is uncertain. And when he is pretty darn sure of what the result is going to be, he is in some doubt. We have found it of paramount importance that in order to progress we must recognize the ignorance and leave room for doubt. Scientific knowledge is a body of statements of varying degrees of certainty—some most unsure, some nearly unsure, none absolutely *certain.*

—Richard P. Feynmann [190, p. 146]

In the previous chapter, we considered several scientific domains where techniques from data mining and statistical analysis are being applied to find useful information in data. Even a cursory read through the chapter indicates that there are several themes that recur in these application areas. In this chapter, I discuss these themes in more detail; I will use them in Chapter 4 to define an end-to-end process of scientific data mining.

This chapter is organized as follows. First, in Section 3.1, I describe the various types of data encountered in scientific applications. Next, in Section 3.2, I discuss the characteristics of scientific data which are common across several different application domains. It is these characteristics which motivate the tasks in the various steps of the scientific data mining process, especially the processing done in the first few steps. There are some common themes in the types of analyses done with scientific data as well; these are discussed in Section 3.3. Finally, following a summary of the chapter in Section 3.4, I provide suggestions for further reading in Section 3.5.

3.1 Types of scientific data

The tasks in scientific data mining are driven both by the type of data, as well as the problem being solved. From the applications described in Chapter 2, it is clear that scientific data come in many different flavors. These range from simple data in the form of tables or time series to more complex data such as the data from an unstructured mesh in a computer simulation.

3.1.1 Table data

The simplest scientific data type is a collection of floating-point numbers which is obtained, for example, by different sensors connected to an experiment or during the examination of patients in a clinical trial. These variables may represent quantities such as the temperature and pressure at some location in the experiment, or the blood sugar and weight of a patient. If the quantities are measured over time, we can consider each quantity to be represented by a time series. The quantities do not all have to be floating-point values; some can be categorical, such as the gender of a patient, while others could be binary, for example, a variable indicating if a certain option was used in the experiment or not.

These data can often be represented in the form of a table, where the rows correspond to the objects or data items, such as a single run of a physics experiment or a patient in a clinical trial, and the columns represent the variables collected for each object. In many problems, the position of the rows and columns does not have any special significance and they can be interchanged without changing the meaning of the data. However, in time series data, the rows often represent increasing time and are therefore constrained to be in a certain order. This is not the case with image data, which I describe next, where the data too can be represented in the form of a table, but the rows and columns have to be in a certain order.

3.1.2 Image data

More complex data types occur when the data are in the form of images, for example, a satellite image in remote sensing or an experimental image from a fluid-mix problem. An image can be considered to be a two- or three-dimensional array of values defined at regular intervals. These values can be integer or floating point. Further, the integer values are usually nonnegative and represented using 8-bit, 11-bit, 12-bit, or 16-bit data. For example, Figure 3.1, shows two images, one an 8-bit integer image and the other a floating-point image, along with the pixel intensities corresponding to the upper right corner of the image. The two images appear very similar, though their values are rather different. This is because tools for displaying images first convert them into 8-bit values, which are usually scaled to lie between 0 and 255. This can be very misleading if the image values are not originally in the range $[0, 255]$. As a result, tools which not only display images, but also allow a user to look at the actual intensity values can be invaluable in the context of floating-point images (see Section 12.2). In addition, we must take care to ensure that any processing, such as cropping or subsetting of an image, which is performed directly on the displayed version, retains the range of the original data.

Scientific images which have 16-bit integer or floating-point values allow a larger range of pixel intensities. However, many of the typical image processing algorithms focus on commercial quality images in 8-bit format, where the values range from 0 to 255. While it is possible to rescale images with a larger range of values to the range $[0, 255]$ and apply the traditional image processing software, a preferred option is to apply floating-point versions of the image processing algorithms and convert the values to $[0, 255]$ only at the end, or if necessary for display. This ensures that the larger range of values is fully exploited during the analysis. It unfortunately also implies that much of the image processing software that is available for 8-bit data cannot be applied to scientific images directly.

In many scientific domains with image data, in addition to the pixel values, additional information is associated with each image, such as the time of day when the image was

(a) (b)

250	251	252	253	254
249	250	251	252	253
248	249	250	251	252
247	248	249	250	251
246	247	248	249	250

(c)

350.0	351.0	352.0	353.0	354.0
349.0	350.0	351.0	352.0	353.0
348.0	349.0	350.0	351.0	352.0
347.0	348.0	349.0	350.0	351.0
346.0	347.0	348.0	349.0	350.0

(d)

Figure 3.1. *Sample images in* (a) 8*-bit integer format and* (b) *floating-point format.* (c) *and* (d) *are the values of the pixels at the top right corners of the* 8*-bit integer and floating-point images, respectively. Note that the images appear very similar as they are converted to* 8 *bits, and scaled to a range from* 0 *to* 255*, for display. However, the floating-point image has pixel values greater than* 255.

obtained, the instrument used, the settings of the instrument, and so on. These metadata, that is, data describing the data, are useful in interpreting the results obtained from the analysis of the image.

Image data are often available as two-dimensional data. However, in some fields, such as medical imaging, three-dimensional data are also becoming available, for example, a computer-aided tomography (CAT) scan of the human brain composed of multiple slices, where each slice is a two-dimensional image.

3.1.3 Mesh data

Data from computer simulations are more complex than image data. In computer simulations, the domain of interest is overlaid with a grid (also called a mesh) and appropriate partial differential equations solved at these discrete points on the grid. In the simplest case, if the grid is a Cartesian mesh, it can be considered as an image, where the floating-point values of the variables being output in the simulation are available at each grid point.

However, it is rarely the case that simulations are performed on rectangular domains (in the case of two dimensions), with equally spaced grid points, with equal spacing in the two dimensions. More frequently, the grid points are irregularly spaced, giving rise

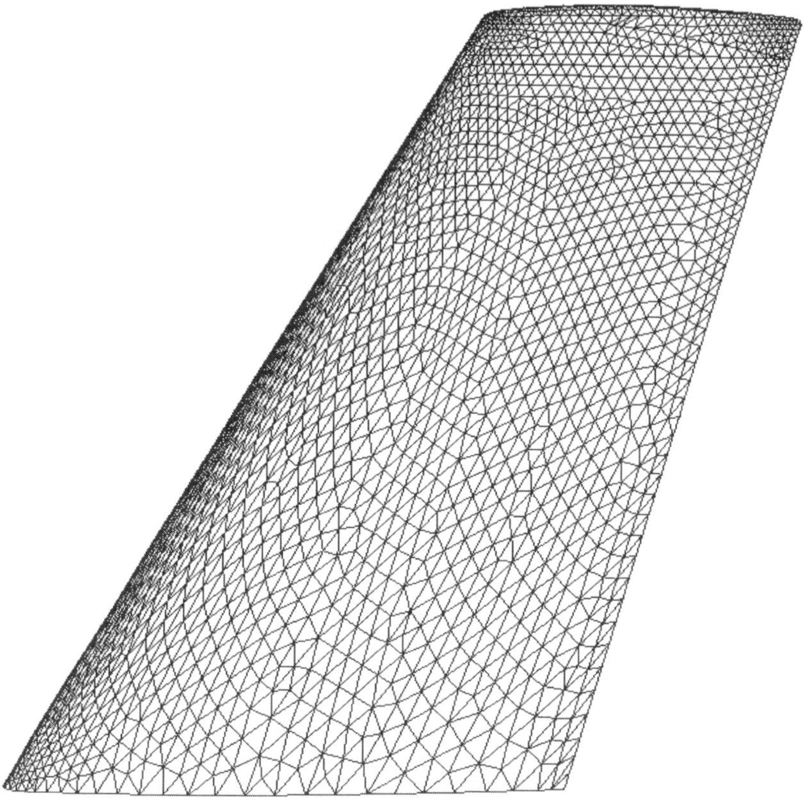

Figure 3.2. *Example of a three-dimensional unstructured grid using tetrahedral meshes around the upper surface of a wing. Image from* [207]*; also available at http://tetruss.larc.nasa.gov/usm3d_v52/IMG/Wf2-4a/winggrid.gif.*

to unstructured meshes. The elements used in such meshes can be homogeneous; for example, all elements in the three-dimensional mesh shown in Figure 3.2 are tetrahedral; or heterogeneous, where there is a mix of elements of different types, such as triangular and quadrilateral elements in a two-dimensional mesh. And then, there is a hybrid mesh structure, which is locally structured, but globally unstructured (see Figure 3.3). Such meshes typically arise in adaptive mesh refinement schemes, where instead of running the simulation on a mesh that has a uniformly fine resolution over the whole domain, it is run on a mesh which has fine resolution only in the regions which require such resolution, and a coarse resolution mesh elsewhere. In such adaptive techniques, if a region of the mesh requires additional resolution, it is refined further, with the increased number of points leading to more computation, but only in limited regions of the computational domain. When this finer resolution is no longer needed, the mesh is coarsened. This approach lowers the total computational costs of running the simulation compared to an approach which uses a fine mesh throughout the computational domain. However, the code is now more complex due to the refinement and the coarsening of the mesh.

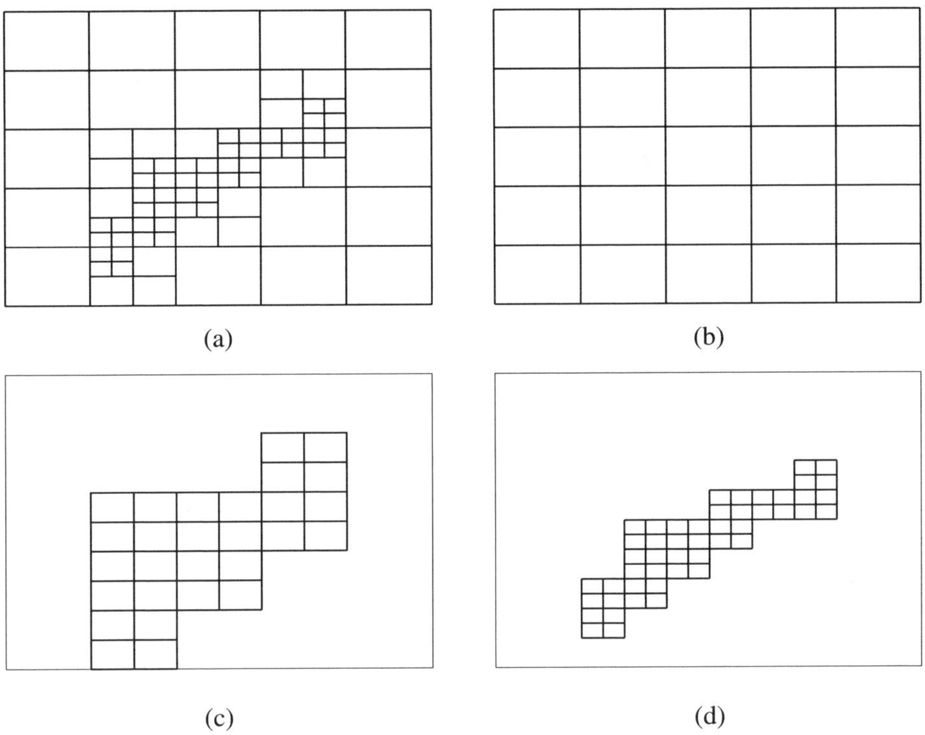

Figure 3.3. *A grid illustrating the concept of adaptive mesh refinement.* (a) *The composite grid which is globally unstructured, but locally structured.* (b) *through* (d)*: the individual grids which make up the composite, from the coarsest to the finest scale. These are shown relative to the boundary of the domain.*

Mesh data are available in both two and three dimensions. It has a time component as the goal of the computer simulations is to study a phenomenon as it evolves over time. At each time step in the simulation, several variables may be output. These are often the variables of interest, such as pressure, density, the velocity components in the different dimensions, and so on. But, additional information, in the form of checkpointing files, can also be output at specific time steps in the simulation, enabling the simulation to be continued from that time step onward. This checkpointing is very useful in long-running simulations on parallel machines, where the machine may not be available to run the complete simulation at one time.

A detailed description of how complex scientific phenomena are simulated using computational techniques is beyond the scope of this book. Readers interested in the general area of computational science and scientific computing are referred to the texts by Heath [265], Heinbockel [268], and A. B. Shiflet and G. W. Shiflet [542], while those interested in details of techniques such as finite-difference and finite-element methods can find more information in the texts by LeVeque [373] and Hughes [283], as well as specialized texts on the subject. While many problems in the analysis of simulation data do not require an in-depth understanding of how the data were generated, in some cases, for example, if we need to calculate the gradients of variables at grid points, we may need to use the same computational scheme as was used in the simulation code to ensure that the analysis is done to the same accuracy as the simulation.

3.2 Characteristics of scientific data

I next describe various characteristics of scientific data and discuss how they influence the analysis process. These are the common themes we have seen in many of the applications discussed in Chapter 2 when we focus on the data being analyzed rather than the analysis task. The issues raised in this section often determine the tasks which should be included in the analysis, the choice of algorithms for a task, and any considerations which may influence the interpretation of the results of the analysis.

3.2.1 Multispectral, multisensor, multimodal data

In some domains, such as remote sensing, a sensor can observe a scene using different frequencies. Thus, instead of a single image, a multispectral image is obtained. This consists of a separate image for each of the different frequencies, for example, one image for near-infrared, and one each for red, green, and blue in the visible spectrum. In hyperspectral imagery, the number of bands can be in the hundreds.

There are several formats in which such multiband or hyperspectral images are stored. The three common ones are band interleaved by pixel (BIP), band interleaved by line (BIL), and band sequential (BSQ) [527]. In both the BIL and BIP formats, the data are stored one line at a time. BIP stores all the bands for the first pixel in a line, followed by all the bands for the second pixel in a line, and so on. BIL stores band 1 for the first pixel, followed by band 1 for the second pixel, and so on. In BSQ, each band is stored separately; within a band, the pixels of the first line are stored in order, followed by the pixels of the second line, and so on. The different storage formats can result in computational inefficiencies if an algorithm requires access to data values which are not stored contiguously in memory. Further, in remotely sensed imagery, when there are more than three bands, we need to select a subset of three to map to the R, G, B colors in the display. The image may appear different based on which of the three bands are chosen for display.

In multispectral imagery, all the spectral bands are usually at the same resolution. However, sometimes, the images can be at different resolutions. Further, in some applications, such as medical imaging or astronomy, the different sensors used in imaging can focus on different organs or different characteristics of the scene, with the resulting images appearing very different. Often, a scientist may want to combine or fuse the information from different images of the same scene, to exploit the complementary information in the images.

A similar problem occurs when data in the form of text, images, plots, and tables have to be mined together, as in the case of mining biological literature [351]. The problem becomes more complex if we also consider the data in documents pointed to by the current document.

3.2.2 Spatiotemporal data

In many scientific domains, the data have both a spatial and a temporal nature. Mesh data from simulations, by their very nature, have a temporal component as the goal of the simulation is to understand a phenomenon as it evolves over time. In other application areas, such as medical imaging and remote sensing, there is a need to detect changes over time, for example, analyzing images of the brain before and after treatment for a brain tumor. The time intervals between the collection of data may be variable. Such problems require the data to be registered; that is, there should be a mapping of the points from the data at one time instant to the data at the next time instant. This allows us to incorporate changes in the data due to factors such as changes in the position of a patient from one MRI scan to the next, or changes in the viewing angle of a remotely sensed image. This collection of data at different time instances can also introduce problems such as differences in the lengths of shadows in remotely sensed images taken at different times of the day.

In other problems, spatiotemporal data may be in the form of time series data, where the data are collected at regular time intervals. In such situations, the traditional ideas developed in time series analysis can be applied [102, 472, 337].

3.2.3 Compressed data

In some domains, the data sizes are so large that they are stored in a compressed form. This is often true of simulation data. In such cases, we can uncompress the data and then apply the traditional analysis algorithms. However, one can also consider algorithms which can be directly applied to compressed data, as discussed by Machiraju et al. [404].

3.2.4 Streaming data

Many scientific applications involve the ongoing collection of data. These include problems in astronomy, experimental physics, and surveillance. Often, the data being collected must be analyzed on the fly to identify either interesting events, which can then be studied further, or untoward incidents (such as damage to an experimental setup), which must be prevented. This requires the mining of data as they are being generated or collected. Given the real-time requirements, often only simple analysis can be done. Such analyses must be amenable to incremental algorithms, where models describing the data are built incrementally and refined as additional data become available. In many applications, the real-time analysis is usually combined with more detailed off-line analyses; such off-line analyses can also be used to generate the models for real-time use.

3.2.5 Massive data

As we have observed, in many scientific applications, the data collected or generated can be quite large, often being measured in terabytes or more. This characteristic is not unique to science and engineering data sets, but is also true for data arising from commercial applications. In scientific applications, the large size of the data could be due to either a large number of small- to moderate-sized data, or a small number of very large data sets, or a large number of very large data sets.

The large size of scientific data sets often implies that much of the data are never looked at, much to the consternation of scientists who wonder about the science being left unexplored in the data. Exploratory data analysis techniques are often needed to identify regions of interest in data, thus reducing their size prior to detailed analysis. Sampling can also be used to process reduced amounts of data, at least in the initial iterations of analysis. Alternatively, we can move some of the processing closer to the generation or collection of the data, creating reduced representations of the full data set as it is being generated or collected. In any approach to reducing the size of the data, we must be careful to recognize that the data not being included could have a substantial effect on the results of the analysis, especially when the data reduction techniques are motivated by what we expect to see in the data. In addition, the reduced representations may be sensitive to the algorithms and parameter settings used to create them, making it unclear if the reduced representations accurately reflect the original data.

3.2.6 Distributed data

Data in scientific applications can be distributed either globally, as in data collected from different telescopes, or locally, as in data generated by a computer simulation on a parallel machine. Both situations can be challenging for data mining algorithms.

One approach to handling globally distributed data is to work with reduced data sets from each source. For example, astronomers create catalogs for each astronomical survey and mine the data across these catalogs. Efforts such as the National Virtual Observatory [597, 173] provide the infrastructure for such analyses, as well as the ability to access the original data as needed. A slightly different situation arises when the data are in one place and many scientists want access to it. Again, one approach would be for each scientist to work with a reduced summary version of the data. A different approach would be to move the analysis software to the data, which is the approach taken at NASA's Goddard Earth Sciences Distributed Active Archive Center [402].

In computer simulations run on large parallel machines, the data are available in several files, each containing a subset of results over part of the computational domain. Putting the entire data set back together may not always be an option, especially if the computational domain was very large, necessitating its decomposition in the first place. This can lead to some interesting challenges. Parallel versions of many algorithms are needed, especially for the initial prepreprocessing of the data. The identification of objects can be a problem when the objects are split across one or more data files. This issue gets aggravated when the problem is one of tracking objects, especially when the objects split and merge over time.

When the data from a computer simulation are analyzed as they are being generated, we can use parallel implementations of various analysis algorithms. This, however, can lead to load imbalance as the size of the data reduces over time, going from values at each mesh point, to objects, to feature vectors. If the data are analyzed after they have been written to files, we have the option of applying the analyses algorithms in parallel, processing a file at a time, along with any data needed from neighboring files. In this case, additional logic will be needed to handle objects split across files. However, it may preclude the use of certain algorithms which require iterating through the data, unless the data are first read back into a parallel machine.

3.2.7 Different data formats

In a single application domain, it is very rare for a single format to be used for storing data available as, say, images. In fact, it almost appears that there are as many different output formats as there are scientists! For example, in climate simulations, data can be output in network Common Data Format (netCDF) format [449], in GriB (Gridded Binary) format [231], or in HDF format [263]. Similarly, in astronomy, there is the Flexible Image Transport System (FITS) format [194], which can store floating-point image data, unlike the more traditional image formats such as *tif* or *pgm*.

In several of the older image formats, there is typically a header associated with each output file (or image) which describes the image. This can include information on how many bits are used to represent the values in the image, if the values are scaled, the corresponding scale factors, and so on. The header can also include information on the instrument used to collect the image, the time the image was obtained, and the settings on the instrument. Much of this metadata had to appear in a fixed order in the file; any enhancements to the data formats to account for advances in computer science, such as newer languages and data structures, had to ensure that changes made to the formats were backwards compatible. This allowed data collected by an instrument over several years to be read by a single piece of software. The newer data formats, however, tend to be "self-describing" and allow greater flexibility in how variables are stored in a file.

The use of these different formats in a single application domain does have practical implications for a data miner. Often, a nonnegligible percentage of time is spent in just dealing with specifics of the data formats used in an analysis problem and in ensuring that the data have been read correctly. If standard data formats are used, such as *tif* or *pgm* for images, conversion tools, such as the "convert" command in ImageMagick software [289], can be used to convert the data into a single format for analysis.

3.2.8 Different output schemes

Another time-consuming aspect of scientific data mining arises from the different ways in which scientists, while using the same data formats, output the data. This problem occurs more frequently in simulation data. For example, in climate simulations, one scientist may output results for each month in a separate file, while another may output the results for an entire year in a single file. If a data miner is trying to perform the same analysis on these two different simulation outputs, they will need two different code segments just to read in the data. Of course, this does not take into account a common problem in simulation output, where all variables for a simulation are dumped out at each time step, while analysis is often done using one, or a few, variable(s) at all time steps. If the data are stored on tape in long-term storage, this implies that a lot of data may need to be accessed, though only part of the data are actually useful in the analysis.

A similar problem occurs when all variables at one time step are output to a single file or they are output to different files with either a single directory for all time steps or separate directories for each time step. If the simulation is run for many time steps, it is possible to run into the problem that common commands for listing the files in a directory no longer work due to the large number of files in a directory.

As a result of all these issues, often a nontrivial amount of the time spent in the initial exploratory analysis is in just reading in the data and bringing them to a consistent format.

3.2.9 Noisy, missing, and uncertain data

Scientific data, especially data resulting from observations and experiments, tend to be noisy. The noise may be random or structured. For example, data from the FIRST astronomical survey have noise in the form of lines at 60 degrees corresponding to the placement of the telescopes of the VLA [174] (see Figure 2.2), while data from Planar Laser-Induced Fluorescence (PLIF) imagery [321] have noise in the form of vertical lines through the image. Such noise in image data must be reduced prior to further processing. This must be done carefully, without destroying the signal. The amount of noise in an image can also be a strong indicator of how much a scientist can trust the analysis done on the image.

Missing data are another common occurrence in observations and experiments. This is frequently due to sensor malfunction. In science data, missing values cannot simply be interpolated from neighboring values, especially if large areas of an image are missing or the neighbors are too far away for interpolation to make sense.

Another related issue is one of outliers. While they are often removed in the initial data exploration step, they could also indicate previously unobserved phenomena, a bug in the simulation code, or an incorrect experimental setup. Hence, outliers in scientific data must be handled with care and addressed appropriately.

In some problems, especially when we measure or observe a quantity using two different sensors, or measure what we expect to be the same quantity at two different times, we may obtain two different values. This results in an uncertainty associated with the data, where we might have a range of values for a variable, rather than a single value. Analyzing data in the presence of uncertainty is very challenging and is increasingly becoming an important topic in scientific data mining.

3.2.10 Low-level data, higher-level objects

A key characteristic of scientific data is that it is rarely in a form that can be directly input to a pattern recognition algorithm. While it is possible to take image pixel values or variables at a mesh point and input them to a classification algorithm [53], often, the pattern of interest is not at the level of a pixel or grid point. For example, we may be interested in classifying galaxies with a bent-double morphology in astronomical image data [315], or in identifying "mushroom"-shaped "objects" in simulation data [314]. These galaxies or objects must first be extracted from the pixels in an image, or grid points in a computational domain, before they can be identified as bent-doubles or mushrooms. Or, we may be interested in comparing a simulation to an experiment, where the two data sets are of different sizes and quality; this would preclude a direct pixel-to-pixel comparison. In such cases, we need to identify objects in the data and compare them based on their location and characteristics.

This extraction of higher-level information, in the form of objects, from lower-level data, which is in the form of pixels or variables at grid points, is a critical step in scientific data mining. It is also a challenging and time-consuming step for the reasons I outline next.

In problems where the data represent natural objects, such as galaxies or the coherent structures observed in edge turbulence in plasma, there is a tremendous variety in terms of the shape, the sizes, and the intensities of the objects. Thus, the techniques used to extract the objects cannot depend on the specific characteristics of the objects to extract them from the data. In contrast, for man-made objects such as semiconductor masks or buildings, it is possible to use characteristics such as straight lines to extract the objects from the images.

The objects in an image can often be difficult to extract due to the noise in the images. This frequently occurs in data from experiments and observations. However, what is noise is often dependent on the problem being solved. For example, if one is interested in identifying buildings in a remotely sensed image, the clouds in the image would be part of the noise. However, if one were interested in classifying the types of clouds in an image, the clouds would be the signal.

In many image data sets, the objects of interest could be partially occluded by other objects as a result of the image being a two-dimensional projection of a three-dimensional scene. For example, in a remotely sensed image, the side of a rectangular building may be partially occluded by a tree growing close to the building, resulting in an image where only three sides of the building can be easily extracted. Or, in medical imagery, a bone may occlude the tissue of interest. A partial solution to such problems is to use three-dimensional imagery or obtain images of the scene from different angles.

Objects of interest in scientific data can be at different scales. This could be due to the inherent resolution at which the data are collected, for example, the same building may appear at different scales in images at different resolutions. Or, it could be the result of the variation in the scales of the objects; for example, a remote sensing image may have both large and small buildings in the same image. The difference in scale may also be the result of the way in which the image was obtained as some objects may appear much smaller than others if they are further away from the camera. It is also possible for the structure of objects to vary over time. For example, coherent structures, known as bubbles and spikes, which form during Rayleigh–Taylor instability in the process of fluid mixing, are a few grid points in size at the beginning of the mixing process, but grow to occupy several hundred grid points as the simulation progresses [318]. This variation in scale may make it difficult to use a single algorithm, with a fixed set of parameters, to extract the objects in the data.

In the case of spatiotemporal data, the objects of interest may not persist over the entire time the data are being collected. Often, objects are born, exist for a while, split or merge with other objects, and then die [318]. In this process, they may change shape substantially, but still be recognizable as the objects of interest by scientists. Identifying such objects can be a challenging problem.

3.2.11 Representation of objects in the data

Once the objects of interest have been extracted from the data, they must be represented by features characterizing them. Again, we need to represent the objects by higher-level features instead of actual pixel values or values at mesh points to account for the fact that the patterns of interest are at a higher level. Further, by representing the objects using certain common characteristics, we can focus on what makes the objects similar or dissimilar, instead of being distracted by minor variations in the objects. These characteristics are usually determined by the problem to be solved. For example, if we are interested in categorizing objects in a data set by their shape, then shape descriptors are relevant. However, if in the same data set, we are interested in clustering objects by their texture, then, texture descriptors, not shape descriptors, would be more relevant.

The word "feature" is somewhat of an overused word and has come to mean different things in different communities. In the context of this book, I consider a feature to be any measurement which can be extracted for an object from the data; that is, one can write a piece of software to extract the feature from the data. Such features could include the mean value of the intensities, the Fourier descriptor for the shape of an object, the texture values of the object, and so on. In some domains, the word "feature" can mean the higher-level object itself, for example, in simulation data, a scientist may be interested in finding mushroom-shaped "features." In the terminology used in this book, this would imply first identifying all objects in the data, then extracting the low-level characteristics, such as the shape and texture, and finally using these features to identify (using perhaps a classification algorithm) which of the objects meet the requirements of being shaped like mushrooms.

The features extracted from the objects in scientific data must usually be invariant to scale, rotation, and translation. After all, a bent-double galaxy remains a bent-double even if it is shrunk or enlarged, rotated, or moved to another location in the image. Further, the features must be robust to small changes in the data. This is to account for the fact that the data may be noisy. If small changes in the data lead to large changes in the values of the features, then such features are unlikely to be reliable.

For spatiotemporal data, we may also need to include a time component as one of the features. While such data are usually collected at regular time intervals, this may not always be the case. For example, in computer simulations, data may be output more frequently when rapid changes are occurring in the data. Such differences in time intervals must be handled appropriately.

3.2.12 High-dimensional data

Once the objects have been identified, and features representing them extracted from the data, the original data have essentially been reduced to a matrix, where each row corresponds to an object or data item and the columns represent the features for the object. Thus, each object is represented by a feature vector.

Typically, many features are extracted for each object, such as statistical features represented by the histogram of intensity values, shape features represented by Fourier descriptors or coefficients of angular radial transform [418], or texture features represented by summary statistics of the wavelet transform. Each of these features can have many components; for example, a 16-bin histogram will have 16 components. Further, if the data are multispectral or several variables are being considered, then each spectral band or variable will contribute a 16-bin histogram to the feature vector. As a result, feature vectors in scientific data can be rather long, with hundreds of features. This does not include features which are extracted as metadata from the original data. Such features could include the name of the image from which the object was extracted or the software directory where the file containing the image is located. They are necessary to help trace back the object to its original source and are useful in analyzing and interpreting the results of data mining.

A key problem with having high-dimensional feature vectors, where the number of features is the dimensionality of the problem, is the "curse of dimensionality." This manifests itself in different ways, ranging from poorly defined distance measures to the need for many samples to effectively represent the distribution of the data in the high-dimensional space [86]. This can affect any subsequent processing of the feature vectors to extract patterns in the data. As a result, feature selection and dimension reduction techniques are often used to reduce the number of features necessary for characterizing an object.

3.2.13 Size and quality of labeled data

In classification problems, there is a need for labeled data, or objects which have a label associated with them. For example, given a training set of galaxies which have been identified as either bent-double or non-bent-double, we can build a classifier, which, given an unlabeled galaxy, can associate a label with it. In scientific problems, such labeled training sets are usually generated manually. This is in contrast to many commercial applications, where training sets can be generated historically, for example, by keeping track of which customers buy a certain product.

The manual generation of training sets can be tedious, subjective, and inconsistent. If an object is difficult to label, it is very likely that when a scientist is shown the same object at two different times, it will be assigned two different labels. This can be due to an inherent difficulty in labeling the hard-to-label cases, but can also arise due to concept drift in cases where the scientist is labeling many examples. The labeling approach can sometimes be ad hoc, with each scientist using a different metric for assigning a label to an object. In these cases, it is helpful to ask if the scientists would like an object to be labeled as a positive example, instead of asking them if it is a positive example. The latter option requires the scientist to be certain, while the former option leaves open room for doubt and lets the scientist determine if further studies are necessary to correctly identify the object.

Another issue with labeling data is that of ground truth. In commercial data, it is easy to determine if a customer did or did not buy a product. In scientific data, it is much harder to ascertain ground truth. In some cases, for example in labeling volcanoes on Venus [71], it is nearly impossible to determine if an object in an image is indeed a volcano. In such cases, expert labeling takes the place of ground truth. Such expert labeling may also be used in cases where it is difficult, not just impossible, to collect ground truth. For example, in remote sensing, an expert may label an area of an image as being an oil spill. Ground truth labeling would require going to the area, collecting a sample of the material, and verifying that it is indeed oil. Needless to say, when we use expert labeling, we need to remember that experts do not always agree and may not always be right!

In some scientific domains, we also need to realize that the data we see may not represent all aspects of an object. For example, in astronomy images, we are seeing a two-dimensional projection of three-dimensional data. Thus, a real bent-double galaxy, when viewed from the wrong angle, may appear to be non–bent-double. In such cases, there is no way we can deduce from a visual analysis of a single image whether the galaxy is really a bent-double galaxy. In other situations, such as video surveillance, it is possible to use multiple cameras to resolve such issues.

Further, in scientific applications, as in some commercial applications, training sets can be unbalanced, with a larger proportion of one type of objects than the other. Thus, a training set may have a few positive examples which are very similar to each other, and many negative examples, each of which is different from the positive examples in its own way.

These issues with the quality and quantity of the training data can significantly influence the results of analysis in classification problems.

3.3 Characteristics of scientific data analysis

In addition to the common characteristics of the data, there are also common themes in the types of analysis which are done using the data. These are determined more by what the scientists want to do with the data, rather than the data themselves; they too determine the tasks which should be included in the analysis process.

In the simplest problems, the scientists may be interested in extracting statistics on the data, such as the mean and standard deviation, to determine if the data are roughly what they expect them to be. Such analysis is often done as a prelude to more complex analysis to ensure that the data meet the basic requirements for their quality. It also allows obvious problems, such as mislabeled variables, missing values, and variables with scientifically meaningless values, to be addressed.

In some problems, the detection of outliers is done to check the data quality, while in others, it is the main goal of the analysis. For example, scientists may be interested in analyzing time series data from different sensors to identify anomalies. Depending on how they interpret the anomaly, they can use the analysis results to prevent an untoward incident or devote more resources to observe a rare event.

In other problems, the task may be to summarize the data. For example, we may want to count the number of buildings in a satellite image, or the number of deformed cells in a blood sample. This may be nontrivial as we first need to define what constitutes a building or a deformed cell. We would also need to separate all cells from noncellular matter in the blood sample. If the data have a time-varying component, we may need to generate these summaries over time, observe how they change, and perhaps summarize the change. The summaries may also involve the generation of distributions for various quantities of interest, such as the sizes of buildings, or the heights and weights of children participating in a health survey.

Sometimes, the scientists conduct an experimental or an observational study with the intent of finding certain types of objects, such as bent-double galaxies, or certain behavior, such as the presence of edge harmonic oscillations in tokamak plasma. In these problems, we have examples of the objects or behavior of interest, and the task is to analyze all the data to determine all occurrences of these objects or behavior. In other words, given a set of objects of a certain type, we want to predict if a new object is of the same type as well.

A similar predictive task arises when a scientist wants to conduct an experiment, but given the high cost of running the experiment, first performs many computer simulations to understand what might happen in the experiment. This situation gives rise to many analysis tasks. We may want to validate the simulations against the experiments, predict what would be the outcome if the experiment was run under certain conditions, or based on the experimental observations, refine the theory behind the simulations.

A more challenging problem arises when the analysis of a data set is done for reasons other than its collection, or the scientist is just interested in finding what else is there in the data without having a specific hypothesis in mind. So, they may want to know if there is a natural grouping of the objects in the data, or if there are any outliers, which may indicate something unusual.

One of the most difficult types of analysis involves understanding causality, where the scientist is interested in determining if one fact, such as adding a particular element to steel, results in another fact, such as improving the strength of the steel. The challenge arises as many factors may contribute to the strength of steel, and careful analysis would be required to ascertain that it is indeed the addition of a certain element at a certain stage of processing that results in stronger steel.

There are two characteristics of analysis which are encountered more frequently in the context of scientific data:

- **Uncertainty quantification:** In scientific problems, technical decisions are often made based on the results of data mining. The consequences of an inaccurate decision can range from a loss of productivity to the life threatening. For example, an astronomer may decide to allocate telescope time to studying a galaxy identified as a possible bent-double galaxy. If the classification label is incorrect, it results in a poor utilization of scarce telescope resources. On the other hand, if classification determines a tumor to be benign when, in fact, it is not, the consequences can be far more serious. Or, if a model built using streaming data from a physics instrument determines that the incoming data are not predictive of an untoward incident, when in fact, they are, then it may result in damage to the instrument.

Consequently, in many scientific domains, it is important to be able to evaluate how much one can trust the results of data mining. This quantification of margins and uncertainty can be rather difficult, especially as all steps in the processing of the data must be taken into account. Though it is an important step, it is rarely done, even in cases where the data quality is good. When the original data is of poor quality, with associated uncertainties in the data themselves, the quantification of margins and uncertainty becomes even more important and challenging.

- **Ambiguity in problem definition:** Problems in scientific data mining are not always well defined. This is especially true when the goal of the analysis is scientific discovery or when the phenomenon being analyzed is poorly understood. Sometimes, the scientist may not know what will be found in the data. In other cases, the scientist may have a theory or hypothesis but does not want it to influence what is found in the data. This often leads to the analysis problems being poorly understood, and therefore, poorly defined, with the scientist unable to provide answers with certainty when specific questions arise in the process of the analysis. This characteristic is not typically observed in commercial data, where there is no ambiguity about whether a person is a customer or not and no doubt about whether a customer bought a certain item or not.

  This ambiguity in problem definition, and the lack of certainty in validating the results of the application of an analysis algorithm can make scientific data mining very challenging. It essentially implies that uncertainty and ambiguity are likely to be a key component of the analysis process. Consequently, we must ensure that the analysis is the result of a careful and considered application of various algorithms and choice of parameters, and that the results not only are scientifically meaningful, but also reflect the data, not the algorithm or the parameters. This often results in the analysis being very iterative and interactive. Many different algorithms may need to be applied and the results compared to each other to ensure consistency. Or, we may need to start with an initial solution approach for a poorly defined problem, and refine it as required during the analysis, perhaps even redefining one of the initial steps in the analysis.

3.4 Summary

In this chapter, I have described some of the common issues and themes which occur in mining data from different scientific applications. Starting with a description of the data types encountered in these applications, I have highlighted the common themes which occur in scientific data analysis. I will use them to define the data mining process in the next chapter, and the tasks which comprise this process will be discussed in further detail in the rest of the book.

3.5 Suggestions for further reading

An excellent source of additional information on the issues discussed in this chapter are the domain scientists themselves. They frequently have the best understanding of the issues which must be considered in analyzing their data. They, along with the information on the Web pages which describe the instruments used, can be an invaluable resource in understanding the data and their characteristics. Further information on possible solutions to the issues discussed are given in subsequent chapters of this book.

Chapter 4

The Scientific Data Mining Process

> *"When* I *use a word," Humpty Dumpty said, in rather a scornful tone, "it means just what I choose it to mean—neither more nor less."*
>
> —Lewis Carroll [87, p. 214]

In Chapter 2, I described various ways in which data mining was being applied to science and engineering problems. Next, in Chapter 3, I identified common themes across these diverse applications. In this chapter, I use these themes to describe an end-to-end data mining process, including various tasks at each step of the process and the order in which these tasks are usually performed to extract useful information from scientific data sets. These tasks will be discussed in more detail in later chapters. I will also comment on some overall characteristics of the data mining process and explain why I choose to define the process in a manner which differs from the more commonly used definition of data mining.

This chapter is organized as follows. Section 4.1 describes the end-to-end process of scientific data mining, motivated by the applications in Chapter 2 and the common themes identified in Chapter 3. Next, in Section 4.2, I make some general observations on the process, followed in Section 4.3 by my rationale for defining the data mining process to be broader than the common definition of the process. Section 4.4 concludes the chapter with a brief summary.

4.1 The tasks in the scientific data mining process

The description of scientific data types in Section 3.1 and the observations about the low-level nature of the raw scientific data discussed in Section 3.2.10 indicate that the raw data cannot be input directly to pattern recognition algorithms. This suggests that the data must first be preprocessed to bring it to a form suitable for pattern recognition. The common themes discussed in Chapter 3 suggest the tasks that might comprise this preprocessing of the data.

Figure 4.1 outlines one way in which these tasks might be incorporated into an end-to-end data mining system for analyzing data from a science or engineering application.

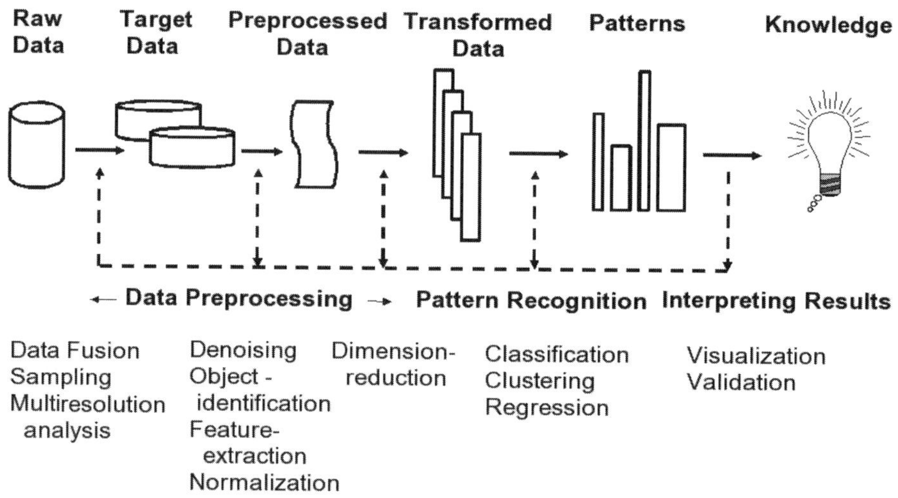

Figure 4.1. *The end-to-end scientific data mining process.*

Starting with the raw data in the form of images or meshes, we successively process these data into more refined forms, enabling further processing of the data and the extraction of relevant information. The terms, such as *Raw data* and *Target data*, used in Figure 4.1 are somewhat generic and also used in the mining of commercial data, where they are used to describe the process of *Knowledge Discovery in Databases* (KDD) [185]. In adapting this process to scientific data sets, I have retained the basic framework, but changed the tasks necessary to transform the data from one form to the next.

First, a few brief observations about the scientific data mining process; I will expand on these observations later, but it is helpful to keep them in mind while reading this section. The data mining process is essentially iterative and interactive—any one step can lead to the refinement of previous steps and the domain scientists should be actively involved in each step to validate the results. Not all of the tasks are needed in every problem and the order in which the steps are performed may change from one problem to another. The tasks listed are the ones I have found to be useful in many scientific applications; I am sure there are tasks I have missed as they may be specific to a particular problem. Each task can be addressed using several algorithms; these algorithms vary in their computational complexity, number of input parameters, suitability to a problem, and robustness to input parameters.

I next describe each of the steps in the scientific data mining process in more detail, followed by some general observations on the end-to-end process. I also discuss the ways in which the approach outlined in this chapter differs from mining of commercial data sets and the more traditional view of data mining as one step of the KDD process.

4.1.1 Transforming raw data into target data

The original or "raw" data which are provided for data mining often need extensive processing before they can be input to a pattern recognition algorithm. These algorithms typically require, as input, the objects (or data items) in the data set, with each object described by a set of features. Thus, we first need to identify the objects in the data and extract features representing each object. In scientific data sets, the data may need to be processed before we can even identify the objects in the data. These processing steps may include tasks such as:

- **Data size reduction:** One task that is very helpful in the initial processing is to reduce the size of the data set. This is especially useful when the data set is very large and we are trying to understand the problem, doing exploratory analysis, or trying out different techniques for the various steps to determine a good solution approach. As we have observed in Chapter 2, scientific applications can result in very large data sets, and it is often helpful to first understand the data by working with a smaller sample and doing some exploratory analysis.

  One approach to reducing the size of the data is through sampling. During the initial analysis, we can select a small sample randomly from the full data set. Later on, as we begin formulating a solution approach, we need to increase the size of the data to reflect the variability which may be present naturally in the data. This would enable us to determine if the solution approach can handle this variability. Finally, most scientific problems require that a full pass be made through the entire data set. If the algorithms used in each task in the solution approach are robust enough, the final step can be fully automated.

  An alternative to sampling is to use multiresolution techniques, where we work with coarse resolution data. This may often be desirable in problems where the data are stored in a compressed form using an approach such as wavelets, which directly gives us a lower-resolution version of the data. Thus, the initial processing during the exploratory data analysis phase can be done on the lower-resolution data, with the full-resolution data being used in later iterations of the analysis.

  Of course, if the data set is massive, one can combine both the sampling and the multiresolution approaches. These techniques for reducing the size of the data are discussed further in Chapter 5.

- **Data fusion:** If the scientific data have been obtained from different sensors, they must be fused before the complementary information in the different data can be exploited. In the case of image data, the step of data fusion may include registration, which is the process of aligning two images of a scene or object, taken at different times or at different resolutions. For example, one frame of a video of a scene may need to be registered with the frame taken the very next instant. This is because even a stationary camera may move a little, resulting in successive frames of the video being misaligned by a pixel or two. Or, satellite images, obtained to understand how the vegetation in a region has changed over the years, must be registered so scientists can map corresponding areas for change detection.

  A different type of data fusion is required when the data are of different modalities, for example, text and images, or video and audio. In this case, fusion may occur at the feature level, where the different modalities of data each contribute features describing an object. For example, a technical document may be represented by text features representing the words in the document, image features representing the images in the document, and other features representing the tables in the document. We can also have fusion at the decision level, where each modality is processed independently, and the final decision made by appropriately combining the decisions from the different modalities. In both these cases, the features or decision may be appropriately weighted to account for the quality of data in each modality.

These issues of data fusion, including data registration, are discussed in more detail in Chapter 6.

- **Image enhancement:** As we have observed, the presence of noise can make it difficult to analyze the data. This is especially true in the case of images if the objects of interest are occluded or not clearly separated from the background due to low contrast; in such cases, it may be difficult to extract the objects from the image.

  Though there are general techniques to reduce noise in images or improve their contrast, many of the techniques used are domain specific. This is because the noise in the data may be due to domain effects, such as the sensor characteristics, or due to external circumstances, such as atmospheric turbulence in astronomical data sets. These noise characteristics may differ across images in a data set, making it difficult to use fully automated techniques to remove the noise. Also, any processing to reduce noise from an image can adversely affect the signal, and appropriate trade-offs may be necessary to reduce the noise or improve the contrast, while minimizing the effect on the signal.

  Techniques for enhancing images are discussed further in Chapter 7.

4.1.2 Transforming target data into preprocessed data

Once the data have been reduced through sampling and/or multiresolution techniques, complementary data sources have been fused, and the data enhanced to make it easier to extract the objects of interest, potential next steps in the scientific data mining process include:

- **Object identification:** Here, the objects of interest are identified in the data, for example, by isolating the pixels which form a galaxy from the background. This step is often time consuming for images and meshes as many different techniques may need to be tried. In addition, depending on the quality of the data, the parameters used in the algorithms may need to be changed, making completely automated analysis difficult.

  In some problems, the identification of objects may be relatively easy, for example, in the case of a physics experiment where we are interested in classifying certain types of events, an "object" may be the time window in which each event occurs. In other problems, the objects may be poorly defined to the extent that defining the objects of interest is one of the main goals of the data mining endeavor. For example, if the physical phenomenon is poorly understood, the scientists may be unsure of what they are looking for in the data. In other problems, the object of interest may evolve over time, splitting and merging with other objects, making it difficult to have a single definition of what constitutes an "object." Or, the data structure used to store the data may preclude easy identification of the object, for example, in an unstructured mesh. In addition, if the data have been generated using a parallel computer, they may be distributed over several files, making it difficult to identify an object split across many files.

Techniques for object identification, especially for images and meshes, are discussed further in Chapter 8.

- **Feature extraction:** As observed earlier, the word "feature" is used to imply different things in different domains. In this book, I use "feature" to mean any low-level attribute or measurement which can be extracted from the data. Features are typically used to represent the objects in the data and are extracted after the objects have been identified. These features are then used in the pattern recognition step.

  The features extracted for an object are usually scale-, rotation-, and translation-invariant. This is because the patterns in the data usually remain unchanged when they are scaled, rotated, or translated. For a feature to be useful, it must be robust, that is, it must be insensitive to small changes in the data. Often, we may need to try different ways of extracting the same feature as some techniques may yield features which are more consistent across the objects of interest.

  In addition to features representing the objects themselves, we frequently need to include "housekeeping" features for each object. These represent the metadata and include information such as the image in which the object was found, the location of the object in the image, the resolution of the image, and so on.

  The features which are extracted for an object are very dependent on the patterns of interest. For example, if the pattern reflects the shape of an object, then various shape-based features should be extracted. Often, it helps to extract features reflecting different properties of an object such as shape, texture, and statistical distributions of intensities, as it may not be obvious at first which type of feature is likely to be most discriminating for the pattern recognition task at hand.

  Different types of features are discussed further in Chapter 9.

- **Normalization and cleanup of features:** Once we have identified the objects in the data, and extracted features representing the objects, as well as features representing the metadata for the objects, then we have a data item (or object) by feature matrix representing the data set. Each data item in this matrix is described by its feature vector. It is often helpful to process these features to ensure they are consistent and accurately represent the data. For example, the features may need to be normalized as the units corresponding to some features may make them appear more important than other features. This normalization may need to take into account the number of features corresponding to each feature type, for example, the number of Gabor texture features may be much larger than the number of features representing the histogram of the intensity of the pixels in the object.

  In addition to normalization, we should check that the features representing an object are valid and complete. For example, features may be missing because a sensor was inoperational. Or, feature values may be outliers and may not have a physical meaning due to inaccuracies in the collection or the input of the data values. Such features must be appropriately processed else they may adversely affect the step of pattern recognition. Techniques for doing this are discussed further in Chapter 9.

4.1.3 Converting preprocessed data into transformed data

Once the initial steps of preprocessing have been applied to the raw data, the objects in the data identified, and features representing them extracted, we have a matrix where the rows represent the data items or objects and the columns represent the features. We could also have considered the rows to be the features and the columns to be the objects, but in this book, we will focus on the data item by feature matrix. In the initial stages of the analysis, we may not consider all the objects, but only those found in the subset of data being analyzed. In addition, we may often extract far more features than necessary to represent each object, as we may not know which features are the most discriminating. These features may number in the tens or even the hundreds or thousands.

Reducing the number of features representing each object is an important step prior to pattern recognition. There are several reasons why this can be helpful:

- Some of the pattern recognition tasks (such as finding nearest neighbors or classification) can be considered to be a search through a high-dimensional space, where the dimension is the number of features. For example, in classification, we need to find the hyperplanes which separate, say the positive examples from the negative ones. If each example, or data item, is represented by many features, trying to determine the appropriate hyperplanes can be difficult. The *curse of dimensionality*, as observed by Bellman [31], indicates that in the absence of simplifying assumptions, the sample size needed to estimate a function of several variables to a given degree of accuracy grows exponentially with the number of variables. In scientific data, where training sets are usually generated manually, large sample sizes may not be an option in classification problems.

- Some of the features extracted to represent an object may be irrelevant (such as the location in the file from which the object was extracted), while others may not be discriminating enough (such as the color feature in a sample of all red objects). While the former may be necessary for identifying where an object came from, we could save on computational time by not extracting features unless they are discriminating enough. However, this assumes that the initial sample of data is representative enough so that a feature, once discarded, does not turn out to be discriminating when the entire data set is considered.

- Reducing the number of features may improve the ability to visualize or interpret the patterns in the data. If the number of features is reduced to a small number, say 3–10, it may be possible to use techniques such as three-dimensional visualization or parallel plots (see Chapter 12) to understand how the data items are grouped together based on the features.

- By reducing the number of features, we can reduce both the computational time and the memory requirements of the pattern recognition algorithms which are applied subsequent to the dimension reduction.

- In some scientific domains, the features extracted from the images or meshes may be stored in databases. Efficient indexing schemes for retrieval of these data are feasible when the number of dimensions is 8–12 (see [209, 177]). Dimension reduction would therefore provide more efficient access to the feature data for analysis.

There are several ways in which the number of features representing an object can be reduced. These include both techniques which transform the data into a lower-dimensional space as well as techniques which select a subset of the features. I will discuss these techniques in more detail in Chapter 10.

4.1.4 Converting transformed data into patterns

Once the data items are represented by a possibly reduced set of features, we can use this transformed data to identify patterns. Depending on the problem, this identification of patterns can take several forms, including:

- **Classification and regression:** If we have a class label associated with each data item, that is, we have a training set, we can use it to build a model of the data which separates one class label from another. There are several classification algorithms, each of which builds a different type of model, and each with its pros and cons. If the label is continuous instead of discrete, the problem can be addressed using regression techniques.

- **Clustering:** If there are no class labels associated with the objects, then we can use clustering techniques to group the objects based on their similarity in feature space. The objects in each cluster are then analyzed to determine what characteristics brought them together.

- **Information retrieval:** A related task is to retrieve objects which are similar to a query object. This can be done by identifying objects which are close to the query in feature space. If the features are an accurate representation of the objects, then objects which are close to the query in feature space are likely to be similar to it.

- **Outlier or anomaly detection:** In some problems, we may be interested in identifying the data items which are anomalous; that is, they do not belong with the rest of the items. Such problems typically arise in streaming data, where near-real-time analysis is required.

- **Association rules:** These techniques are popular in commercial data in the context of market basket analysis where they are used to determine the types of items which are often bought together. Association rules are just making inroads into scientific data mining in domains such as bioinformatics.

- **Tracking:** In spatiotemporal data, the features associated with an object, such as its size and location, can be used to track the object from one time step to the next.

Scientific problems often have characteristics which can make the direct application of various pattern recognition techniques a challenge. I will discuss these in more detail in Chapter 11.

4.1.5 Converting patterns into knowledge

Once the patterns in the data have been identified, they must be visualized and presented to the scientist for validation. This may result in the iterative refinement of one or more of the steps in the data mining process.

This visual display of the results has to be done with care to maintain any relationships between the objects. This can be difficult when the objects are in a high-dimensional space and they have to be projected into two or three dimensions for display. Another problem arises when the size of the data set is so large that a typical display is insufficient.

This visual display of information is not restricted to just the final step in the data mining process. In fact, it is often useful to apply the same techniques for information visualization during the initial exploratory analysis of the data and in evaluating the results from the intermediate steps. Such validation can often yield important insights into the data and suggest appropriate solution approaches for the next steps. I will elaborate on the visualization and validation of results in Chapter 12.

4.2 General observations about the scientific data mining process

In Section 4.1, I briefly described the various tasks which comprise the overall process of scientific data mining. Next, I make some general observations about the process of mining science and engineering data.

- The data flow diagram presented in Figure 4.1 is one that I have found to cover the needs of many of the scientific applications I have encountered. It is by no means the only approach to the analysis of scientific data sets; variations and enhancements may be necessary as required by an application. For example, the order in which the tasks are done may change from one application to another. Depending on the quality of the data, it may be necessary to enhance image data from different sensors before fusing them as the enhancement may make the fusion algorithms less sensitive to the setting of the parameters. Also, not all tasks may be necessary in all applications; for example, denoising data is not a key part of the analysis of data from computer simulations, but is often required for experimental data. Tasks in addition to the ones described in Section 4.1 may be required as appropriate for specific problems, for example, to incorporate domain-specific information.

- The data size shrinks as we move to the right from the raw data to the patterns identified in the data. The raw data may be in the form of images, while the pattern may be a simple model in the form of a decision tree, which can be used to label the objects in the images. If the original data are being processed using a parallel computer, then the shrinking of the data size would imply that it would be inefficient to use the same number of processors throughout the implementation of the entire process. In addition, there could also be issues of load balancing. For example, if we are analyzing image data, each processor could work on a subset of images. The denoising and enhancement of the images, as well as the identification of the objects in the images, could be done in parallel. However, if some images had far more objects than others, the processors working on those images would have more work to do in the feature extraction step than the processors working on images with few objects. Further, if more than one processor is processing a single image, some processors may have no objects in their subimage, while other processors may have many. An object could also be split between the subimages assigned to different processors, requiring appropriate merging of the processed data.

- The scientific data mining process is very interactive. The domain scientists are involved in every step, starting with the initial, perhaps tentative, formulation of the problem to providing information on the noise characteristics of the sensors for use in enhancing the data, validating the objects identified in the data, identifying robust ways of calculating the features for the objects, selecting features which may or may not be relevant to the problem, and most importantly, validating the results obtained at each step.

  The scientific data mining process is also an iterative process. The results of any one step may indicate that a previous step needs to be refined. For example, identification of the objects in image data may indicate that the noise in the images has to be reduced further in order to separate an object from the background. Or, the step of pattern recognition may indicate that some features which are key to discrimination are not rotation invariant and therefore objects which are rotated versions of objects in the training set are not being labeled correctly. Or, the error rate of the pattern recognition step could be high, indicating that the features extracted are not representative enough of the patterns being considered or that the quality of the training data could be improved.

  If the data set is very large, a practical approach to analyzing it might require working initially with a small subset of the data, processing these data using the initial steps of Figure 4.1, and when a set of algorithms is found to work, processing additional data to see if the quality of the results still hold and if it is possible to proceed further along the analysis pipeline. If the initial sample is not representative enough, the new data added may indicate a different choice of algorithms or parameter setting for the algorithms.

- The tasks in Figure 4.1 can be implemented using several algorithms. For example, there are several ways of reducing noise in image data and several different classification algorithms to create a model given the training data. These algorithms differ in their suitability to a problem, the assumptions they make about the data, their computational complexity, the accuracy of the results, their robustness to input parameters, and their interpretability. Often, several algorithms may need to be tried before one suitable for the data and problem is found. In some cases, we may even need to design algorithms which are tuned to the characteristics of the data or problem.

- Each task in the process described in Figure 4.1 generates some output which can be directly input to the next step. However, it is often advantageous to save the results which are output at each step. It allows us to experiment with different algorithms for a step without having to repeat all the previous steps. Also, if new data become available, they can easily be incorporated into the process by repeating only the steps which are absolutely necessary, without processing all the data twice. Saving intermediate data essentially makes it easier to handle the iterative nature of the scientific data mining process.

4.3 Defining scientific data mining: The rationale

As I discussed in Chapter 1, there are several definitions of data mining, with some considering it to be one of the steps in the Knowledge Discovery in Databases (KDD) process in which patterns are extracted from data, and others considering it to be the overall process of extracting useful information from data. When this data arises from science and engineering applications, it is clear that we cannot analyze the data by directly applying algorithms which extract patterns from this data.

As we have seen in Chapter 2, the data in many scientific applications are in the form of images or mesh data from simulations, where the values of the variables are available at each pixel or mesh point. However, we are interested in patterns among the objects in the data, objects which are at a higher level, for example a galaxy in an astronomical image, where the galaxy is a collection of pixels. Going from the data at the level of pixels, or mesh points, to data which are suitable for pattern recognition, is a very time-consuming process, involving trial and error. As I have described in this chapter, there are several tasks which must be performed before the data are ready for pattern recognition. In many problems, these tasks typically take 80–90% of the total time spent in analyzing the data. Often, the quality of the final results is very dependent on the quality of the initial preprocessing. Further, these tasks provide the means of obtaining input from the domain experts, input which is key not only to understanding the problem, but also solving it. For these reasons, I consider data mining to be the overall process of extracting information from data, not just one step in the KDD process. Of course, one might also say that for many scientific applications, the data are not stored in databases, and the term KDD might not be applicable in the strictest sense of the term.

4.4 Summary

In this chapter, I have described the tasks involved in a typical end-to-end process for the analysis of scientific data sets, where one starts with the raw data in the form of images or meshes and extracts useful information. The process may involve all or some of the following tasks: sampling, multiresolution, data fusion, object identification, feature extraction, dimension reduction, pattern recognition, visualization, and validation. I shall describe how we can accomplish these tasks in the next several chapters. I start by describing ways in which we can reduce the size of the data to make the initial exploratory analysis feasible for massive data sets.

Chapter 5

Reducing the Size of the Data

I have no data yet. It is a capital mistake to theorise before one has data. Insensibly one begins to twist facts to suit theories, instead of theories to suit facts.

—Sherlock Holmes [160, p. 3]

In many of the early steps of the scientific data mining process, we may be faced with the prospect of a very large data set, where it may not be possible to view the entire data set all at once. This occurs frequently in the case of three-dimensional simulation data, where many variables are output for each grid point, or in analysis of images, say from remote sensing, where the high-resolution, hyperspectral data sets can be very large. This can be overwhelming as it is difficult to decide how best to proceed with the analysis. One solution to this challenge is to work with data that are a subset of the full data yet reflect almost all of the characteristics of the full data. Such a subset can then be used to determine an appropriate solution approach or to choose the parameters for an algorithm. Alternatively, one can consider a "coarse" version of the data, where the fine-scale structure has been lost, but the data size has been reduced to make it easier to manipulate. I next discuss both these approaches to reducing the size of the data.

This chapter is organized as follows. First, in Section 5.1, I describe ways in which we can sample a data set to identify a smaller subset. There are different ways in which this can be done, depending on our goals. Next, in Section 5.2, I describe multiresolution techniques which are rather effective in images for viewing the data at a coarser resolution. Finally, Section 5.3 provides a summary of the chapter and Section 5.4 provides suggestions for further reading.

5.1 Sampling

One approach to reducing the size of the data is to work with a smaller sample of them. How one samples the data is often dependent on how the sample will be used. In the following, we assume that the data have already been generated; the issue of sampling to generate the data is beyond the scope of this book. However, at the end of this section, I briefly mention some techniques which are useful for increasing the coverage of the data generated and thus can be used in analysis for increasing the size of the sample being analyzed.

There are several stages in the scientific data mining process where we may need to sample the data. These include:

- **Sampling for exploratory analysis:** When we first encounter a data analysis problem, one of the initial tasks is to look at the data to try to understand both the data and the problem which needs to be addressed. This exploratory analysis can be done using the techniques for visualization described in Chapter 12. However, such exploratory analysis is not restricted to just the beginning of the scientific data mining process. It can be used after any step in the process. Suppose we have completed the step of image denoising; we then need to look at the output of this step to determine which class of algorithms would be the most suitable for the next step of identifying objects in the denoised images. Exploratory analysis could indicate if gradient-based methods are appropriate or if region-growing techniques would be a better choice. It could also help confirm that the results from the denoising step are what we expected. At this stage, we are not quite ready to start applying various algorithms and selecting parameters; rather, the intent is to become more familiar with the data and identify possible solution approaches.

  In problems where the data set is small, say a handful of moderate-sized images, or the domain scientist has already isolated an initial representative subset, then there is no need to do any sampling in support of the exploratory analysis. However, if the data set is very large, then it makes sense to select a smaller sample for this analysis.

  If nothing much is known about the data set, as may sometimes be the case, one approach is to randomly select a smaller subset. For example, if the data set is a large number of small- to moderate-sized images, we can start by first selecting some images randomly and then gradually increasing the size of the sample until we have some idea of what is in the data, and we can proceed to the next step of proposing a solution approach. If, however, the data are a collection of rather large images, we may start by selecting one at random, splitting it into smaller subimages which can be viewed easily, then selecting another image at random, splitting it into subimages, and so on.

  If the data have both a spatial and a temporal component to them, then the sampling should be done in both space and time. A straightforward approach would be to use systematic sampling where we select samples which are equispaced in time. However, in some domains, where scientists have some idea of the rate at which a phenomenon evolves, we may use samples which are equispaced in log time or use other alternatives suggested by the scientist.

- **Sampling to identify a solution approach:** Once we have had a chance to look at a sample of the data, the next task is to select a sample to use in trying out different solution approaches to the problem. Our goal is to identify algorithms and parameter settings which will work well across the full data set. This means that our sample must reflect the diversity of the data. For example, if some, but not all, of the images in our data set have low contrast, then, ideally, such low-contrast images should be in our sample so that we know that we need to include contrast enhancement as one of the steps in the analysis. Or, if our task is to classify objects into one of five classes, then, we must have a few examples from each of the classes so that we can try to identify features which can be used to discriminate among the classes.

If the sample of the data used earlier in the exploratory analysis phase is diverse enough, it may be the starting point for the data subset used to test the solution approaches. It is, of course, difficult to tell when our sample reflects the true diversity of the data, especially, when we first start analyzing the data.

One approach is to keep increasing the size of the sample, and evaluating our algorithms on the sample, until we have an indication that some of the items in the current sample have characteristics which we had not accounted for earlier. For example, suppose the first sample from a set of images is such that all have good contrast, so we proceed to apply a gradient-based edge detector to identify the objects in the images. However, when we increase the sample size, we find that the results of the edge detector in some of the images is less than satisfactory. Upon investigation, we find that these images have low contrast. To address this, we can include a contrast enhancement step prior to the edge detection. If the results are still not satisfactory, we may need to not only include a contrast enhancement step, but also select a different algorithm for identifying the objects as well. It is issues like this which make the process of scientific data mining very iterative and interactive, with the steps in the process being refined as ever larger subsets of the data are analyzed.

Iterative techniques for selecting a representative sample size can also be used in some of the later steps. For example, Fayyad and Smyth [187] discuss how random sampling to generate models in clustering can cause problems in the detection of new classes which occur with low probability in the data. They suggest an iterative sampling scheme which progressively refines the sample selected by considering the outliers.

Sampling to select a representative set is never trivial. However, in scientific data mining, especially when the data are poorly understood, or the data were collected for the purpose of scientific discovery, one must be exceptionally careful in using sampling approaches, lest the item of interest be ignored altogether. Often, the outliers contribute more to the analysis; they enable a better choice of algorithms and parameters and also enhance the scientific discovery process by indicating the presence of something unusual in the data.

- **Sampling in classification:** Another area where sampling plays a role is in the creation of a training set for use in classification algorithms (see Section 11.2). If the training set is generated manually, it is likely to be quite small. In such cases, the issue of sampling may not arise, though the considerations used in sampling are helpful in increasing the size of the training set.

  Suppose we are fortunate to have a large set of labeled examples and, for the sake of efficiency, we need to select a sample for use in classification. One approach is to select the sample randomly. However, to ensure that we have enough examples from each of the classes, especially the minority class, we may prefer stratified sampling. Here, the items are divided into groups according to their class label and we select an equal number of items randomly from each of the classes. This ensures that all the classes are adequately represented. However, it may be the case that the full data set does not have enough items of the minority class, or there is not enough variation in these items to provide a good representation of the minority class in the training data.

Such problems have to be addressed in other ways, for example, by working with the domain scientists to identify additional examples of the minority class.

Another issue in creating the training set is the size of the sample. This can be a difficult question to address as the answer is usually dependent on the data set, the accuracy required from the classifier, how widely separated the items are from the different classes, how many features are used to represent each item, and so on. A rough rule of thumb is to have 5–10 times more items per class than features [613]. Since scientific data are often of a high dimension, and the training sets generated manually, it may be difficult to meet this requirement.

The evaluation of classification algorithms is usually done by calculating the cross-validation error using the training, evaluation, and test data sets. Section 11.2.7 describes the issues which must be considered when we sample the data set to form the training set, the evaluation set, and the test set.

Sampling appears to be a relatively simple task, but if not done correctly, it can have a profound effect on the results of subsequent processing. Hence, the results of analysis done using a sample of the data must be interpreted with care; one must always be aware that the results may reflect the sample or the sampling technique rather than the actual population from which the sample was obtained.

Sampling techniques also play a role in the generation of the data themselves and form an integral part of the field referred to as the design of experiments or experimental design. These experiments can be either physical or computational [437, 180]. Though the generation of data is outside the scope of this book, the approaches used to ensure a good coverage of the data during its generation are also relevant in selecting a good representative sample. For example, if we are collecting soil samples from a certain region, it does not make sense to collect many samples from a small subregion, where the samples are likely to be very similar and not necessarily representative of the whole region. Common sense would dictate that we collect samples from several subregions; the more diverse, the better. The process of deciding the next sample to be collected, based on the samples collected thus far, is called adaptive sampling [584]. The idea is to use the characteristics of the samples collected so far, and our knowledge of the population, to collect the next set of samples so that the combined set of samples is more representative of the population as a whole. As a side benefit, this may also give us a better idea of the variables in the population other than the variable of interest. The idea of adaptive sampling is also related to active learning in classification (see Section 11.2.7) and to adaptive mesh refinement in computational science, where a mesh is refined more in regions where the function being studied is changing rapidly. Both these techniques try to address the question of what is the next point we should consider so that the addition of this point maximizes what we learn about the function being modeled.

5.2 Multiresolution techniques

A different approach to reducing the size of the data is through multiresolution techniques. The scale, or resolution, of an image is a concept which is exploited in several steps of the scientific data mining process. It is easier to explain the concept in terms of an image, though the idea also occurs in one-dimensional signals as well as two- and three-dimensional simulation data. When we view a scene, we tend to perceive it in several different ways.

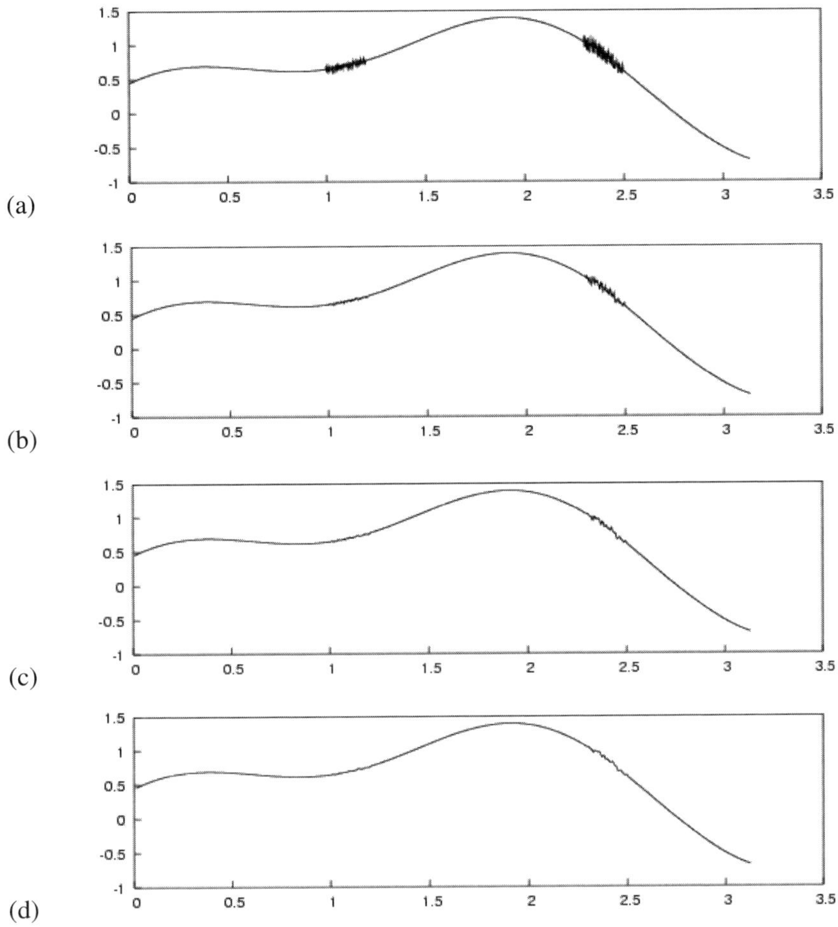

Figure 5.1. *Notion of scale in a one-dimensional signal:* (a) *Original signal with fine-scale structure in two regions.* (b) *through* (d) *The original curve smoothed using Gaussians of* $\sigma = 1.0$, 2.0, *and* 3.0. *The larger the value of* σ, *the larger is the structure which is smoothed away.*

First, we may see a forest, with several trees, and as we get closer, we may recognize different types of trees, and finally, see the characteristic shapes of the leaves in each tree, and maybe some insects on the leaves. If our task is to classify the trees in the forest based on, say, the shape of the leaves and other characteristics, then a certain level of fine-scale detail is necessary. The very fine details, such as the cellular structure of the leaves, are irrelevant, and the insects on the leaves can be considered as noise. If, on the other hand, we are interested in the approximate size of the forest, then the fine-scale details of the leaves, or even the type of trees, are unimportant. This level of detail in the data is referred to as the scale or the resolution of the data.

Consider the one-dimensional signal shown in Figure 5.1(a). It is a smooth curve, with two regions where there is some fine-scale structure, with the region on the left having a finer-scale structure than the region on the right. When this curve is smoothed with a

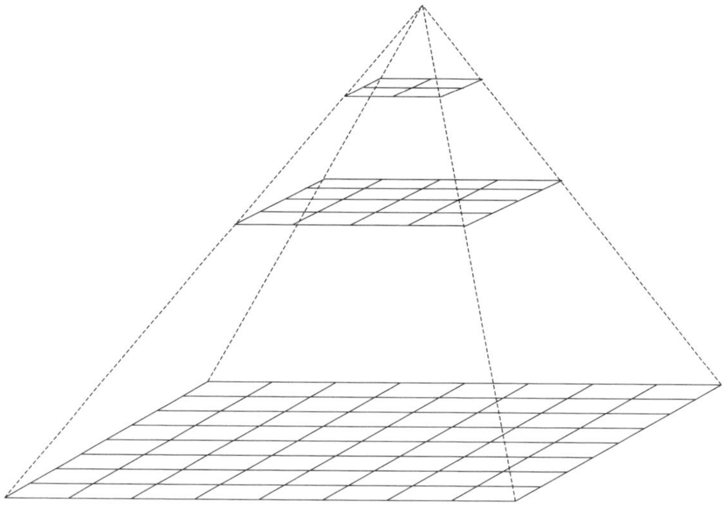

Figure 5.2. *A pyramid structure, with the finest scale image of size* 8×8 *shown at the bottom. At each level, the image is reduced in size by a factor of* 2 *in each dimension, with a* 4×4 *image at the second level, a* 2×2 *image at the third level, and a single pixel at the top of the pyramid.*

Gaussian (see Section 7.2.1) of $\sigma = 1.0$, it yields the curve (b); smoothing with Gaussians of $\sigma = 2.0$ and $\sigma = 3.0$ yields the curves in panels (c) and (d), respectively. As the curve is smoothed, the fine-scale structure is removed from it. The larger the σ, the larger the scale of the structure which is smoothed away. Note that in the curve (d), the very–fine-scale structure in the left region is smoothed away, while some remnants of the fine-scale structure in the right region remain.

One of the key steps in exploiting the different scales present in an image is to first decompose it into these scales. There are several different types of multiresolution decompositions. These essentially build a pyramid structure, which has the finest scale image at the bottom and successively coarser images at the higher levels of the pyramid, as shown in Figure 5.2. As subsampling is often used to represent the coarser levels, when the images are stacked one on top of the other, they have a pyramid structure—hence the name. There are three commonly used multiscale decompositions for one-, two-, and three-dimensional images (or data with regular spacing), as discussed next.

- **Gaussian pyramid:** Here, each coarser resolution level is obtained by first convolving the image with a Gaussian (see Section 7.2.1) and then subsampling by a factor of two in each dimension. The Gaussian smoothing essentially acts as a low-pass filter, smoothing the fine-scale structure, that is, the high-frequency components, in the image. The width of the Gaussian determines the scale of the structures which are smoothed—the larger the width, the larger the size of the structures which are smoothed away, as shown in Figure 5.1. By reducing the size of the image by two in each dimension, and then applying the constant width Gaussian to the resulting image, we are essentially smoothing away structures which were twice as large in the previous level of the pyramid.

The idea behind the Gaussian pyramid is useful in several tasks in the scientific data mining process. For example, it is used in the Marr–Hildreth and the Canny edge detection techniques (see Section 8.1) where the image is first smoothed using Gaussians at several different widths. The intent is to smooth away the high-frequency components in the image, which are typically the noise or the very–fine-scale structure. As a result, the edge image is not cluttered by the small edges corresponding to these structures. This allows us to identify the "salient" edges, or the edges which persist across several scales and are thus considered as the ones which are important in the image. The Gaussian pyramid is also important in scale-space theory [382] where it is used to focus on the salient parts of the image. The idea of using smoothing by Gaussians of different widths is also part of the Retinex technique for removing multiplicative noise from images (see Section 7.2.4).

- **Laplacian pyramid:** The Laplacian pyramid [76] is closely related to the Gaussian pyramid. Instead of considering the Gaussian-smoothed and subsampled image at each level, we consider the difference between the image from the previous level and the Gaussian-smoothed version at the current level. Thus, if I_0 is the original image and I_1 is the Gaussian-smoothed image, then the base of the pyramid is the image $(I_0 - I_1)$. This difference image is referred to as the *detail image* at level 0 as it contains the high-frequency or detail components. Next, I_1 is subsampled to create I_1', which is the coarse image at level 1. The next level in the pyramid is then created by obtaining the detail image at level 1, which is $(I_1' - I_2')$, where I_2' is I_1' convolved with the Gaussian. I_2' is then subsampled to yield the coarse image corresponding to level 2, and so on. Note that given the coarse image at level 1 and the detail image at level 0, we can reconstruct the original image using interpolation (upsampling followed by a low-pass filter) on the coarse image and adding the result to the detail image. This approach can be applied recursively to the coarse image at the top level of the pyramid and the detail image at the next lower level to yield the coarse image at the lower level, until we finally reconstruct the original image.

  The Laplacian pyramid is a simple approach to compressing images without visible distortion. It is also useful in progressive transmission of images, where an initial coarse version of the image is transmitted first (using the coarsest level of the pyramid). The image is successively refined by transmitting the information in the lower levels of the pyramid, which is used to reconstruct an improved image by using interpolation. The transmission can be stopped at any time if the user recognizes the image or realizes that it is not of interest. This idea of progressive transmission is important in remote viewing of large-size images as it allows a user to explore large data sets without using expensive bandwidth transferring bits, which may not really be useful.

  Further, since the coarse version retains all the significant characteristics of the original image, we can use the Laplacian pyramid as a way of reducing the size of the data during the process of visual exploration. The computations involved in generating the pyramid are simple and fast, making the Laplacian pyramid well suited for browsing through image data sets quickly by allowing views of the data at a coarse resolution.

  The idea behind the Laplacian pyramid is also exploited in the creation of the Scale-Invariant Feature Transform (SIFT) regions (see Section 8.3.3).

- **Wavelets:** Wavelets are more sophisticated decomposition schemes. Like the Laplacian pyramid, the image is decomposed into a series of subimages at ever coarser resolution. However, unlike the Laplacian, which represents an image of size $n \times m$ using $\approx 4/3\ n \times m$ pixels, the wavelet transform is not an overcomplete transform [439].

The theory behind wavelets is rather mathematical. Several authors have provided simple and intuitive approaches to understanding the core ideas behind wavelets, including Hubbard [280]; Stollnitz, DeRose, and Salesin [568]; and Burrus, Gopinath, and Guo [75]. A good reference for practical implementation issues is the text by Wickerhauser [625]. Wavelets are also closely related to filter banks in signal processing literature [599]. I next briefly describe the algorithm behind wavelets using a one-dimensional signal.

There are two aspects to the implementation of wavelets—the decomposition of the signal to give the coarse representation, also referred to as the analysis step, and the reconstruction of the signal from the coarse representation, also referred to as the synthesis step. In the analysis step, the signal is first decomposed into its high- and low-frequency components using a high-pass and a low-pass filter, H_H and H_L, respectively. Each of these components is then subsampled by a factor of two to yield the next coarse level in the wavelet representation. The values in the high-frequency component are referred to as the detail coefficients, while the values in the low-frequency component are referred to as the smooth coefficients. The next level of the wavelet transform is generated by applying the high- and low-pass filters to the smooth coefficients from the previous level.

The synthesis step reverses the computations and allows us to reconstruct the signal from its decomposition. First, the low- and high-frequency components at level i are upsampled. Then, the corresponding low- and high-frequency synthesis filters are applied to each and the resulting signals summed to get the component at the next level of finer resolution, $i-1$.

There are several standard analysis and synthesis filters, ranging from the very simple Haar wavelets to the more complex Daublets, symmlets, and Coiflets [67]. For example, the unnormalized low-pass and high-pass analysis filters for the Haar wavelet are

$$H_L = \frac{1}{2}[1 \quad 1] \quad \text{and} \quad H_H = \frac{1}{2}[1 \quad -1]. \tag{5.1}$$

The corresponding low-pass and high-pass synthesis filters are

$$H_L = [1 \quad 1] \quad \text{and} \quad H_H = [1 \quad -1]. \tag{5.2}$$

The normalized filter coefficients for the analysis and synthesis filters are identical:

$$H_L = \frac{1}{\sqrt{2}}[1 \quad 1] \quad \text{and} \quad H_H = \frac{1}{\sqrt{2}}[1 \quad -1]. \tag{5.3}$$

Figure 5.3 shows a continuous dotted curve represented using 64 discrete intervals. Coarser representations using 32, 16, and 8 intervals are shown on the left, while the corresponding detail coefficients are shown in the right column. These were obtained using the unnormalized Haar low-pass and high-pass analysis filters, respectively.

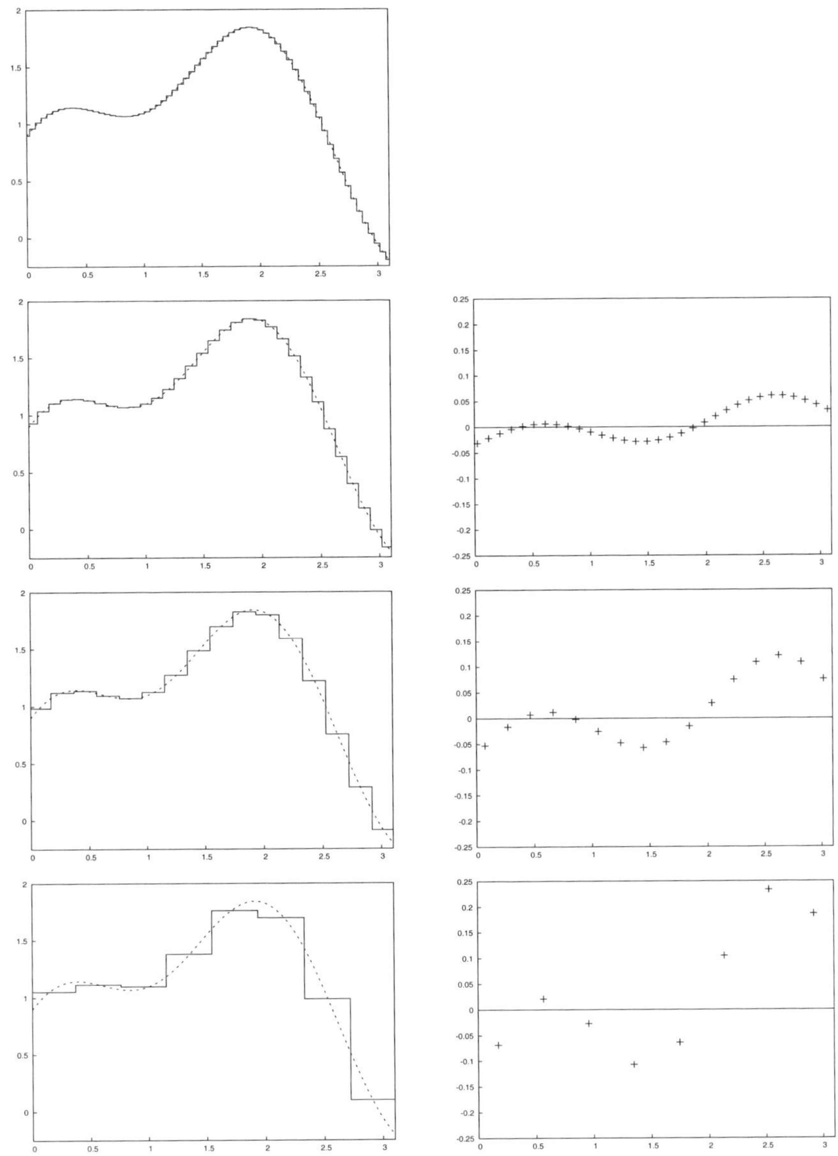

Figure 5.3. *Representing a curve using the unnormalized Haar wavelet. Top row: the original curve (dashed line) and its discrete representation using 64 values. The left column in rows two through four shows the smooth representation using 32, 16, and 8 values. The right column shows the corresponding detail coefficients. All the smooth coefficients are shown using the same scale, which is different from the scale used for showing all the detail coefficients. Note that the detail coefficients are larger in value at the coarser resolutions than at the finer resolutions.*

All the detail coefficients are shown using the same scale; this scale is different from the single scale used for displaying the smooth coefficients. In areas of the curve which are nearly flat, the detail coefficients are small, as would be expected based on the high-pass filter. This would imply that if we drop the detail coefficients with small values by setting them to zero, we can reconstruct an approximation to the original curve with relatively little error.

When the wavelet analysis filters are applied to two- or three-dimensional images, the computations are very similar. For a two-dimensional image, we apply the filters first along the rows of the image, followed by subsampling of the rows, and then along the columns, followed by subsampling of the columns. With the low- and high-pass filters, this gives rise to four subimages, corresponding to filtering by H_{LL}, H_{HH}, H_{LH}, and H_{HL}, where the first subscript indicates the filter applied along the rows, and the second subscript indicates the filter applied along the columns. These four subband images correspond, respectively, to the coarse image, the diagonal detail, the vertical detail, and the horizontal detail in the image. As in the one-dimensional case, the filters can be recursively applied to the coarse image at any resolution to give the smooth image at the next coarser level and the corresponding detail images.

The wavelet transform, though a complete representation of the data, is not translation invariant. If, however, we skip the step of subsampling, we obtain the undecimated transform, which results in $k \times n \times m$ coefficients for an $n \times m$ image and k multiresolution levels. Further details regarding the translation invariant, undecimated transforms are available in [540, 66]. The representation with subsampling is also referred to as the decimated wavelet transform.

Wavelets are useful in several tasks in the scientific data mining process. Their multiresolution properties allow them to be used for initial exploration of the data through the viewing of coarse versions of large images. They have also been used in progressive transmission of astronomical images [624], in image registration [377] (see Section 6.4), in image denoising (see Section 7.2.1), in representing texture features (see Section 9.4), and in reducing the dimensionality of time series data (see Section 10.4).

Based on the discussion so far, it is tempting to create a multiresolution pyramid by simply averaging pixel values over a 2×2 neighborhood of the pixel values at the finer scale. However, this is not recommended as it can lead to severe aliasing problems [382].

5.3 Summary

In this chapter, I focused on ways we can reduce the size of the data to make the analysis more manageable. Of course, eventually, we do have to analyze all the data. However, when faced with analyzing a massive data set measuring in terabytes, it is often helpful to make the problem tractable by reducing its size, especially in the initial exploratory phase. We can either use sampling techniques and work with a smaller subset of the data or use multiresolution techniques to create a coarser-resolution version of the data. These approaches are useful in other steps of the scientific data mining process as well. For example, sampling is used in the creation of a training set for classification, or a sample of a large data set is used in clustering to make the process more computationally efficient. Further, multiresolution techniques are used in denoising, feature extraction, and scale-space analysis.

In the next chapter, I will discuss another step in the initial phase of scientific data mining, namely, data fusion. If we have data from several different sensors, at different resolutions, it is often useful to try to combine the data available so that we can exploit the complementary information in the various data sources.

5.4 Suggestions for further reading

Sampling is a subject which has been studied extensively in statistics; texts of further interest include the books by Lohr [388] and Thompson [583]. In addition, as massive data sets become the norm, practical approaches to sampling are discussed in various journals in computer science, data management, and data mining. In the area of multiresolution techniques, the subject of wavelets and related techniques is one of active research. An easy-to-understand introduction to wavelets is given in the books by Hubbard [280] and by Stollnitz, DeRose, and Salesin [568] in the context of computer graphics. The latest developments in the field can be found in various journals in image and signal processing.

Chapter 6

Fusing Different Data Modalities

Live with the data before you plunge into modeling.

—Leo Breiman [60, p. 201]

In the previous chapter, I described ways we might reduce the size of the data through techniques such as sampling and multiresolution analysis. These approaches can be used in several of the steps in the scientific data mining process, but they are especially helpful in the early exploratory phases when the data sets are extremely large. In this chapter, we consider a different aspect of scientific data mining which too must be addressed at the beginning of the analysis process, namely, the availability of data from several different sources and the appropriate ways in which we can use these data to improve the analysis.

In many scientific problems, we have access to different modalities of data. For example, in medicine, we may have both magnetic resonance imaging (MRI) scans as well as positron emission tomography (PET) scans for a patient. Or, in remote sensing, we may have multispectral satellite imagery, where each image has four bands covering the red, green, blue, and near-infrared frequencies. Given the diversity of sensors which are becoming available, such multimodal sources of data are rapidly becoming the norm in many scientific domains. In such cases, it makes sense to exploit the complementary information in all these different modalities to improve both the analysis and the decision-making process. This process of combining data from different sensors is referred to as data fusion.

This chapter is organized as follows. First, in Section 6.1, I describe the many reasons why we might want to combine data from different sensors. Next, in Section 6.2, I describe the different levels at which we can combine information from these sensors. These are the sensor level, the feature level, and the decision level. Sections 6.3 through 6.5 describe the techniques used in these levels in more detail. Section 6.6 provides a summary of the chapter and finally, in Section 6.7, I provide suggestions for further reading.

First, I present an observation—data fusion, like scientific data mining, is a multidisciplinary topic. Its techniques are drawn from a variety of fields, including computer vision, signal and image processing, pattern recognition, statistics, and so on. I describe many of these techniques in the later chapters of this book, so the placement of this chapter early on may seem inappropriate. However, when we have the opportunity to exploit several different modalities of data, we should think about how we will combine the information in these different sources from the very beginning. In addition, as we shall see in this chapter, the fusion of data can be done during one or more of the steps of the data mining process. We can exploit the complementary information either at the level of the raw sensor data, or the level of the objects and features extracted from the data, or at the level of the decisions made from the data. If data fusion is done at the sensor level, it is one of the first steps in the scientific data mining process. Hence, this chapter is placed before the other chapters; as appropriate, I include pointers to techniques which are described in later chapters.

6.1 The need for data fusion

Data fusion is the process of combining data from different sensors to improve the results of the analysis. This improvement could manifest itself in several different ways, including increased confidence in the results, increased robustness of the techniques used in the analysis, improved coverage in both space and time of the phenomena being observed, reduced ambiguity, and improved detection of objects in the data.

Data fusion techniques have long been used in military applications, where the benefits of using multiple sensors was recognized in problems such as military surveillance and battlefield management systems. Much of the early literature in data fusion focuses mainly on the needs of such applications, though more recently, several nonmilitary domains such as remote sensing, medicine, intelligent transportation, and robotics have also started to exploit the complementary information available via multiple sensors.

In some problems, the sensors used are identical to each other, for example when multiple cameras are used to increase the spatial coverage of a scene or to provide different views of the scene. In the latter case, we can use the information from different cameras to reduce any ambiguity in the analysis. In other problems, the sensors may be different, providing complementary information. For example, in medical imaging, an MRI would focus on the soft tissue, while computed tomography would provide information on the bones, and a PET scan would provide functional information. Similarly, in nondestructive evaluation, a visual examination could complement information from eddy-current analysis and ultrasonics. In the area of remote sensing, there is often a trade-off between the resolution, the signal-to-noise ratio, and the spectral coverage. This trade-off exists when there are competing constraints in designing the sensors, making it necessary to use multiple sensors if we desire high resolution, improved signal-to-noise ratio, and a broad spectral coverage. Data fusion is also important in problems such as biometrics or mine detection, where the error rates using a single sensor may be too high, making it necessary to use more than one sensor.

Data fusion is useful even in commercial applications, where we want to exploit information collected in different surveys. These surveys may have focused on different aspects of a problem, resulting in some information available for a subset of customers, but not all customers. Or, in some surveys, aggregate information at a higher granularity is available, say, by zip codes. In such problems, the idea is to use these complementary sources of information to fill in the gaps in the databases. It goes without saying that this must be done with care, lest the missing data be replaced with incorrect data.

A more recent demand for data fusion techniques comes from applications involving networks of autonomous sensors, where it is important for each sensor to take into account what is observed by other sensors, so it can improve its decision-making process and that of the network as a whole. Multiple sensors are also used in the monitoring of complex machinery, where they can be used to predict failures, and in experimental physics, where they are used to measure various quantities of interest relevant to understanding the phenomena being observed.

While using more than one sensor can be beneficial, using multiple sensors may not always be beneficial. It is possible to have catastrophic fusion (see [429, Section 1.5]), where the results of using a multisensor system are poorer than using a single sensor or a few of the individual sensors. For example, in a problem involving the identification of volatile organic compounds, it was found that the improvement in classification accuracy obtained using the data from three sets of sensors was not statistically significant compared to the results obtained using two sets of sensors [470]. However, using two sets of sensors yielded a better performance than using a single set of sensors. This situation arises when data from a sensor do not provide any discriminatory information. In such cases, just adding data from a sensor because it is available can lead to problems with the curse of dimensionality (see Chapter 10).

6.2 Levels of data fusion

The fusion of data from multiple sensors can be done at different levels. Reflecting the early driving force of military applications, one of the most widely used models for categorizing the functions related to data fusion is the US Joint Directors of Laboratories (JDL) model. This model, originally proposed in 1985, provides a distinction among data fusion processes relating to the refinement of "objects," "situations," "threats," and "processes" [241, 242]. Though the recent refinement of the model has reduced some of the military focus of the terminology, the tasks associated with the processes tend to be related to military applications, with an emphasis on tracking moving targets.

A different approach to categorizing the functions in data fusion was proposed by Dasarathy [132], who focused on the type of inputs and outputs considered by each function. In this functional model, the type could be either data, features, or objects. For example, if the input was of type data and the output was of type features, the functions would be feature extraction functions. If instead, both the input and output were of type features, then the functions used would be feature-refinement functions.

Another commonly used model of data fusion considers three levels of fusion—the sensor level, the feature level, and the decision level. This fits in well with the overall process of scientific data mining. We can exploit the multiple data sources in the form of raw data by combining the data obtained from the sensors. Or, fusion can occur after the objects have been identified and the features extracted from them; here the task would be one of combining the features. Or, we can combine the information at a high level, after the pattern recognition step, when the models have been built and the step of fusion is used to combine the decisions from the models.

There are specific situations where each of the three different levels of data fusion are appropriate:

- **Sensor-level data fusion:** In some problems, it makes sense to combine the raw data obtained from the sensors. This is especially true when the sensors are commensurate; that is, they measure the same quantity. For example, images from multiple cameras viewing a scene can be fused at the image level. Or, satellite images of a region taken with similar sensors can be fused. When the data are in the form of images, this is also referred to as pixel-level fusion. A key challenge in this case is that the images must be registered, or aligned correctly, before we can combine them. In addition, if these images are at different resolutions, they must first be transformed into the same resolution. However, such processing can degrade the data during the registration or the transformation.

Sensor-level data fusion can be appealing as it allows us to exploit the multiple sources of data in their raw form. However, it can be time consuming, as raw data are often larger than processed data. More importantly, the process of registration, and any transformations involved such as resampling or warping of the images, can degrade the data, resulting in errors in registration. This could adversely affect the results of further processing.

- **Feature-level data fusion:** When the sensors used in an application are noncommensurate, that is, they are measuring very different quantities, then the fusion has to be done at a higher level. In feature-level fusion, the raw data are processed to identify the objects and extract relevant features for them. Then, fusion is done by simply combining the features for each object.

  Feature-level fusion reduces the need for registration. However, we must ensure that the identical object has been identified in the different data sources prior to combining the feature vectors for the object. This can be a problem if the resolutions of the data are such that an object is present in one data source, but not in another. Also, as we have observed before, combining the features results in a longer feature vector and all the associated problems of high dimensionality of the feature space.

- **Decision-level data fusion:** At the highest level, the data can be fused by combining the decisions made by several sensors. This, too, can be appealing, especially if the sensors are distributed with a relatively low bandwidth among them, making it prohibitively expensive to communicate either the raw data or the feature vectors. In this approach, as with feature-level data fusion, we can incorporate information about the reliability of different sensors by appropriately weighting the decisions. A key challenge, however, is the optimal approach to combining the decisions.

Data can also be fused at multiple levels to exploit the different characteristics of the different data sources. For example, in [323], we describe an approach to detect human settlements in satellite images and show how we can effectively exploit low-resolution, multispectral images and higher-resolution, panchromatic images. The information in the four-band (red, green, blue, and near-infrared) multispectral image is first used for pixel-level classification. The pixels with the associated class labels are then used to identify regions where there is a high likelihood of human settlements. Next, the higher-resolution panchromatic images are used to confirm if these regions do indeed have human settlements. The pixel-level processing done on the lower-resolution images enables us to reduce the regions processed in the higher-resolution images, making the approach computationally efficient. In our application, we combined the complementary information in the multispectral images with the panchromatic images, albeit at different stages of the decision-making process.

I next discuss the different techniques which can be used in the different levels of data fusion.

6.3 Sensor-level data fusion

The tasks involved in fusing the raw data collected by different sensors are very dependent on the type of data and the context in which we plan to use them.

The simplest case arises when the data obtained from different sensors are just scalar measurements taken over time. For example, in a physics experiment, we may have several sensors measuring the values of different variables, such as pressure and temperature, over time; or measuring the same variable at different locations. The intent may be to understand how the variables are changing when a specific phenomenon of interest occurs in the experiment. Or, we may be interested in understanding what causes a certain phenomenon so that we can use the values of the variables to try to predict when the phenomenon will occur. In these cases, data fusion may be as simple as representing the experiment as a series of time intervals, each characterized by several variables. These variables would essentially be the mean or median values of the sensors over each time interval. In other words, data fusion would be done by simply concatenating the different sensor measurements into a feature vector for each time interval.

A similar approach can be used in problems where multiple sensors are used to measure different properties of an object such as its radar signature, size, and speed, with the intent of using these measurements to identify the object as belonging to one of several categories. Data fusion is again done by concatenating the measurements from these sensors to form a feature vector which is then used in an appropriate pattern recognition algorithm.

Another simple example occurs in multispectral imaging, when relatively few bands are used. Here, one can display three bands at a time by assigning each of them to the red, green, and blue channels of the display, thus displaying the three bands simultaneously. Of course, when there are many spectral bands in the image, this simplistic approach is inadequate.

A more complex problem arises when the sensor measurements are obtained for moving objects which are being tracked over time. The problem now becomes one of using the data from all the sensors to correctly assign or associate an object with a track. This problem occurs frequently in military applications; I discuss the solution approaches briefly in Section 6.3.1.

If the data are in the form of images, data fusion becomes much more complex. A key challenge is that the images which need to be fused must first be aligned appropriately. This process is referred to as image registration. It is a common task not only when two images need to be merged, but also when two images of a scene, taken at different times, need to be compared to detect changes. Registration is also important in problems where, instead of aligning one image to another, both images have to be transformed into a single standard coordinate system. Such coordinate systems exist, for example, in remote sensing, where an image might be rectified so it is aligned to a planar map, or it is orthorectified so that each pixel is corrected for topographic distortion as well [527]. A standard coordinate system exists even for the human brain, motivated by the fact that there is a lot of variation across the brains of normal patients, making it difficult to compare, say, the MRI scan from the brain of a patient with a disease to a normal, disease-free brain. This standardized, anatomically based, coordinate system is referred to as an atlas [582], and all images are registered to it prior to analysis. I describe the techniques for data registration in Section 6.3.2.

6.3.1 Multiple target tracking

In military and other surveillance applications, one of the key tasks of interest is tracking moving objects. This can be rather difficult if there are several moving objects in a scene, some of which may be partially or completely occluded. Further, if the data are noisy and there is little to distinguish one object from another in the images, the task of correctly assigning objects to tracks becomes even more complicated. In such cases, an obvious solution is to use multiple sensors to extract additional characteristics on the moving objects so that these characteristics may be used in the accurate association of objects to tracks. This, of course, raises the obvious question, which one of the many measurements should be used to update the track? The task of assigning a single measurement to one of many tracks can itself be seen as data fusion, as we are finding the correspondence between two different types of data.

Section 11.5 briefly describes some of the issues which arise in tracking and the common techniques used for tracking multiple objects. Specifically, the association of objects to tracks can be considered to be a bipartite graph-matching problem, where the objects form one set of nodes, the tracks form the other set, and the goal is to correctly match each object to a track. The number of links between these two sets can often be reduced by exploiting the characteristics of the objects. For example, if a radar signature indicates an object is of a particular class, and a velocity measurement indicates how far it could have moved since the previous time, we can remove links to tracks associated with objects of a different class and far away from the current object. This can, however, be nontrivial given the errors in accurately assigning an object to a class based on its radar signature, or in measuring the velocity of an object.

More recently, two techniques have been developed to handle the issue of multiple measurements. The probabilistic data association approach [44], when applied to a single target, assumes that all measurements which have been observed and validated can be assigned to a track—however, with different weights or probabilities. This is in comparison to selecting only one of the validated measurements and ignoring the rest. The problem becomes more difficult when we have multiple moving objects because now, not only do the tracks compete for the measurements, but the measurements can compete for the tracks. This contention is handled by the joint probabilistic data association algorithm [44] which determines the measurement to track association probabilities and combines them to find the state estimate.

A more complex approach, the multiple hypothesis tracker, evaluates the probability that there is a target given the sequence of measurements. Essentially, a series of hypotheses is maintained, representing the different assignments of measurements to objects, with a track associated with each object. Based on this, each hypothesis can predict where its associated object will be at the next frame in the sequence. When new measurements come in, they are compared with the predictions, and each hypothesis is extended in several ways depending on how many of the new measurements are valid or feasible for the current hypothesis. With the addition of these "child" hypotheses at each frame, the hypothesis tree can grow quite large, representing all possible hypothesis of all previous measurements. This tree is finally pruned to remove unlikely branches. The multiple hypothesis algorithm can be computationally exponential in both time and memory; usually, approximations to it are implemented [124].

Multiple target tracking is a rather difficult problem, especially if the data are noisy, there are several moving objects, and several sensors are collecting measurements. A brief summary of various techniques is given in Chapter 11 of [429], the survey paper by Smith and Singh [550], the book by Blackman and Popoli [44], as well as several chapters in the *Handbook of Multisensor Data Fusion* [242] which describe the more advanced techniques in detail. In particular, Chapter 21, aptly titled "Dirty Secrets in Multisensor Data Fusion," discusses the pitfalls and problem areas which remain to be addressed in building a robust and accurate system for multiple target tracking. Finally, the detection of moving objects in video sequences, a topic related to tracking, is discussed further in our Section 8.4.

6.3.2 Image registration

Registration is the process of alignment of data, such as images, in order to relate the information in one image (called the target image) to information in another image (called the reference image). As mentioned earlier, there are two main tasks which require registration. The first is change detection, where the data have been acquired over time and must first be aligned before any changes can be detected, and the second is data fusion, where data from different sources must be combined to exploit the complementary information.

Registration is a multistep process. First, characteristic features are identified in the target and the reference images. These features are called "ground control points" in remote sensing and "landmarks" in medical imaging. The term "fiducial points" is also used when the control points are selected before the image is obtained, for example, by using an object of known dimensions at a specific location in the scene to be imaged. In the next step, the features are paired by correspondence matching, which essentially matches a feature in one image to the corresponding feature in the other image. If the features are defined as known objects, for example the point of intersection of two specific structures in the brain, then this correspondence is implicit in the definition of the control points. Next, the matched control points are used to find a transform function which warps the target image to match it to the reference image. Finally, the transform that brings the features into alignment is applied to the full image.

There are many options for each of these steps in image registration. In a broad sense, a registration technique can be thought of as comprising the following four components [65]:

- **A feature space:** This is the set of characteristic features used to perform the matching. These features can include pixels which satisfy certain constraints such as the pixel with the highest intensity in an object; the centroids of specific objects such as certain organs in a brain image; edges and corners; intersection of edges such as the location of the intersection of roads; points of high curvature; and so on. One can also think of these features as the salient points in an image (see Section 8.3). Depending on the problem, these features or markers can be very general, such as the intersection of roads; or they can be very specific and unique, such as the intersection of two specific linear structures in the brain. If the features selected are objects or regions in the images, then we need to select a characteristic point (say the centroid) in the region as it is computationally more efficient to do registration using the transformation of points rather than the transformation of entire regions.

  Techniques for extracting these features are discussed further in Chapters 8 and 9. Though the extraction of the control points or features is the first step in registration, which in turn, is the first step in fusing image data, a nontrivial amount of processing may be required to identify and extract these features. For example, if we plan to use the intersection of roads as control points in registering two images from remote sensing, then, we first need to identify the roads before we can obtain the intersection points. Doing this in an automated way can be quite difficult.

It is important to accurately identify the location of the control points, as any errors will manifest themselves as errors in the registration. This can, however, be a challenge if the images are of low quality due to noise or sensor errors. Therefore, prior to the extraction of the control points, the images must be preprocessed to reduce noise or any blurring, and to incorporate any corrections to account for the motion of the sensor or orbital distortions due to the satellite.

The control points must also be distributed as uniformly as possible across the image so that all points in the image can be registered correctly. Concentrating the control points in one part of the image may give good results in that region, but poor registration elsewhere [509]. It may not always be possible to have uniform distribution of the control points, especially if we select features which are not uniformly distributed in the image. For example, we may select road intersections as control points, but this would be a poor choice if only a small part of the image has roads. In such cases, we may want to consider other features in the part of the image where there are no roads to ensure a uniform distribution of features.

When the data are in the form of three-dimensional images, as is increasingly the case in medical imaging, the problem of extracting the landmark points becomes even more difficult; instead of single points randomly scattered, points on specific surfaces can be used in the registration [195].

- **A search space:** The search space defines the set of potential transforms that can be used to establish the correspondence between the images, for example, rotation, translation, scaling, polynomials, and so on. These transforms impose restrictions on the types of geometric distortions which can occur during registration. The most common transformations assume a rigid body model, which asserts that the object can undergo global rotations, scaling, and translations, but the relative distances and internal angles within the image cannot change during registration. In other words, the objects retain their relative shape and size. In what appears to be an issue of terminology, we observe that in medical imaging, a rigid body transformation does not include scaling [195], while in remote sensing, it does [65]. Representing scaling, translations, and rotations from one coordinate system to another is very simple. For example,

$$x' = x + p, \tag{6.1}$$

$$y' = y + q, \tag{6.2}$$

$$x' = \alpha x, \tag{6.3}$$

$$y' = \alpha y, \tag{6.4}$$

$$\begin{bmatrix} x' \\ y' \end{bmatrix} = \begin{bmatrix} \cos\theta & \sin\theta \\ -\sin\theta & \cos\theta \end{bmatrix} * \begin{bmatrix} x \\ y \end{bmatrix} \tag{6.5}$$

represent, respectively, translations in x and y coordinates, scaling by α in the x and y coordinates, and a clockwise rotation through the angle θ of the point (x, y), resulting in the point (x', y').

Other simple transformations include horizontal and vertical shear, a change in aspect ratio, and polynomials. More complex transforms based on thin-plate splines, multiquadrics, or piecewise linear functions are also possible [641], and nonrigid transformations must be used in medical imaging for registering organs such as the heart or the brain. For problems such as interpatient registration and registration to an atlas, curved transforms based on solutions to the equations of continuum mechanics are also becoming popular [582]. These equations describe the elastic and fluid properties of the anatomy being registered and are solved using computational techniques from partial differential equations. Also, in some problems, we may need to consider local transformations if a single global transformation does not give us the necessary accuracy in registration.

- **A similarity metric:** This is a metric used for evaluating the match between images for a given transformation in the search space. It often depends on how the registration problem is formulated. For example, if the fiducial markers are at anatomically known points in a medical image, we already know the correspondence between the markers in the target and reference images. In this case, the similarity metric may be just the sum of the squared error between the marker location in the reference image and the transformed location of the corresponding marker from the target image. The sum is taken over all the markers and the error for each may be weighted to reduce the contributions of the less reliable markers.

  When the images to be registered are three dimensional, instead of matching points from one image to another, we may want to match surfaces [195]. The similarity metric is defined in an analogous way by considering the "error" between one surface and the transformed version of the other surface. However, when we transform a point from one surface, there is no guarantee it will lie on the other surface. In this situation, we consider the closest point on the other surface to correspond to the point being transformed. The calculation of this closest point, however, can be computationally expensive.

  In some problems, we can use an intensity-based similarity metric. For example, a small region around each control point in one image may be extracted as a window and correlated with the points in the second image. This idea is very similar to the block-matching approach for identifying moving objects in video sequences, a topic discussed in Chapter 8. The correlation measure can be something simple, such as the absolute differences between the pixel intensities in the two windows, or more sophisticated normalized cross-correlation, or a cross-correlation coefficient [65, 527]. Correlation measures can also be used for edge images if we are using edges as the feature for registration.

  It is also possible to combine different features in a similarity metric to make the registration more robust and accurate. For example, in three-dimensional medical imaging, we could consider the sum of the errors at the fiducial points and at points on the surface to be registered.

- **A search strategy:** The search strategy determines how we compute the parameters of the transform. This involves transforming the control points from the target image and evaluating the similarity metric based on how close each point is to the corresponding point in the reference image. However, we first need to ensure that we have a correspondence between the landmark points in the target image and the points in the reference image. In some problems, this correspondence is known, for example, in medical imaging where the landmarks may have been selected at anatomically specific locations. In other problems, especially when the control points are determined automatically, the correspondence is unknown.

There are several ways in which we can find the correspondence. We can use characteristics of the control points such as the curvature or the intensities in a small window around the point. Alternatively, we can use clustering techniques (see Section 11.1), match the minimal spanning trees of the control points, or match the convex hull of the control points [229].

Once we have a correspondence between the control points in the target and reference images, we need to determine the parameters of the transform. In some cases, the parameters can be obtained in closed form using a linear least-squares formulation. For example, the coefficients of a polynomial transformation can be obtained by minimizing the sum of the squared error between points in the reference image and the transformed version of the corresponding points from the target image. If, instead, we use a transform composed of scaling, translation, and rotation in two dimensions, then the four parameters (the scale factor, the x and y translation, and the angle of rotation) can be obtained by considering two sets of matched points, which gives us four equations in four unknowns [539].

In more complex registration problems, we may need to use a nonlinear similarity metric to obtain accurate results [195, 652]. For example, we may want to minimize the energy functional in transforming the three-dimensional surface representing the transformed target image to the reference image. Or, we may want to maximize the mutual information between the reference image and the transformed target image. In such problems, an exhaustive search for the parameters of the transformation is always an option, albeit an expensive one. More commonly, iterative techniques are used including those from multivariate optimization such as gradient descent and the Levenberg–Marquardt technique, as well as the more stochastic approaches such as simulated annealing and evolutionary algorithms [414], which avoid getting stuck in a local minima (see Section 11.7.2).

The search for the optimum parameters for the transform can be computationally expensive. If we are using intensity-based similarity metrics, we can exploit multiresolution approaches based on wavelets or Gaussian pyramids (see Section 5.2) to make the task of registration more computationally efficient. Typically, a not-so-accurate registration is obtained using the data at a coarser resolution. This narrows the search space at the finer resolution, where a more accurate registration is done. The basic idea behind this approach is discussed further in Section 8.4.2 in the context of detecting moving objects, as well as in [363] in the context of registration.

Finally, once the transformation has been obtained based on the control points, we need to apply it to the rest of the points in the image. If the transformation is applied to all pixels in the target image, that is, it is a forward transformation, it can produce holes or overlapping due to the discretization and rounding which occurs after the mapping. Hence, the reverse approach is chosen and each point in the reference image is mapped back to a point in the target image using the inverse transform. The rounding and interpolation is done using the values from the target image [539]. When the mapping back from the reference pixel to the target is not exact, nearest neighbor resampling or bilinear interpolation is used, though more complex approaches can also be used [652].

After the images have been registered, they can be fused at the pixel level by simply averaging the two images, pixel by pixel. However, as we shall see in Section 6.4, this may not be the preferable approach.

Registration can be a difficult task, given the diversity of sensors that generate data, the presence of noise in the data, the absence of fiducial markers, the sheer volume of data even from a single sensor, and the technical challenges associated with the identification of characteristic features and optimal transforms. Oftentimes, the approaches used are specific to the application domain and the problem being solved.

There are several general reviews of the techniques used in image registration, including the early survey by Brown [65], the more recent survey by Zitová and Flusser [652], and the excellent text by Goshtasby [229]. In addition, many innovative approaches are proposed within the context of specific application areas. These exploit the characteristics of both the objects being imaged and the sensors used to image them.

Medical imaging, with its diversity of sensors, the use of three-dimensional imaging, and objects ranging from soft tissue to bone in a single image, presents some of the more challenging problems in registration. The solutions to these can be found in several survey papers [423, 406, 195] as well as in several chapters in the *Handbook of Medical Imaging*, edited by Bankman [20], and the books by Modersitzki [433] and Hajnal, Hill, and Hawkes [238].

Remote sensing, too, is an application area where registration plays a key role, not just in image fusion, but also in change detection and mosaicking, where two or more images are merged to create a composite single image [229]. With the advent of very high resolution and hyperspectral imagery, there is an increasing need for subpixel registration algorithms as well as techniques which incorporate sensor-related transformations and knowledge of the application and the content of the images [168]. An interesting review of data fusion techniques in remote sensing, with a focus on pixel-based approaches, is given by Pohl and van Genderen [480].

Registration is also important in computer vision problems such as three-dimensional shape reconstruction from multiple views [647, 183, 258]. Inferring a three-dimensional object from its two-dimensional projections requires identifying corresponding points in the multiple views. This is usually done using feature-based techniques or optical flow techniques. The latter is commonly used in estimating the motion of the objects in successive frames of a video sequence (see Section 9.5). We have earlier observed the similarity between registration and block-matching techniques used in the detection of moving objects in video. This is another example where techniques from motion estimation are used to find the transformation between two images. In fact, the original article by Lucas and Kanade [397], describing the popular optical flow technique named after them, is actually on registration and not optical flow.

Another area related to registration is that of shape analysis [162] (see also Section 9.3). Often the similarity between two shapes is obtained by finding landmark points on them and evaluating the cost of transforming one shape to the other based on the landmarks.

6.4 Feature-level data fusion

Data fusion at the feature level is usually considered when either the data are available as features or the data are from noncommensurate sensors. In the latter case, it may not make sense to fuse the raw data from the sensors.

The most commonly used approach to fusing data at the feature level is to first identify the objects in the different types of data to be fused and extract features representing them. Then, the features are concatenated to form a single feature vector representing the object. In this process, we need to ensure that the objects being fused in the two data sets are the same. This is done by spatially associating the centroids of the two objects. However, in some problems, this can be a challenge. For example, if the two data sets to be fused are at different resolutions, an object could be clearly identified in one data set but could be missing in the other if it is smaller than the resolution of the data set. This can be the case with astronomical data when we want to combine information from several different surveys.

The concatenated feature vector, representing the contributions of several different data sources, can then be used in an appropriate pattern recognition algorithm. For example, we extended our similarity-based object retrieval system [314] to work with multivariate data from computer simulations. We exploited the fact that in a computer simulation, many variables, such as pressure, density, and velocity, are output at each grid point. Instead of using just one of these variables in the retrieval system, we could use all of them. In our approach, retrieval was done using tiles (rectangular regions in the image) as objects. We combined the different variables by first extracting the same set of features for each of them. So, for each tile in an image, we would extract the texture features for the density variable, the pressure variable, and so on. The features were then concatenated to form a long feature vector which was used in retrieval based on a nearest-neighbor approach in feature space.

Combining feature vectors has also been used in landmine detection, an application where data fusion is important to improve accuracy and reduce false alarms. In one study [562], data from a metal detector and ground-penetrating radar were combined, while in another study [235] the data from an infrared camera were also included. Both studies experimented with feature-level and decision-level techniques. Based on the data sets used in these studies, the results indicated that using multiple data sources is beneficial in reducing the false-alarm rate, with the feature-level techniques performing slightly better than the decision-level techniques.

Data fusion at the feature level can also be used to estimate missing features. For example, in the context of a nonscientific data mining problem, we improved the accuracy of retrieval results in a problem where we combined features representing faces in an image with text features from the associated caption [15]. If one of these sets of features is missing, we could estimate the missing features by using relevance feedback on the results retrieved using the known set of features.

Features in image data can also be used to improve data fusion done at the sensor level. If two registered images are fused by averaging the images pixel by pixel, then the contrast of features uniquely present in either image is reduced. One solution to this problem is to fuse the data using pyramid-based schemes which create a multiresolution decomposition of the two images (see Section 5.2). Then, a composite decomposition is created by appropriately merging the two decompositions. Li, Manjunath, and Mitra [377] have observed that if a Laplacian pyramid is used for this, the resulting image has blocking artifacts in regions where the two images are very different. They propose using a wavelet decomposition which allows them to identify important features as ones with strong detail coefficients. The composite decomposition is performed by taking the larger of the detail coefficients from the two decompositions. An inverse wavelet transform on the composite decomposition provides the fused image.

Feature-level fusion is also useful when the data to be fused are distributed and large enough that it does not make sense to move the data to a single location. This is especially true if the data are such that only a small part of the data is relevant to making a decision. For example, in medical images, only the regions with the tumors might be of interest. In such cases, we can extract features from these regions for each type of data and send them to a centralized location, where the features would be combined to make a decision. The alternate approach of sending the entire image to the centralized location could be bandwidth intensive, especially if the data are high-resolution, three-dimensional images. If the features are too large in size and the bandwidth constraints are severe, then one can use the distributed features to build a decision tree classifier and use it for the decision making [427].

A key problem with feature-level fusion is that the concatenation of the feature vectors from each type of data can result in each object being described by hundreds of features. This is especially true when the feature itself has many components, as is the case with texture features or histograms. This can result in the curse of dimensionality when we apply pattern recognition algorithms, such as classification and clustering techniques, in a high-dimensional feature space. Dimension reduction techniques, such as those discussed in Chapter 10, are recommended for the identification of the key features prior to the step of pattern recognition.

6.5 Decision-level data fusion

In problems where the data from each sensor are too large to be sent to a centralized place for fusion, or there are issues of privacy which prevent the open sharing of data, fusion at the decision level is an alternative. The basic idea is rather simple—each sensor makes a decision based on what it observes and the decisions are combined in some way to represent the combined data from all the sensors.

The difficult part in decision-level fusion is the combination of the decisions from each of the sensors. Several authors have observed the similarity between this problem and the problems of ensemble learning and incremental learning (see Section 11.2). In ensemble learning, we build many classifiers from a single data set, with some randomization so that each classifier is different in some way. Then, the results from the classifiers are combined either by a simple majority vote, or by using some weighted voting scheme where decisions from classifiers with higher accuracy carry more weight than decisions from classifiers with lower accuracy. In incremental learning, the data for building the classifier are obtained incrementally, so the goal is to learn the new information as the new data becomes available while retaining the information learned from the old data, which are now inaccessible.

The idea of using incremental learning in data fusion is explored further in [470], where the authors train an ensemble of classifiers, each on different data sources and each represented by a set of features for a data item. The features are different across the different data sources. Within the data from each source, a weight is assigned to each data item to indicate how likely they are to be included in the training set for the next classifier to be built in the ensemble, and the ensembles themselves are assigned a weight based on their training performance. The data sources are also assigned weights to indicate their reliability. A final decision is then made, taking into account both the weights of each classifier in the ensemble for a data source and the weight of the data source. For the data sets considered in [470], this approach is shown to give more accurate results than an ensemble of classifiers trained using the concatenated feature vector.

When we use decision-level data fusion, we may run into problems if not all sensors observe an object, or the sensors produce conflicting data, or there is uncertainty in the data. In such situations, probabilistic techniques, such as those based on the Dempster–Shafer theory of evidence, are used to fuse the decisions [538, 85, 533].

6.6 Summary

In this chapter, I discussed the topic of data fusion, or the process of combining data from different sensors, to exploit the complementary information available from each sensor. The expectation is that this will improve the results of the analysis by increasing our confidence in the results and reducing the ambiguity. I described three different levels at which we can fuse the data—the sensor level, the feature level, and the decision level. Each of these has its challenges, and while data fusion does have its benefits, caution is urged lest we have catastrophic fusion, resulting in worse results than using a single sensor.

Multiple sensors are becoming the rule rather than the exception, and sensor networks are gaining acceptance in a variety of problems, increasing the need for data fusion techniques. Further, as distributed data become more common and privacy issues influence what data we can or cannot share, the techniques being used in data fusion will be used in other areas as well, which in turn will result in new approaches which can benefit the more traditional data fusion applications.

6.7 Suggestions for further reading

There are several books which cover the general topic of data fusion. The two edited books by Hall and Llinas [242] and by Blum and Liu [49] provide a good overview of different techniques and applications, while the texts by Mitchell [429] and Hall and McMullen [243] focus more on the techniques.

Chapter 7

Enhancing Image Data

It has been my experience, as well as many others who have looked, data is generally much less accurate than it is advertised to be.

... I learned never *to process any data until I had first examined it carefully for errors. There have been complaints that I would take too long, but almost always I found errors and when I showed the errors to them they had to admit that I was wise in taking the precautions I did. No matter how sacred the data and urgent the answer, I have learned to pretest it for consistency and outliers at a minimum.*

—Richard Hamming [248, pp. 313 and 316]

In the previous chapter, I described ways we can use data fusion techniques to exploit the complementary information available in problems where the data are collected by different sensors. These data are often in the form of images, for example, observational images from an astronomical survey, experimental images from a physics experiment, or, in the case of simulation data, output which is displayed as an image. These images, especially those from observations and experiments, tend to be noisy, often with poor contrast and nonuniform illumination. Cleaning and enhancing these images is a crucial first step in their analysis.

In this chapter, I will discuss ways we can improve the quality of image data. Though the focus is mainly on images, similar techniques can be applied to one-dimensional signals as well as temporal data, such as video images. These image enhancement techniques are mainly motivated by the need to improve the quality of experimental and observational data; however, they can also be applied to simulation data to smooth the fine-scale structure and focus on the coarser features.

The techniques discussed in this chapter range from the very simple to the complex. The efficacy of a method often depends on the noise characteristics and the specific parameters used in the enhancement algorithm. Often, we may need to apply more than one algorithm in succession to achieve the desired result.

This chapter is organized as follows. First, in Section 7.1, I describe why enhancing image data is an important step in scientific data mining. Next, in Section 7.2, I explain various approaches for reducing the noise in images while preserving the signal. These techniques range from very generic to more specialized. In Section 7.3, I describe ways in which the contrast in the images can be enhanced to clearly separate objects of interest from the background. Section 7.4 describes the nonlinear methods of mathematical morphology which are used in many tasks in image analysis, including de-noising. Finally, Section 7.5 summarizes the chapter and Section 7.6 provides suggestions for further reading.

7.1 The need for image enhancement

Enhancing the raw input data to clearly identify the objects of interest is an important step in scientific data mining. If we can reduce the noise in the data or enhance the contrast of an image, we can greatly simplify subsequent steps in the processing of the data. For example, the presence of noise may interfere with the identification of objects in the image. If the objects are detected using a gradient-based technique (see Chapter 8), we may find that the regions of high gradient occur not only at the object boundaries, but also in the regions with noise [322]. Selective denoising of the images may be required before the objects can be identified. In some problems, we may find that the contrast in one part of an image is not as strong as in another part of the image. If a simple thresholding approach is used to identify the objects of interest, we may completely miss the objects in the lower-contrast regions. Or, we may need to use locally adaptive methods, which, depending on the image and the variation in contrast, may not give satisfactory results. In such situations, enhancing the contrast may be beneficial.

Cleaning the data may be useful even when the data are not in the form of images. For example, if we are interested in the problem of finding peaks in a one-dimensional signal, it may be helpful to smooth the data first so that small variations due to noise in the data do not adversely affect the results. This example also illustrates another benefit of cleaning the data—it makes the subsequent steps of processing less sensitive to the parameters used, making the analysis more amenable to being automated.

Data enhancement algorithms can also be applied temporally. For example, techniques used to remove noise in images can be used to reduce small camera motion in video data [112]. Here, the approach is applied across the frames of a video sequence to smooth out the temporal variation caused, for example, by the slight motion of a stationary camera due to wind.

The enhancement of the data is often done after a subset of the data has been extracted through sampling (see Section 5.1). Depending on the specifics of a problem, several steps may precede or succeed the step of image enhancement. For example, if we are working with data from different sources and the data need to be fused, we may do some initial cleaning of the data, then fuse them, and then enhance them further as necessary. Also, several different types of enhancement techniques may be used in a single problem. For example, we may first reduce the noise and then apply contrast enhancement to clearly identify the boundaries of the objects of interest so that it is easy to extract them in the object identification step (see Chapter 8). Depending on the problem, we could reduce the noise by either a single application of a technique or successive applications of one or more techniques.

The approach one uses to enhance the data often depends on the quality of the data and the source of noise in the data. In experimental and observational data, the quality of the data may be degraded by sensor noise, presence of dust particles, or atmospheric turbulence. Other factors that could also contribute to the poor quality of data include nonfunctional sensors, leading to missing data such as missing pixels in a CCD camera image, as well as blurred images due to camera movement in a video sequence. Frequently, we may need to try several different denoising and enhancement techniques before we can find a satisfactory solution to a problem.

7.2 Image denoising

The noise in two-dimensional signals can be of various types. Additive noise occurs when noise is added to the value of each pixel. This noise can be impulse noise (also called salt-and-pepper noise), when it affects some of the pixels. It can also be Gaussian noise when a value drawn from a zero-mean Gaussian probability density function is added to the true value of each pixel. Noise can also be multiplicative, as in the case of variable illumination or speckle noise in SAR imagery [459]. The effects of variable illumination can be suppressed by considering the image as a product of two factors—an illumination function and a reflectance function—and taking its logarithm, making the two functions additive instead of multiplicative. The varying illumination can then be subtracted out, as discussed in Section 7.2.4.

I next describe various techniques for reducing noise in data without significantly affecting the signal. I focus mainly on two-dimensional image data, though several of the techniques can be applied to one-dimensional signals as well. Note that it is seldom possible to leave the signal unaffected while reducing the noise. A good application of a denoising technique will leave the signal largely unchanged while reducing the noise. However, in some cases, this may not be possible, and care must be taken to check that the signal has not changed substantially before the image is processed further.

7.2.1 Filter-based approaches

The idea behind filter-based approaches is to consider the pixel values in a small window around the current pixel and replace the current pixel by some linear or nonlinear combination of these values [596]. Many of these approaches use discrete convolution, which is the weighted sum of one discrete function, the image, with respect to another discrete function, the spatial mask, which has first been reflected about the origin and then variably displaced. Let I be the image and M be the mask. Then, the convolution C of I with M at the location (x,y) in the image is defined as

$$C(x.y) = M(x,y) * I(x,y) = \sum_{n'=-\infty}^{\infty} \sum_{m'=-\infty}^{\infty} I(x-m', y-n') M(m', n'). \tag{7.1}$$

Usually, the mask is defined over a small window centered at the origin and is zero elsewhere. Hence, the summation can be done over the small window. Further, if the mask is symmetric, as is often the case, it does not have to be reflected about the origin.

By using an appropriate mask, or filter, we can enhance an image. For example, a commonly used mask is the averaging filter, or the mean filter, where all values in the filter window are weighted equally, that is,

$$M = \frac{1}{4}\begin{pmatrix} 1 & 1 \\ 1 & 1 \end{pmatrix} \tag{7.2}$$

for a 2×2 mask, with the origin at the top left, and

$$M = \frac{1}{9}\begin{bmatrix} 1 & 1 & 1 \\ 1 & 1 & 1 \\ 1 & 1 & 1 \end{bmatrix} \tag{7.3}$$

for a 3×3 mask, with the origin at the center. Essentially, the mean filter replaces each pixel by the mean of the values in a window around it. It acts as a low-pass filter, reducing the higher-frequency components in the image. The mean filter is frequently used to smooth or reduce additive Gaussian noise. There is a trade-off between the size of the filter and the reduction in noise. Larger filters smooth the image more, but, if too large a filter size is used, then pixel values far from the current pixel influence the new value of the pixel, which may be undesirable. Application of a larger filter is also more computationally expensive. An alternative is to use a Gaussian mask, where the values in the mask are smaller as we go further out from the origin of the mask:

$$G_\sigma(x,y) = \frac{1}{2\pi\sigma^2}\exp\left(-\frac{|x|^2+|y|^2}{2\sigma^2}\right), \tag{7.4}$$

where σ is the standard deviation of the Gaussian.

For example, Figure 7.1 shows the effect of a simple 3×3 Gaussian filter on a synthetic image with 20% random noise. The addition of the noise makes a region that originally had uniformly grey intensity, appear to be speckled, with pixels of varying intensity. The zoomed-in region clearly shows that the application of the Gaussian blurs the sharp edges between regions with different intensities. Further, in contrast to the noisy image, the smoothing reduces the intensity variation in the speckled regions, though it is not as uniform as in the original image.

It should be mentioned here that convolution with a Gaussian can also be used for scale-space analysis [382]. The idea in scale-space representation is to generate a one-parameter family of derived signals in which the fine-scale structure is successively suppressed. One way this can be done is by successive convolutions with a Gaussian filter with increasing larger σ. Recall also that Gaussian convolution is used in creating a multiresolution pyramid, as discussed in Section 5.2.

Unlike the Gaussian or mean filters, some filters, such as the minimum mean-squared error (MMSE) filter, are more adaptive and use the local image characteristics to determine their behavior [596]. This filter works best with Gaussian or uniform noise. The pixel values in the denoised image are given by

$$I(x,y) - \frac{\sigma_n^2}{\sigma_w^2}\{I(x,y) - M_w(x,y)\}, \tag{7.5}$$

where σ_n^2 is the noise variance, σ_w^2 is the variance in the image in a small window around the current pixel (x,y), and $M_w(x,y)$ is the mean value of the pixels in the window. If there is no noise in the image, the noise variance is zero, and the image remains unchanged. In areas of the image where the pixel intensity is fairly constant, σ_w^2 will equal σ_n^2, and the MMSE filter will function as a mean filter. In the areas of the image where the local variance is much higher than the noise variance, the influence of the second term in (7.5) is small, and the denoised pixel value is close to the original pixel value. As observed in [596], areas with high local variance are typically regions with edges or other details, and a good denoising filter should try to preserve these details.

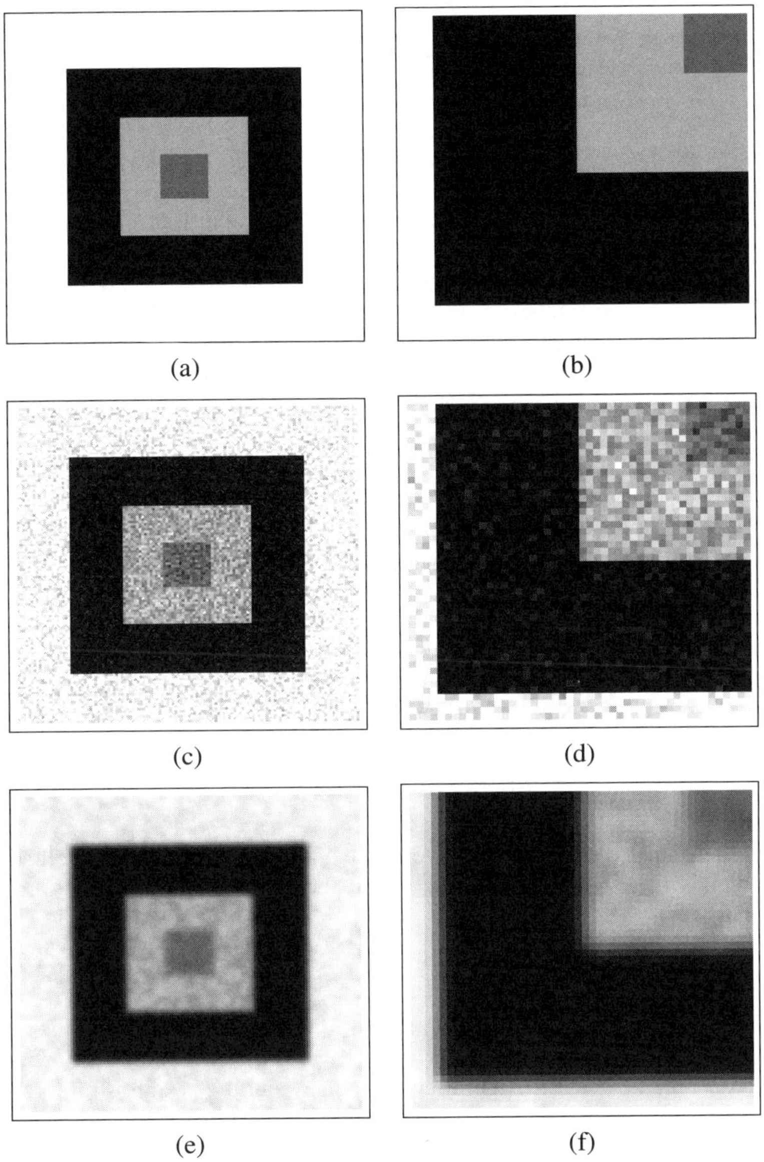

Figure 7.1. *Using the Gaussian filter to denoise an image.* (a) *A synthetic image;* (b) *a zoomed-in region near the lower left corner of* (a); (c) *a noisy version of image* (a), *with 20% random noise;* (d) *the corresponding zoomed-in region of* (c) *showing regions which had uniform intensity in* (b) *appearing speckled due to the addition of the noise;* (e) *the noisy image after smoothing using a* 3×3 *Gaussian filter;* (f) *the corresponding zoomed-in region of* (e). *Note that the sharp transition between different grey levels is now blurred and regions with a uniform grey level now have a variation in the intensity.*

There are several ways in which the noise variance in an image can be estimated. One of the more popular noise estimators is the median absolute deviation (MAD), which defines the standard deviation of noise as

$$\text{median}(|I(x,y) - m|)/0.6745, \tag{7.6}$$

where m is the median of the image and 0.6745 is a factor to make the estimate unbiased for the normal distribution.

In addition to these simple averaging filters, there are also more sophisticated filters, such as the bilateral filter [585] which computes the weights in the filter based on both spatial proximity and similarity of intensity values. The bilateral filter has been shown to be related to isotropic diffusion approaches to denoising, which we discuss later in this section. The idea behind the bilateral filter is to apply in the range of an image what traditional averaging filters do in its domain. Since images typically vary slowly over space, the pixels near each other have similar values. As the noise in the nearby pixels is mutually less correlated than the signal values, when we use an averaging filter, the noise is averaged away, while the signal is preserved. The bilateral filter also applies a similar idea in the range of the image by using a range filter which averages pixel values with weights that decay with dissimilarity. In other words, the weights are indicative of the intensity difference (in a grey-scale image) between the center pixel and its neighboring samples. Such a filter is nonlinear as the weights depend on the image intensity.

In the bilateral filter proposed by Tomasi and Manduchi [585], the weights of the range and domain filters are combined using simple multiplication of the weights at each point in the stencil. The weights are determined using a Gaussian function. Specifically, the domain weights between pixels p_1 and p_2 are given by

$$w_d(p_1, p_2) = \exp\Big(-\frac{\| p_1 - p_2 \|^2}{2\sigma_d^2}\Big), \tag{7.7}$$

where σ_d is the standard deviation in the domain and $\| p_1 - p_2 \|$ is the Euclidean distance between the pixels p_1 and p_2. The range weights between pixels p_1 and p_2 are defined as

$$w_r(p_1, p_2) = \exp\Big(-\frac{\{\delta(I(p_1) - I(p_2))\}^2}{2\sigma_r^2}\Big), \tag{7.8}$$

where σ_r is the standard deviation in the range and $\delta(I(p_1) - I(p_2))$ is a measure of the distance between the intensities at pixels p_1 and p_2. If the two pixels vary widely in their intensities, as is the case at an edge, then the corresponding weight in the range filter is small, thus reducing the smoothing. Each weight in the filter is the normalized product of the corresponding weights w_d and w_r. The bilateral filter is thus described by three parameters: σ_r, σ_d, and the filter width.

A different approach to calculating the weights of the filter is the gradient inverse weighted filter [608] which forms a weighted mean of the local pixels with the weights depending on the difference between the value of the central pixel and the value of the local pixels. A small 3×3 window around the central pixel is considered and the filter can be applied iteratively. This filter has been shown to reduce Gaussian noise well while preserving the image structure.

Another filter which performs denoising by taking the weighted average of the values in a region around the current pixel is the one based on the Univalue Segment Assimilating

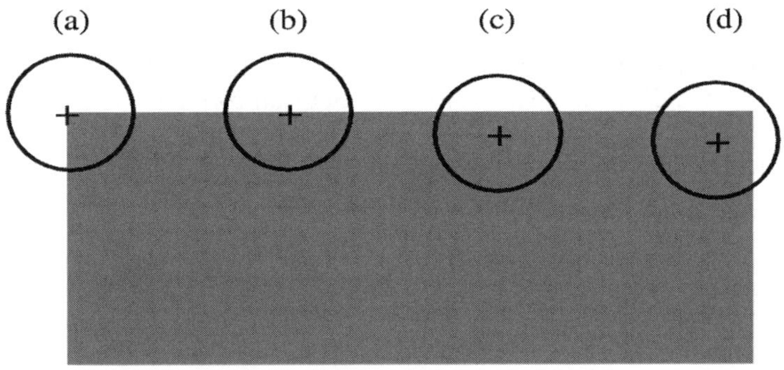

Figure 7.2. *The definition of Univalue Segment Assimilating Nucleus (USAN), illustrating how the USAN varies based on the location of the pixel. The object of interest is in grey on a white background. Four different placements of a circular mask are shown. The nucleus is indicated by +. Within each mask, the area in grey indicates the USAN, where the pixels have similar intensity to the pixel at the location of the nucleus.*

Nucleus (USAN) principle [553]. The idea behind the USAN is rather simple and elegant. Consider a circular window or a mask whose center, referred to as the nucleus, is placed on a pixel. When the intensity of each pixel within the mask is compared with the intensity of the pixel at the location of the nucleus, we can define an area of the mask which has the same or similar intensity as the nucleus. This area is called the USAN. For example, consider the grey rectangle which is the object of interest in Figure 7.2. The figure shows four placements of the circular mask whose nucleus is indicated by +. Note that when the nucleus is near an edge, the USAN is about half the area of the mask. Near a corner, it is less than half the area, and if the mask is completely within the object, the USAN is at the maximum and equal to the area of the mask.

The filter then uses a weighting scheme similar to the gradient inverse weighted filter by considering the difference between the intensity of the pixel at the nucleus and the pixels in the USAN. The equation used for weighting, as well as related thresholds, are described in more detail in [553]. One of the more interesting applications of the USAN principle is in detecting corners and edges, a topic which I discuss further in Chapter 8.

Another simple nonlinear filter is the median filter, which is frequently used to remove salt-and-pepper noise. It essentially replaces each pixel by the median of the pixel values in a window around the current pixel.

Filters such as the median filter can also be applied to temporal data. For example, it has been applied across the frames in a video sequence to reduce small camera motion and to remove the effects of falling snow [111]. The filters can also be applied in one and three dimensions to smooth data and reduce the noise.

7.2.2 Wavelet-based approaches

Denoising data by the thresholding of wavelet coefficients has been extensively studied by several authors. The early work in this area focused on one-dimensional signals, though the approach has been extended to images as well [612, 140, 155, 156, 93, 563, 98]. The basic idea behind the use of wavelets in denoising is that the noise in an image is usually at a higher frequency than the background, but not as high as the edges in the image. Thus, if we thresholded the detail coefficients in a wavelet-transformed image, we would reduce the noise while preserving the edges.

The statistical approach to denoising data considers the observed data $I'(x,y)$ as a representation of a noisy version of the true image $I(x,y)$:

$$I'(x,y) = I(x,y) + \epsilon(x,y), \quad x = 1,\ldots,N; \quad y = 1,\ldots,M, \tag{7.9}$$

where the image is of size $N \times M$. The noise, $\epsilon(x,y)$, is assumed to be an independent and identically distributed Gaussian with zero mean and standard deviation of σ_n, i.e., $\mathcal{N}(0,\sigma_n^2)$, independent of the signal. The goal in denoising is to obtain an estimate $\hat{I}(x,y)$ of the original signal $I(x,y)$ from the realization $I'(x,y)$ with minimal mean-squared error.

If we denote the observed data, the noiseless data, and the error matrices in equation (7.9) by $\mathbf{I}'$, $\mathbf{I}$, and $\mathbf{E}$, respectively, then the three steps in wavelet-based denoising can be summarized as follows:

1. obtain the wavelet coefficient matrix $\mathbf{C}$ by applying a wavelet transform $\mathbf{W}$ to the data:

$$\mathbf{C} = \mathbf{W}\,\mathbf{I}' = \mathbf{W}\mathbf{I} + \mathbf{W}\mathbf{E}; \tag{7.10}$$

2. threshold the detail coefficients of $\mathbf{C}$ to obtain the estimate $\hat{\mathbf{C}}$ of the wavelet coefficients of $\mathbf{I}$:

$$\mathbf{C} \rightarrow \hat{\mathbf{C}}; \tag{7.11}$$

3. inverse-transform the thresholded coefficients to obtain the denoised estimate:

$$\hat{\mathbf{I}} = \mathbf{W}^{-1}\,\hat{\mathbf{C}}. \tag{7.12}$$

Under our assumptions, the application of the orthogonal transform $\mathbf{W}$ preserves the normality of the errors $\mathbf{WE}$ in the new wavelet basis. Thus, the empirical wavelet coefficients $C(x,y)$ of the observed signal can be written as noisy realizations of the true wavelet coefficients $\{\mathbf{WI}\}(x,y) = \mu(x,y)$ of the unknown signal:

$$C(x,y) \sim \mathcal{N}(\mu(x,y),\sigma_n^2), \quad x = 1,\ldots,N; \quad y = 1,\ldots,M. \tag{7.13}$$

I next briefly discuss the various parameters in each of the three steps in wavelet denoising.

First, we need to select a wavelet $\mathbf{W}$ for the forward and inverse transformations. We can choose from either the orthogonal wavelets, which include the Haar, the Daubechies family (daublets), the least asymmetric wavelet family (symmlets), and the Coiflets, or from the biorthogonal families of b-spline and v-spline wavelets [134]. The latter are symmetric and provide perfect reconstruction properties. We also have a choice between using decimated or undecimated transforms (see Section 5.2). Denoising using the coefficients of the undecimated transforms generally leads to fewer visual artifacts than denoising based on the corresponding decimated transform [117, 446], but requires more computational resources, in terms of both compute time and memory.

In addition to the wavelet, we need to select the number of levels of multiresolution and the boundary conditions necessary to handle the application of the wavelets at the boundaries of the image. It can be difficult to make appropriate choices for all these parameters, so some observations might be helpful. It has been argued that biorthogonal wavelets are preferable over the orthogonal ones as symmetry leads to visually more pleasing results in images [93]. However, a couple of studies [99, 198] have found that the choice of wavelet, the number of levels, and the boundary treatment rule has little effect on the results. So, a first option may be to keep the computational costs low by selecting a symmetric wavelet with a smaller support, a small number of multiresolution levels, and an easy-to-implement boundary condition. Based on the results, the options may then be varied appropriately.

For the thresholding step, we need to determine both a threshold value and a thresholding or shrinkage function, which determines how the threshold is applied to the wavelet coefficients. The thresholding function must be selected first as some threshold values depend on it.

There are four ways in which we can apply the threshold.

(1) The *hard* wavelet shrinkage function depends on a single threshold parameter. It keeps a wavelet coefficient if it is larger in absolute value than the positive threshold λ and sets it to zero otherwise:

$$f(x) = \begin{cases} 0 & \text{if } |x| \le \lambda, \\ x & \text{otherwise.} \end{cases} \tag{7.14}$$

(2) The *soft* wavelet shrinkage function also depends on a single threshold value. A coefficient is shrunk toward zero if its absolute value is larger than the positive threshold and is set to zero otherwise:

$$f(x) = \begin{cases} 0 & \text{if } |x| \le \lambda, \\ \operatorname{sgn}(x)(|x| - \lambda) & \text{otherwise.} \end{cases} \tag{7.15}$$

(3) The *garrote* wavelet shrinkage function also depends on a single threshold value. A coefficient is shrunk toward zero if its absolute value is larger than the positive threshold and is set to zero otherwise:

$$f(x) = \begin{cases} 0 & \text{if } |x| \le \lambda, \\ \left(|x| - \frac{\lambda^2}{x}\right) & \text{otherwise.} \end{cases} \tag{7.16}$$

(4) The *semisoft* shrinkage function is more general than either the hard or the soft shrinkage functions, and it includes both of them as special cases. The semisoft function depends on two positive thresholds, λ_1 and λ_2, with $\lambda_1 \le \lambda_2$:

$$f(x) = \begin{cases} 0, & |x| \le \lambda_1, \\ \operatorname{sgn}(x)\frac{\lambda_2(|x| - \lambda_1)}{\lambda_2 - \lambda_1}, & \lambda_1 < |x| \le \lambda_2, \\ x, & |x| > \lambda_2. \end{cases} \tag{7.17}$$

There are several options available for the choice of thresholds in wavelet denoising [198]. Some thresholds, such as the universal threshold, are independent of the thresholding function. Others, such as the Stein's Unbiased Risk Estimator (SURE), provide thresholds that depend on the shrinkage function, resulting in different thresholds for the different shrinkage functions. Also, certain thresholds assume a unit noise scale for the wavelet coefficients; others do not. This noise variance can be estimated, for example, by using the MAD estimator (equation (7.6)).

The *universal* threshold for a normal signal of size N is simply $\lambda = \sqrt{2 \log N}\sigma_n$, where σ_n is the standard deviation of the noise and can be estimated using equation (7.6) [155]. The universal threshold is determined independently of the thresholding function.

The *top* method for determining the threshold requires as input the percentage p of the wavelet coefficients to keep [66]. It does not need an estimate of the noise variance as it determines the thresholds directly from the data. Given p, the fraction of largest coefficients to keep, the threshold λ is simply set to be the $(1-p)$th quantile of the absolute values of the wavelet coefficients of interest. For the thresholding functions requiring a single threshold, coefficients above the threshold are assumed to comprise the signal, while those below are assumed to be part of the noise. Though this threshold does not have any theoretical optimality properties, it is simple enough to allow the user to experiment with any given data set by selecting different percentages of the wavelet coefficients to represent the signal. Note that the threshold is independent of the shrinkage function. For the semisoft shrinkage function, we obtain the lower and the upper thresholds by using two different values of p.

There are other, more complex, thresholds such as SURE [156] which depend on both the thresholding function and the multiresolution level, and the BayesShrink [97] which is appropriate for soft thresholding. While these do provide better denoising than the simpler methods, they are more complex to implement. A comparison of wavelet-based denoising with spatial filtering methods indicates that the latter are comparable in performance based on the mean-squared error metric, but that the wavelet-based approaches result in images which are less grainy [198].

Finally, an observation regarding wavelet-based approaches for denoising—they can sometimes create noticeable artifacts which degrade the image substantially [197]. This makes them unsuitable for use when a large number of images have to be processed in an automated way. However, more recent extensions to wavelets, such as ridgelets and curvelets, are possible alternatives which may address these issues [564, 157].

7.2.3 Partial differential equation–based approaches

The use of a Gaussian filter to denoise an image results in smoothing both the signal and the noise. To reduce the smoothing of an image in areas of interest such as edges, various researchers have turned to methods based on the solution of partial differential equations (PDEs), such as the diffusion equation. Convolving an image with a Gaussian filter G_σ,

$$G_\sigma(x,y) = \frac{1}{2\pi\sigma^2} \exp\left(- \frac{|x|^2 + |y|^2}{2\sigma^2} \right) \tag{7.18}$$

with standard deviation σ, is equivalent to the solution of the diffusion equation in two dimensions

$$\frac{\partial}{\partial t} I(x,y,t) = \nabla I(x,y,t) = \frac{\partial^2}{\partial x^2} I(x,y,t) + \frac{\partial^2}{\partial y^2} I(x,y,t), \tag{7.19}$$

where $I(x,y,t)$ is the two-dimensional image $I(x,y)$ at time $t = 0.5\sigma^2$, with initial conditions $I(x,y,0) = I_0(x,y)$, where I_0 is the original image. In a general form, this can be written as

$$\frac{\partial}{\partial t} I(x,y,t) = \nabla \cdot \Big(c(x,y,t)\nabla I(x,y,t)\Big),$$
$$I(x,y,0) = I_0(x,y), \tag{7.20}$$

where $c(x,y,t)$ is the diffusion conductance or diffusivity of the equation, ∇ is the gradient operator, and $\nabla\cdot$ is the divergence operator. When c is a constant, independent of x, y, or t, it leads to a linear diffusion equation, with a homogeneous diffusivity. In this case, all locations in the image, including the edges, are smoothed equally. As this is undesirable, a simple improvement is to change c with the location x and y in the image, thus converting the equation into a linear diffusion equation with nonhomogeneous diffusivity. When the function c is image dependent, the linear diffusion equation becomes a nonlinear diffusion equation. This approach was proposed by Perona and Malik in their 1987 work [473] (also see [474]), where they considered two values for the function c:

$$c(x,y,t) = 1 \Big/ \left(1 + \frac{|\nabla I|^2}{K^2}\right) \tag{7.21}$$

and

$$c(x,y,t) = \exp\Big(-|\nabla I|^2/2K^2\Big). \tag{7.22}$$

Here, c varies as a function of the image data, and it is small when the gradient of the image is large, resulting in lower diffusivity near the edges. The conductance parameter, K, enables backward diffusion when it is smaller than the gradient ∇I, thus enhancing the edges.

Note that if c is calculated based on the gradient of the original image I_0, it is independent of the time t, resulting in a linear nonhomogeneous diffusion equation. In the work of Perona and Malik, the gradient of c is taken of the image at time t, making it a nonlinear, nonhomogeneous diffusion equation. While Perona and Malik, as well as several other authors, use the term "anisotropic" to refer to the case where the diffusivity is a scalar function varying with the location, I reserve this term for the case where the diffusivity is a tensor, varying with both the location and the direction. Instead, following the terminology used in the PDE literature, I use the term "nonhomogeneous" to refer to the case where the scalar diffusivity varies with location.

The first formulation of the Perona–Malik approach resulted in an ill-posed problem, where images close to each other could produce divergent solutions and very different edges [523]. A common solution to this problem is to use a regularized (or smoothed) version of the image to calculate the gradient in equations (7.21) and (7.22); that is, the gradient ∇I is replaced by

$$\nabla I_G = \nabla(G_{\sigma_r} * I(x,y,t)), \tag{7.23}$$

where G_{σ_r} is another Gaussian with standard deviation σ_r. In other words, the gradient is taken after the image at time t is smoothed by convolution with a Gaussian. σ_r can be considered as a regularization parameter which describes a uniform smoothing used in the calculation of the gradient of the image at time t. The subscript r indicates that the σ is for the Gaussian used in regularization. This σ_r should not be confused with the σ of equation (7.18) which characterizes the time parameter of the smoothing through diffusion.

There are several options for the choice of the diffusivity function, $c(x,y,t)$. Following the original two suggestions by Perona and Malik (equations (7.21) and (7.22), and their regularized versions), several authors have suggested variations or enhancements. For example, Charbonnier et al. [101, 45] suggested using the diffusivity

$$c(x,y,t) = \left(1+\frac{|\nabla I_G|^2}{K^2}\right)^{-1/2}, \tag{7.24}$$

while Weickert [617] suggested the use of

$$c(x,y,t) = \begin{cases} 1 & \text{if } |\nabla I_G| = 0, \\ 1-\exp\Big(-C_m/(|\nabla I_G|^2/K^2)^m\Big) & \text{if } |\nabla I_G| > 0, \end{cases} \tag{7.25}$$

where the coefficient C_m, for a given value of m, is derived using

$$1 = \exp(-C_m)(1+2C_m m). \tag{7.26}$$

For $m = 2$, 3, and 4, $C_m = 2.33666$, 2.9183, and 3.31488, respectively. This value of the diffusivity function is designed to better preserve edges, with the effect most enhanced for larger values of m.

In addition to the form of the diffusivity, we also need to choose the value of σ_r used in the regularization of the image to calculate the gradient (see equation (7.23)). Although σ_r is often kept constant throughout the evolution of the PDE, this may not always be desirable. As the image is smoothed, and the unwanted intensity variations diminish faster than the signal, the gradient measurements become more reliable. To account for this, the parameter σ_r can be reduced as the equation evolves [622].

Another parameter to be determined in the diffusivity is the conductance parameter K. This is typically done experimentally, with a single value of K used during the entire evolution of the PDE. However, Li and Chen [378] have suggested that this might not be appropriate. When the gradient magnitude in the image exceeds the value K, the corresponding edge is enhanced. This assumes that the gradient magnitude of the noise is typically smaller than that of the edges in the image. But, in some images, this may not be the case, and as a result, both the edges and the noise (with gradient magnitude greater than K) are enhanced. Further, by fixing the value of K for the entire process, edges that lie beyond the range of edge enhancement can never be enhanced. Both these problems can be addressed by making K a function of the time t. If K is set high at the beginning, it can reduce the noise significantly, while enhancing only the edges with very high gradients. As the image is smoothed, the intensity variation of the noise reduces more than that of the signal. Since the strength of the noise has been reduced, the value of K can also be reduced without any adverse effects. This would allow edges with lower gradients to be enhanced, without correspondingly enhancing the noise. Thus, by changing K with time, edges with different gradients are enhanced at different times in the evolution of the PDE.

Once the diffusivity function $c(x,y,t)$, the regularization parameter σ_r, and the conductance parameter $K(t)$ have been chosen, the diffusion equation must be solved on the discrete grid represented by the image. Here, one can borrow extensively from the work that has already been done in the area of the solution of PDEs through techniques such as finite-difference and finite-element methods (see the references in Section 3.1.3). There are several options which need to be considered in the solution of the PDE itself, including the choice of a solution approach (e.g., finite difference or finite element), the discretization scheme (explicit or semi-implicit), and the solution of the resulting linear system of equations. Various solution approaches are discussed in [620, 23, 614].

The nonhomogeneous isotropic diffusion described so far limits the smoothing of an image near pixels with a large gradient magnitude, which are essentially the edge pixels. As the diffusion near an edge is minimal, the noise reduction near the edge is also small. To address this, anisotropic diffusion was proposed to allow the diffusion to be different along different directions defined by the local geometry of the image [617, 130]. Thus, diffusion across the edges of objects in an image can be prevented while allowing smoothing of the noise along the edge. This prevents the edge from being smoothed during the denoising.

The anisotropic form of the diffusion equation can be written as

$$\frac{\partial}{\partial t} I(x,y,t) = \nabla \cdot \Big(D(x,y,t) \nabla I(x,y,t) \Big), \tag{7.27}$$

where $D(x,y,t)$ is a symmetric, positive-definite diffusion tensor. This 2×2 matrix can be written in terms of its eigenvectors $\mathbf{v}_1$ and $\mathbf{v}_2$, and eigenvalues λ_1 and λ_2, as follows:

$$D = \begin{bmatrix} \mathbf{v}_1 & \mathbf{v}_2 \end{bmatrix} \begin{bmatrix} \lambda_1 & 0 \\ 0 & \lambda_2 \end{bmatrix} \begin{bmatrix} \mathbf{v}_1^T \\ \mathbf{v}_2^T \end{bmatrix}. \tag{7.28}$$

By appropriately choosing the eigenvalues and eigenvectors, different diffusivity tensors can be obtained. For example, in edge-enhancing diffusion, the eigenvectors are defined as follows:

$$\mathbf{v}_1 \parallel \nabla I_G \tag{7.29}$$

and

$$\mathbf{v}_2 \perp \nabla I_G. \tag{7.30}$$

Here, I_G is the regularized (or smoothed) version of the image, that is, the image convolved with a Gaussian filter with standard deviation σ_r (see equation (7.23)). Usually we choose $\lambda_2 = 1$ to allow smoothing in the $\mathbf{v}_2$ direction. λ_1 can be chosen to be any of the diffusivity functions from the traditional nonhomogeneous isotropic diffusion equation. This limits the diffusion in the direction of the gradient.

As with the isotropic diffusion equation, the anisotropic diffusion equation can be solved using techniques from PDEs [619]. Again, several choices must be made for various parameters as well as options for the PDE, the discretization scheme, the linear solver, the number of iterations, and so on. While the results of these techniques are competitive, they are nontrivial to implement, can be time consuming, and are sensitive to the choice of parameters [614, 615].

7.2.4 Removing multiplicative noise

Another problem which can degrade the quality of an image is that of variable illumination. This problem can be addressed by observing that an image $I(x,y)$ is the product of two factors, an illumination function $I_i(x,y)$ and a reflectance function $I_r(x,y)$:

$$I(x,y) = I_i(x,y)I_r(x,y). \tag{7.31}$$

The illumination typically comprises the low-frequency components of the Fourier transform of the image. The reflectance, on the other hand, is intrinsic to the imaged surface. Objects of different materials, when imaged next to each other, cause a sharp change in the reflectance, which is therefore associated with the higher-frequency components. To remove the effects of the nonuniform illumination, we first need to find an approximation to the low-frequency components and then "divide" them out from equation (7.31). Alternatively, working with natural logarithms, we can subtract out the component responsible for the variable illumination:

$$\ln(I(x,y)) = \ln(I_i(x,y)) + \ln(I_r(x,y)). \tag{7.32}$$

One approach to addressing the problem of variable illumination is the Retinex algorithm [357, 497], which is a member of a class of center/surround functions where the center is defined as each pixel value and the surround is a Gaussian function. The single-scale, monochromatic Retinex [306] is defined by

$$R(x,y) = \alpha\,[\ln\{I(x,y)\} - \ln\{I(x,y) * G(x,y)\}\,] - \beta, \tag{7.33}$$

where $I(x,y)$ is the original image, $G(x,y)$ is a Gaussian filter (see equation (7.4)), $R(x,y)$ is the Retinex output, and α and β are the scaling factor and offset which transform and control the output of the logarithm function. Here, the illumination is approximated by convolving the image with a Gaussian to extract the low-frequency components. The scale is determined by the σ of the Gaussian. The multiscale version of the Retinex method [305] is essentially a weighted sum of the single-scale Retinex at different scales:

$$R_M(x,y) = \sum_i \alpha_i\,[\,\ln\{I(x,y)\} - \ln\{I(x,y) * G_i(x,y)\}\,] - \beta, \tag{7.34}$$

where the different scales are obtained by convolving the image with Gaussians with different standard deviations. The sum of the weights is equal to 1.0. In a multiband image, the Retinex algorithm is applied to each band separately.

A technique similar to the Retinex method is that of homomorphic filtering [297] which is used in removing multiplicative speckle noise from synthetic radar imagery. However, comparisons indicate that it is outperformed by the Retinex approach on color images [496].

7.2.5 Problem-specific denoising

In some application domains, the data collected may have specific noise characteristics. For example, in Planar Laser-Induced Fluorescence (PLIF) images, such as the ones arising in experimental physics, the noise in the images appears as vertical streaks [321]. These are typically darker streaks on a lighter background and sometimes may be of the same intensity as parts of the objects of interest in the image. Similar "noise" in the form of vertical lines or streaks also appears in old movies where the film has been corrupted with the passage of time. In this case, we need to differentiate between a vertical streak due to noise and a vertical "streak" due to an object in the movie frame such as a rope hanging from a ceiling [68]. A general approach to denoising is usually not very successful in these cases and we may need to resort to techniques which are specific to the problem.

One commonly used approach is to build a model of what constitutes the noise and use the model to reduce the noise while leaving the signal unaffected. For example, in PLIF imagery, we can use an edge detector to determine edges in the objects, including the edges of the darker noise regions, which stand out from the lighter background. Then, we can exploit the fact that the edges due to the noise must be long straight-line segments, while the edges due to the signal are often curved or shorter straight-line segments. This characteristic of the noise enables us to selectively apply the more traditional denoising techniques only in the noisy regions of an image. However, in regions where the noise and signal characteristics are similar, such as signals which are short line segments near a noisy region which also appears as a short line segment, it may not be possible to completely remove the noise in the data [322].

A similar approach to removing problem-specific noise is used for removing line scratches from archived motion pictures [347, 348]. In this approach, tentative locations of lines are first identified using a Hough transform. Then, stochastic techniques are used to determine the lines which are more likely to fit a predefined model. A generalization of this approach, based on the hypothesis that scratches are not purely additive, is described in [68].

7.3 Contrast enhancement

The quality of image data may also be degraded due to poor contrast. This can be a problem when some areas of an image have good contrast, making it easy to identify the objects in these areas, while other regions have poor contrast, making it difficult to separate the objects of interest from the background. If we use, for example, a gradient-based method for identifying the objects (see Chapter 8), then, the gradient will be small in regions of low contrast and the objects in these regions are likely to be missed. Improving the contrast may not only make the follow-on processing feasible, it may also make it less sensitive to the parameters used.

There are several ways of improving the contrast in an image. The simplest is contrast stretching, where a linear transform is used to stretch the range of the intensity values in an image from its original narrow range to the full range allowed by the type of data used to represent the image. So, if the image is represented as 8-bit data, its range would be from 0 to 255. Thus, if I_{max} and I_{min} are the maximum and minimum intensity levels in the original image, and max and min are the maximum and minimum values possible for the data type, then the new values are obtained as

$$I_{new}(x,y) = (I(x,y) - I_{min})\frac{\max - \min}{I_{max} - I_{min}} + \min. \tag{7.35}$$

However, if the original image has a single outlier pixel with intensity much greater or smaller than the others, it can result in poor scaling of the entire image. A more robust approach to contrast stretching is to take a histogram of the image and then select the I_{min} and I_{max} values at the 5th and 95th percentiles in the histogram. Note that this linear scaling can also be used when converting from floating-point images to 8-bit images for the purpose of displaying the image.

A more sophisticated way of improving the contrast is to modify the dynamic range of the image by modifying the histogram of its intensity values. A common approach is to use histogram equalization which is essentially a nonlinear mapping which reassigns the pixel values such that the histogram of the new image is flat; that is, it has a uniform distribution of intensity values. It can be shown that a mapping which achieves this goal is one where the grey level of a pixel is replaced by the cumulative distribution of the intensities from the minimum grey level up to the grey level of the pixel. In other words, the mapping function is the cumulative probability distribution function of the image [227].

A global histogram equalization may not always be effective, especially when the contrast characteristics vary across the image. An adaptive histogram equalization approach addresses this problem by generating the mapping for each pixel by using the histogram of a window centered at the pixel [284]. As this is done for each pixel in the image, this can be computationally expensive. Alternatives such as dividing the image into regions and using the histogram of a region to determine the mapping for all pixels in the regions can lead to poor quality images due to the blocking effect at the boundary of the regions.

Several authors have proposed alternatives to the traditional definition of histogram equalization. Pizer et al. [478] have proposed a clipped version of adaptive histogram equalization, where the contrast which is allowed through histogram equalization is clipped or limited. This addresses regions of the mapping function where the slope is large and, as a result, two nearby pixel values are mapped to very different pixel values. An efficient implementation of this approach is described in [508] for real-time image enhancement. Yu and Bajaj [639] have proposed an adaptive mapping which uses the local statistics of the pixel such as the local maximum, the local minimum, and the local average intensity in a window around a pixel. They use a fast propagation scheme to calculate these quantities effectively.

Another approach which is used to enhance images is that of sharpening. When transitions between details in an image, such as an edge between a dark and a light region, occur within one pixel, the image captured is such that the intensity at that pixel corresponds to neither the dark nor the light intensity. To enhance such edges and other high-frequency components, as well as to adjust the contrast, we can use a sharpening filter. This filter is also referred to as an unsharp-masking filter, as it identifies the high-frequency components by subtracting out an unsharp, or smoothed, version of the image from the original. This high-frequency component is then scaled and added back to the original image, with the scaling factor determining the amount of sharpening. The unsharp image can be obtained by smoothing using a mean or Gaussian filter. Alternately, we can directly obtain the high-frequency component by using one of the following approximations to the negative Laplacian:

$$\begin{matrix} & -1 & \\ -1 & 4 & -1 \\ & -1 & \end{matrix} \qquad \begin{matrix} -1 & -1 & -1 \\ -1 & 8 & -1 \\ -1 & -1 & -1 \end{matrix} \qquad \begin{matrix} 1 & -2 & 1 \\ -2 & 4 & -2 \\ 1 & -2 & 1 \end{matrix}\,. \tag{7.36}$$

A different approach, which is popular in medical imaging, in particular for the analysis of mammograms, is the use of multiscale contrast enhancement. Here, the contrast enhancement techniques are applied to a multiresolution version of the image obtained by using either the wavelet transform [356, 654] or the Laplacian pyramid [76, 149]. While the wavelet transform can better enhance very small details, it can create artifacts in the large structures in the image. The Laplacian pyramid, on the other hand, appears to be more suitable, providing a balanced image impression.

In some images, the quality of the image can be improved through deblurring. In a single image, blurring is caused by convolution with a point spread function (PSF). The PSF describes how a point of light is transformed as it passes through the imaging system. The blur can be linear in one direction, or circularly symmetric, or a combination of different directional types of blur. One way of determining the PSF is by sending a point of light through the system and then examining what comes out. However, if the system is not available, the PSF can be estimated by seeing how known point sources in the image have been transformed. This would give us an idea of both the size of the mask and the distribution of weights in the mask which is causing the blur.

When the PSF of the blur is known, a typical solution approach is to filter the image in frequency domain [596]. The degraded image $I_d(x,y)$ can be considered to be the original clean image $I(x,y)$, convolved with the degradation function $p(x,y)$, which is the PSF, plus the noise $n(x,y)$ (in the case of additive noise):

$$I_d(x,y) = p(x,y) * I(x,y) + n(x,y). \tag{7.37}$$

In the Fourier domain, this becomes

$$I_D(u,v) = P(u,v) \times I(u,v) + N(u,v), \tag{7.38}$$

where the convolution is replaced by multiplication. The restored image can be obtained by applying a suitable filter to the Fourier-transformed degraded image $I_D(u,v)$ and then taking the inverse Fourier transform. The simplest filter is the inverse filter, which assumes the noise $n(x,y)$ to be zero and the filter to be an inverse of the PSF mask. The inversion is done point by point, where each value in the PSF mask is replaced by its inverse. This can, however, cause problems if the mask value is zero or if the degraded image has additive noise. The Wiener filter addresses these problems by attempting to model the error in the restored image and applying the filter such that this error is minimized. The Wiener filter is thus sometimes referred to as the minimum mean-square estimator. The Weiner filter is

$$\frac{P^*(u,v)}{|P(u,v)|^2 + \frac{S_n(u,v)}{S_I(u,v)}}, \tag{7.39}$$

where $P^*(u,v)$ is the complex conjugate of the Fourier transform $P(u,v)$ of the PSF, and the quantities

$$S_n(u,v) = |N(u,v)|^2 \quad \text{and} \quad S_I(u,v) = |I(u,v)|^2 \tag{7.40}$$

are the power spectra of the noise and the original image, respectively. Note that when the noise is zero, this reduces to the inverse filter. When the noise term is large, the filter value is small, thus attenuating the signal.

In practice, as the original signal is unavailable, the second term in the denominator of equation (7.39) is replaced by a parameter, which must be experimentally determined [596].

When the PSF of the imaging system is not known or cannot be determined, it is possible to deblur the image using blind deconvolution techniques [115]. More recently, a fast blind deconvolution approach based on fast Fourier transforms has been proposed which works well on several images with real blur [84].

7.4 Morphological techniques

Mathematical morphology methods are a class of techniques which can be used for many tasks in image analysis ranging from denoising and enhancement to segmentation and shape analysis. These nonlinear methods have their origin in the study of the geometry of porous media, which can be considered as "binary" data, where a point either belongs to a pore or to the material surrounding the pore. This led to the development of a set-based approach where a point would belong to either an object set (the pores) or its complement (the material surrounding the pores). The data were then processed using simple set operations such as union, intersection, complement, and translation.

The early work in morphological techniques was done by Matheron [422] who formalized the concept of granulometry, a tool used in the domain referred to as stereology, where the goal is to understand three-dimensional structure from cross sections. Matheron used granulometry to understand the distribution of pore sizes in porous media. Later, Serra [534] applied the concepts to solve image analysis problems, and the subject has been extensively studied since then, in the context of both binary and grey-scale images [557, 158].

Morphological techniques are best explained in the context of binary images. The basic idea, as developed in stereology, is to probe an image with a structuring element (also called a template) and to quantify the manner in which the element fits, or does not fit, inside the objects in the image. This approach is very dependent, of course, on the shape and size of the structuring element. Typical elements include crosses, squares, and circles; these can range in size from small, such as 3×3, to larger ones, as required by the problem. Associated with each element is its center, which is translated over each pixel in the image.

There are two primary operations one can perform using the structuring element and a binary image. The operation of "erosion" identifies all pixel locations in the image such that, when the center of the template is placed at the pixel location, the template fits completely inside the objects of interest. In a binary image, the objects of interest are typically all the black regions, with the white regions forming the background. This operation results in the black regions shrinking in size—hence the name, erosion. The amount of shrinkage is determined by the size of the template. Objects smaller than the size of the template will disappear, enabling the use of erosion to remove small objects. The dual of erosion, referred to as "dilation," is obtained by considering all pixel locations such that when the center of the structuring element is placed on these locations, the element hits (or intersects) the object of interest. If we use a 3×3 structuring element, dilation has the effect of changing all background pixels neighboring an object to object pixels (hence the name dilation). It can also be used to close small holes in an object.

In addition to the erosion and dilation operations, there are two secondary operations, opening and closing. "Opening" is the union of all translations of the structuring element which fit inside the objects in the image. Opening is equivalent to erosion followed by dilation. Its complement, "closing," is dilation followed by erosion.

Using these four basic operations, we can implement several image processing algorithms. For example, the inner boundary of an object can be considered as the difference between an object and its eroded version using a circular element, while the outer boundary is the difference between the dilated version and the original object. One can also use two structuring elements simultaneously, one to probe the inside of the object and the other to probe the outside of the object. This is referred to as the hit-or-miss transform and probes the relationship between the objects and their complement relative to the pair of elements. Its most common use is in thinning an object to find its skeleton.

The extension of morphological operations to grey scale essentially involves extending the definition of various set operations such as translation, subset, intersection, and union. For example, intersection and union are replaced by minimum and maximum, respectively. Erosion (dilation) of an image results in assigning to each pixel the minimum (maximum) value found over a neighborhood of the pixel in the image, where the neighborhood is determined by the structuring element.

The application of morphological operations is computationally inexpensive, though care must be taken in the appropriate selection of the structuring elements to achieve the desired results. Additional information on the use of morphological operations for enhancing images by improving the contrast, reducing the noise, and minimizing differences in illumination are discussed in [557, 158] as well as Chapter 11 of the text by Sonka et al. [559].

7.5 Summary

In this chapter, I discussed how image data can be enhanced to make it easier to extract information from them. This enhancement can involve reducing the noise, improving the contrast, and deblurring the images. The techniques discussed can be applied to one-dimensional signals as well as two-dimensional images. They can also be applied across the frames of a video sequence to reduce temporal noise.

The approach used in enhancing image data often depends on the type of enhancement required. Usually, it is necessary to use more than one method or try several different parameter settings in an algorithm before finding a combination which works well. If the algorithm and the parameter settings are not chosen with care, the image enhancement algorithm may result in a poorer quality image than the original. In some cases, for example in denoising, it might not be possible to reduce the noise without adversely affecting the signal. This makes image enhancement one of the more difficult steps to automate in the scientific data mining process.

7.6 Suggestions for further reading

A good source of suggestions for enhancing scientific image data is the domain scientists who may have collected the data and are therefore familiar with the systems which were used to generate the data, the characteristics of the noise in the data, as well as domain-specific techniques used for enhancing the data. In addition, journals in the appropriate field can be an excellent source of information, for example, the *IEEE Transactions on Medical Imaging* can provide valuable insights into the approaches used in the enhancement of medical images.

There are several books which provide additional information. Introductory texts, such as the one by Petrou and Bosdogianni [476] discussing the fundamentals of image processing and the book by Umbaugh [596] discussing computer vision and image processing techniques from a practical viewpoint, provide an easy-to-understand introduction to the subject. These books also provide some intuitive insights into the reasons why the methods work and the issues which must be considered in the practical application of the techniques. Other advanced techniques for enhancing images are described in the books by Jain [297] and Sonka et al. [559].

More sophisticated approaches are usually discussed in books which are specific to a domain or a solution approach. For example, the techniques based on PDEs are discussed in the books by Chan and Shen on variational methods [94]; Sapiro [523] on the use of geometric PDEs in image analysis; and by Sethian [536] in his text on level set methods and fast marching methods. For additional information on denoising techniques based on wavelets and related methods, the text by Jansen [301] on noise reduction by wavelet thresholding provides statistical insights, and the papers by Donoho and coauthors [157] provide details on the recent developments in the field. For domain-specific approaches in medical imaging, the edited handbooks by Beutal, Kundel, and Van Metter [37] and by Bankman [20], as well as the text by Rangayyan on biomedical image analysis [502], are all excellent reference texts. More information on enhancing synthetic aperture radar images is provided in the book by Oliver and Quegan [459]. For video degraded by motion blur, Tekalp [576] discusses ways of restoring the video.

Chapter 8

Finding Objects in the Data

Problems worthy
of attack
prove their worth
by hitting back.

—Piet Hein [267, p. 2]

In Chapter 7, I described how data, especially image data, can be enhanced to make it easier to identify the objects of interest in the data. As discussed in various applications in Chapter 2, these objects could be galaxies in astronomy images, roads or fields in satellite images, tumors in medical images, or vehicles in surveillance data. In this chapter, I describe how we can identify these objects by separating them from the background, a process also known as image segmentation.

There are a variety of techniques for object identification. Some are edge based and focus on the boundary or edge of an object, exploiting the fact that since the object is expected to be different from the background, there should be a gradient in the pixel intensities at the boundary of the object. Other methods, which are region based, focus on the object itself instead of its boundary and are based on the expectation that object pixels are more similar to each other than they are to the background. Both the edge- and region-based methods have their pros and cons, leading some to consider a combination of the two approaches. In other problems, such as tracking in video data, where we are interested in finding the moving objects, we can exploit the motion of the objects to detect them.

This chapter is organized as follows. In Sections 8.1 and 8.2, I describe edge- and region-based techniques. These range from the very simple to the more complex. Section 8.3 describes the concept of salient points or regions, where instead of focusing on objects, we consider landmarks in an image. In Section 8.4, I discuss the problem of detecting moving objects in video, where the motion of an object is used to identify it from the background, which is assumed to be stationary. Section 8.5 describes some domain-specific approaches which are being used or must be developed for problems which do not lend themselves to the more traditional approaches. Since the process of object identification can often result in spurious objects or single objects which are split into many parts, some postprocessing is required to correctly identify the objects. I address these issues in Sections 8.6 and 8.7, focusing on ways of associating image pixels with objects and cleanup techniques, respectively. Section 8.8 describes data structures to represent the objects identified so they can be used in further analysis. Finally, Section 8.9 summarizes the chapter and Section 8.10 provides suggestions for further reading.

8.1 Edge-based techniques

In edge-based techniques for segmentation, we exploit the fact that at the boundary of an object, there is a gradient in the intensity values across the edge. The magnitude of this gradient is dependent on how well the object is separated from the background, that is, the difference in intensities of the object pixels and the background pixels in the region around the edge of the object.

There are several ways in which the gradient of the image can be obtained. These approaches typically involve convolving an image with appropriate filters (see equation (7.1)). These filters, or gradient operators, obtain the directional derivatives in two orthogonal directions at each point in the image. They range from the simplest, which is the Roberts' cross operator [511],

$$\begin{pmatrix} 0 & 1 \\ -1 & 0 \end{pmatrix}, \qquad \begin{pmatrix} 1 & 0 \\ 0 & -1 \end{pmatrix}, \tag{8.1}$$

to the more complex Prewitt operators [484],

$$\begin{pmatrix} -1 & 0 & 1 \\ -1 & 0 & 1 \\ -1 & 0 & 1 \end{pmatrix}, \qquad \begin{pmatrix} 1 & 1 & 1 \\ 0 & 0 & 0 \\ -1 & -1 & -1 \end{pmatrix} \tag{8.2}$$

and the Sobel operators

$$\begin{pmatrix} -1 & 0 & 1 \\ -2 & 0 & 2 \\ -1 & 0 & 1 \end{pmatrix}, \qquad \begin{pmatrix} 1 & 2 & 1 \\ 0 & 0 & 0 \\ -1 & -2 & -1 \end{pmatrix}. \tag{8.3}$$

Note that the Roberts' operators operate on the diagonal elements of a 2×2 window around the current pixel, while the Prewitt and Sobel operators operate along the x and y directions using 3×3 windows. Without loss of generality, we can denote the derivatives in the two orthogonal directions by $\partial I(x,y)/\partial x$ and $\partial I(x,y)/\partial y$, where the two gradients are obtained by convolution with the operators described above. Then, the magnitude of the gradient at the pixel location $I(x,y)$ is given by

$$\sqrt{(\partial I(x,y)/\partial x)^2 + (\partial I(x,y)/\partial y)^2} \tag{8.4}$$

and the direction of the gradient with respect to the x-axis is given by

$$\tan^{-1}\Big(\frac{\partial I(x,y)}{\partial y} \Big/ \frac{\partial I(x,y)}{\partial x}\Big). \tag{8.5}$$

The values in the gradient operators add up to 0, indicating that the gradient is zero in regions with uniform pixel intensities. Both the Prewitt and the Sobel operators obtain the difference of pixel intensities in one direction while smoothing (or taking the average) in the other direction. In the case of the Sobel operators, the weighted average is used.

Once the gradient magnitude has been obtained, it can be thresholded to obtain the edge image. Usually, a simple thresholding is done, where pixels with gradient magnitude less than a certain value are dropped from the edge image. More sophisticated thresholding schemes are also possible, as I describe in the next section.

The gradient-based techniques described thus far tend to work well in images with low noise where the objects of interest have intensity values which are very different from the background, resulting in a high gradient at the edges of the objects. The presence of noise is handled to some extent by the Prewitt and Sobel operators as they perform some limited smoothing of the image. Another approach to this problem, proposed by Marr and Hildreth [421], is to smooth the image first by a Gaussian and then compute its Laplacian, which is equivalent to convolving the image with the Laplacian of a Gaussian:

$$\nabla^2(G_\sigma * I) = (\nabla^2 G_\sigma) * I. \tag{8.6}$$

The Gaussian is defined as

$$G_\sigma(x,y) = \frac{1}{2\pi\sigma^2}\exp\Big(-\frac{|x|^2+|y|^2}{2\sigma^2}\Big), \tag{8.7}$$

where σ is the standard deviation of the Gaussian. The operator ∇^2 produces zero-crossings along the edges in an image; these zero-crossings can be shown to form closed contours. Marr and Hildreth suggest applying equation (8.6) to the image at several different scales, that is, for several different values of σ, and then combining the results to produce the final edge image.

In Figure 8.1, I show the effect of noise in detecting edges in the synthetic image from Figure 7.1. Only the lower left region of the full image is shown for clarity. The images in the right column display the inverse of the magnitude of the gradient obtained using the Sobel operators on the images in the left column. The inverse is used as it is easier to see the effects of edge detection in the presence of noise. A darker value in these edge images indicates a higher gradient magnitude.

This simple example reveals several important challenges to edge detection. In the absence of noise, as in panel (b), the edges can be clearly detected. When a small amount of noise (10%) is added, the edges can still be detected (panel (d)), but there will be nonzero values for the magnitude of the gradient in regions which previously had uniform intensity. This is clearly seen in panel (f), where the image has been corrupted by 20% random noise. Further, noise in regions with a low magnitude of the gradient can make it difficult to identify the edge, as is the case in the edge detected in the upper right corner in panel (f). Also, the gradient magnitude along an edge now varies, so that if a simple thresholding is used to detect the edge pixels, the edge will be irregular with potentially some gaps where the gradient magnitude falls below the threshold.

If we smooth the image prior to edge detection, as shown in panel (g), which is the image in panel (e) after smoothing using a 3×3 Gaussian, we partially address these problems, but create others. Now, the effect of noise in previously uniform-intensity regions is minimized. However, the smoothing diffuses the edge, reducing the value of the magnitude of the gradient, as indicated by the light grey regions in panel (h) in contrast to the darker grey regions in panel (f). If the magnitude of the gradient is small in the original image, this can result in completely missing an edge, as is the case for the edge in the top right corner.

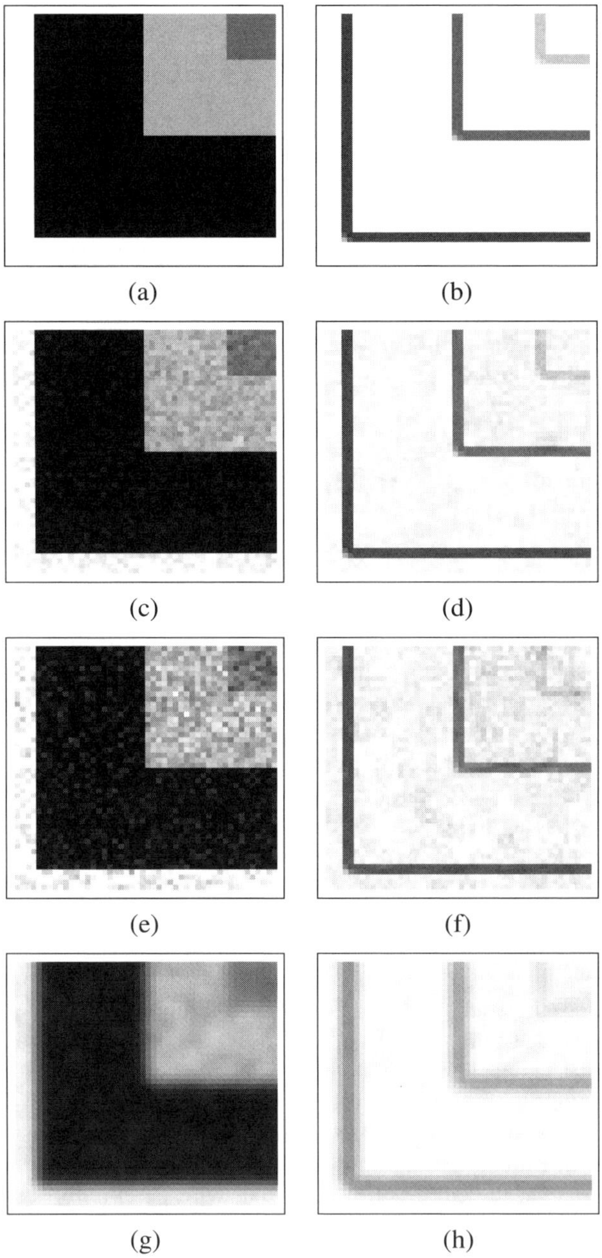

Figure 8.1. *Edge detection in the presence of noise: The images in the right column are the inverse of the magnitude of the gradient obtained using the Sobel operator on the images in the left column.* (a) *The original image.* (c) *The original image with* 10% *random noise.* (e) *The original image with* 20% *random noise.* (g) *The image in* (e) *after the application of a* 3×3 *Gaussian filter.*

Further, the smoothing results in a wider edge as indicated by the blurred darker regions in panel (h).

This illustrates that we should take care in detecting objects in noisy images, as processing the data to address one problem may result in another problem.

8.1.1 The Canny edge detector

Another popular edge detection algorithm is the one proposed by Canny [80] which, like the Marr–Hildreth method, uses Gaussian smoothing at different scales. The Canny edge detector was proposed to meet certain desirable criteria in edge detection including good localization of the edge, a single response to a sharp edge, and few false alarms.

The Canny edge detector first smooths the image using a Gaussian with a specific value of σ. Next, a simple gradient-based edge detector is applied to calculate the magnitude and the direction of the gradient at each pixel. The Sobel operator is typically used, though it is possible to use other methods. Then, instead of applying a simple thresholding operator to the gradient magnitude, the Canny approach first thins the edges obtained and then uses two thresholds to obtain the final edge image.

The thinning of the edges in the Canny edge detector is done by keeping those pixels for which the magnitude of the gradient is locally maximum with respect to the gradient magnitudes along the direction of the gradient. Referring to Figure 8.2, where the gradient direction makes an angle θ with the x-axis, we first calculate the magnitude of the gradients at points A and B on either side of the pixel P in the direction of the gradient. This can be done using linear interpolation. Next, the pixel P considered further only if its gradient magnitude is greater than the gradient magnitudes at points A and B. This process is referred to as nonmaximal suppression, indicating the suppression of values which are not maximal in the gradient direction.

After nonmaximal suppression, the edge image is obtained by thresholding. Canny proposed the use of two thresholds T_{low} and T_{high}. A pixel is retained in the edge image if its gradient magnitude is larger than T_{high}. A pixel is dropped from the edge image if its gradient magnitude is smaller than T_{low}. For pixels with values in between T_{low} and T_{high}, a pixel is retained only if it is adjacent to a pixel which is retained in the edge image. This recursive thresholding, which is also referred to as hysteresis thresholding, extends and fills in the edges determined by T_{high}.

In the original method proposed by Canny, he suggested that different values of σ be used in the Gaussian and the resulting edge images be integrated for the final result. The use of different σ has the effect of determining the scale of the objects detected by the edge detector. When σ is small, the finer details in the image are captured. As σ is increased, the coarser details of the image are selected as being more important. Thus, by varying σ, we can change the level of details which are discernable in an image [382].

8.1.2 Active contours

I next discuss a different category of edge-based segmentation methods that is based on the computation of paths of minimal energy. These techniques, which are referred to as active contours or deformable models, start with an initial guess for the contour of the objects of interest. This contour is then moved by image-driven forces to the boundaries of the desired objects. In this approach, two types of forces are considered: the internal forces,

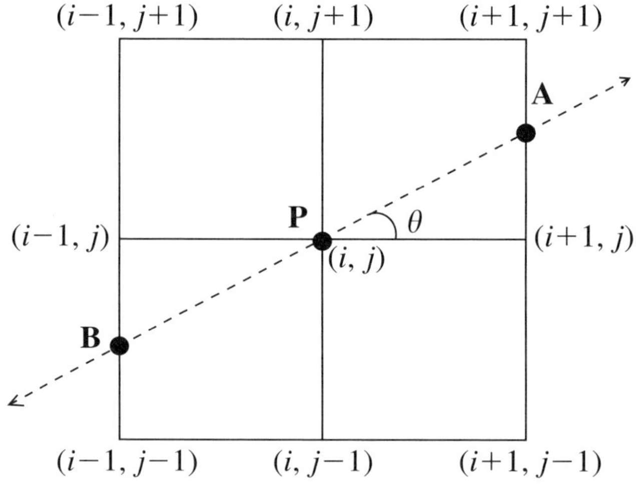

Figure 8.2. *Nonmaximal suppression in the Canny edge detector: The pixel at location $P = (i, j)$ is considered further only if the gradient at P is greater than the gradients at locations A and B which are on either side of P in the direction of the gradient (indicated by the dotted line). The gradients at locations A and B are obtained through interpolation from the gradient values at neighboring pixels.*

which are defined within the contour and are designed to keep the model smooth during the deformation process, and the external forces, which are computed from the underlying image data and are defined to move the model toward an object boundary or other desired features within the image.

There are two forms of deformable models. In the parametric form, also referred to as snakes, an explicit parametric representation of the curve is used [331]. This form is not only compact with an efficient implementation, but also robust to both image noise and boundary gaps as it constrains the extracted boundaries to be smooth. However, there are several difficulties with the parametric form of active contours. They are sensitive to the initial conditions, and the contour must be initialized near the objects of interest. They are not suitable for objects with deep cavities as their formulation prevents them from progressing into boundary concavities. Further, the internal energy constraints of snake models can limit their geometric flexibility and prevent them from representing long tube-like shapes or objects with significant protrusions. Topological changes are also difficult to handle with the traditional implementation of snakes as the formulation severely restricts the degree of topological adaptability of the model, especially if the deformation involves splitting or merging of the contour. All these can cause problems in automated segmentation of images from scientific domains where the number or shape of objects in an image is usually unknown. Several approaches have been proposed to overcome these difficulties with the parametric form of active contours [424, 634, 633] and the topic remains one of active research.

Given these limitations of the parametric deformable models, I will focus on the geometric form of these models, which are also referred to as level set methods or geodesic active contours. As I shall discuss at the end of this section, these methods also have their own drawbacks.

The implicit active contour, or level set, approach was introduced by Osher and Sethian [466] and has since been enhanced by several authors [536, 408, 465, 464]. An easy-to-understand high-level description of the level set method is given in [535]. The basic idea is to start with a closed curve in two dimensions (or a surface in three dimensions) and allow the curve to move perpendicular to itself at a prescribed speed. Instead of explicitly describing the curve using a parametric form as in the snakes approach, the implicit active contour approach takes the original curve and embeds it in a higher-dimensional scalar function, defined over the entire image. The curve is now represented implicitly as the zeroth level set (or contour) of this scalar function. Over the rest of the image space, this level set function ϕ is defined as the signed distance function from the zeroth level set. Specifically, given a closed curve C_0, the function is zero if the pixel lies on the curve itself, otherwise, it is the signed minimum distance from the pixel to the curve. By convention, the distance is regarded as negative for pixels inside C_0 and positive for pixels outside C_0. The function ϕ, which varies with space and time (that is, $\phi = \phi(x,y,t)$ in two dimensions) is then evolved using a partial differential equation (PDE), containing terms that are either hyperbolic or parabolic in nature.

In order to illustrate the origins of this PDE, I next consider the evolution of the function ϕ as it evolves in a direction normal to itself with a known speed F. Here, the normal is oriented with respect to an outside and an inside. Since the evolving front is a zero level set (that is, a contour with value zero) of this function, we require (using a one-dimensional example) that the following condition:

$$\phi(x(t),t) = 0 \tag{8.8}$$

be satisfied for any point $x(t)$ on the zero level set at time t. Using the chain rule, we have

$$\phi_t + \nabla\phi(x(t),t) \cdot x'(t) = 0. \tag{8.9}$$

Since F is the speed in the direction n, normal to the curve, we have

$$x'(t) \cdot n = F, \tag{8.10}$$

where

$$n = \frac{\nabla\phi}{\|\nabla\phi\|}. \tag{8.11}$$

Thus, the evolution of ϕ can be written as

$$\phi_t + F\|\nabla\phi\| = 0, \tag{8.12}$$

where $\phi(x, t = 0)$, that is, the curve at time $t = 0$, is given. This formulation enables us to handle topological changes as the zero level set need not be a single curve, but can easily split and/or merge as t advances.

Equation (8.12) can be solved using appropriate finite difference approximations for the spatial and temporal derivatives [373]. The image pixels are considered to be a discrete grid in the x-y domain with uniform mesh spacing. In order to evolve the level set, we need the specification of an initial closed curve(s), the initialization of the signed distance function ϕ over the rest of the image, the finite difference discretization of (8.12), and the prescription of the propagation speed F. We next discuss each of these issues in detail.

- **Choice of the speed function F:** The speed F depends on many factors including the local properties of the curve, such as the curvature, and the global properties, such as the shape and the position of the front (i.e., the zero level set). It can be used to control the front in several different ways. The original level set method proposed using F as the sum of two terms

$$F = F_0 + F_1(\kappa), \tag{8.13}$$

where F_0 is a constant propagation term and F_1 is a scalar function of the curvature κ,

$$\kappa = \mathrm{div}\Big(\frac{\nabla\phi}{\|\nabla\phi\|}\Big). \tag{8.14}$$

The propagation term F_0, sometimes referred to as the "balloon force," is independent of the geometry of the moving front; the front uniformly expands or contracts with speed F_0, depending on its sign [409]. Note that the F_0 term is hyperbolic, while the F_1 term is parabolic. The key idea is to play one term against the other—the hyperbolic term leads to the formation of shocks from which a representation of the shape can be deduced, while the curvature term smooths the front, which enables us to distinguish the more significant shape features from the less significant ones [409, 545].

Since we are using level sets for image segmentation, we want the zeroth level set to stop evolving at an edge in the image. This can be accomplished by multiplying the speed term by an image-dependent stopping factor $g(x,y)$ given by the following equation:

$$g(x,y) = 1\Big/\Big(1 + \frac{|\nabla I_G|^2}{K^2}\Big), \tag{8.15}$$

where I_G is the image I convolved with a Gaussian with standard deviation σ (see (8.7)).

Equation (8.15) is also the one used by Perona and Malik in their work on smoothing images using the diffusion equation [474] and discussed in more detail in Section 7.2.3. In the original work, this term was used to stop the diffusion process near an edge of an image. When used in the context of level set segmentation, it has the effect of reducing the propagation speed near an edge in the image where the gradient is high. By suitably changing the value of K, we can control the strength of the edges that cause the speed to become zero near an edge. With this definition of the speed F, the level set equation can be written as

$$\frac{\partial\phi}{\partial t} + g(x,y)\,\|\nabla\phi\|\,\Big(\mathrm{div}\Big(\frac{\nabla\phi}{\|\nabla\phi\|}\Big) + F_0\Big) = 0. \tag{8.16}$$

Further modifications to the speed term have been proposed by several authors. For example Yezzi et al. [339, 637] observe that the stopping factor given by equation (8.15) will not force the front to stop at an edge unless it is exactly zero, a situation that rarely occurs in realistic images. They propose adding an additional term of the form $(\nabla g \cdot \nabla \phi)$ to the left-hand side of equation (8.16) resulting in the following equation for the evolution of the level set:

$$\frac{\partial \phi}{\partial t} + g(x,y)\, \|\nabla \phi\| \left(\operatorname{div}\left(\frac{\nabla \phi}{\|\nabla \phi\|}\right) + F_0 \right) + \nabla g \cdot \nabla \phi = 0. \tag{8.17}$$

Since the edge can be defined as a step function, and ∇g involves two derivatives of the image, the term ∇g acts as a "doublet" near an edge. It has the effect of attracting the evolving contour as it approaches an edge and then pushing the contour back out if it should pass the edge. This additional term is called the pullback term or the edge strength stopping force multiplier. With the addition of the doublet term, $(\nabla g \cdot \nabla \phi)$, the level set equation (8.17) becomes equivalent to the geodesic snake model that was proposed simultaneously by Caselles et al. [90] and Kichenassamy et al. [339].

- **The placement of the initial contour:** A key challenge in both implicit active contours and snakes is the placement of the initial contour. Since the contour moves either inward or outward, its initial placement will determine the segmentation that is obtained. For example, if there is a single object in an image, an initial contour placed outside the object and propagated inward will segment the outer boundary of the object. However, if this object has a hole in the middle, it will not be possible to obtain the boundary of this hole unless the initial contour is placed inside the hole and propagated outward. We must be careful to ensure that the sign of the so-called *balloon force* (F_0) is selected to generate the appropriate direction of propagation of the initial contour(s). More than one closed curve can also be used for initialization of the zeroth level set, for example, by starting with several seed contours identified in the image, doing an initial segmentation, and then using region merging (see Section 8.2) to merge small regions [640].

- **Calculation of the distance function:** Once the initial contour has been determined, we need to calculate the signed distance function ϕ, that is, the minimum distance from each pixel in the image to the prescribed initial contour. This can be done by solving the Eikonal equation [536, 572] which is derived from the level set formulation as follows. Suppose the speed function F is greater than zero. As the front moves outward, one way to characterize the position of the front is to compute the arrival time $T(x,y)$ as it crosses each point (x,y). This arrival function is related to the speed by

$$\|\nabla T\| F = 1, \tag{8.18}$$

where $T = 0$ is the initial contour. When the speed F depends only on the position, equation (8.18) is referred to as the Eikonal equation. The solution of this equation for a constant speed of unity gives the distance function. The sign is attached to the function depending on the location of each pixel relative to the original contour. A fast sweeping method to solve the Eikonal equation is described by Zhao [646].

As the front evolves, the signed distance function can often lose the "distance property" [226]. As a result, when the curve stops evolving near an edge, the zero level set is not exactly at the edge. One solution to this problem is to periodically reinitialize the distance function ϕ. However, this can add significantly to the overall computational cost of the level set approach.

- **The discretization of the level set equation:** In order to evolve the level set equation, equation (8.17) must be solved on the discrete grid represented by the image. Using the finite difference approach [373], we can consider the discretized version of the image $I(x, y)$ to correspond to the intensity at the pixels (i, j), at locations (x_i, y_j), where $i = 1, \ldots, N$, and $j = 1, \ldots, M$. The distance between the centers of the pixels, referred to as the grid spacing, is h. The same interpixel distance h is used along the x and y dimensions.

  Following the approach of Sethian [536, page 73], the level set equation (8.17) can be written as

$$\frac{\partial \phi}{\partial t} + g(x, y)\ \|\nabla \phi\|\ \mathrm{div}\left(\frac{\nabla \phi}{\|\nabla \phi\|}\right) + F_0\ g(x, y)\ \|\nabla \phi\| + \nabla g \cdot \nabla \phi = 0. \tag{8.19}$$

  The second term on the left-hand side is the curvature term and is the parabolic contribution to the equation. This can be discretized using a second-order accurate central difference scheme such as the one described by Weickert [618]. The third term is the hyperbolic propagation term and can be discretized using the first-order accurate upwind differencing scheme given in Sethian [536]. The fourth term is a pure advection term and can also be discretized using a similar first-order accurate upwind differencing scheme.

  When equation (8.17) is evolved using an explicit time integration scheme, such as the forward Euler scheme, the time step size needs to be bounded in order to maintain the numerical stability. In general, the stability restrictions imposed by the parabolic terms tend to be significantly more restrictive than those imposed by the hyperbolic terms. To alleviate this overly restrictive stability criteria, the time integration of the parabolic term cna be carried out using an implicit scheme such as additive operator splitting (AOS) [618].

A comparison of level set techniques with other edge detectors such as the Canny edge detector is presented in [616]. In this preliminary study, several observations are made about the pros and cons of the methods considered. The traditional edge detectors, such as the Canny edge detector, are computationally very efficient and require the setting of just a few parameters. In contrast, the implicit active contours are computationally very expensive. Several options have to be determined including the strength of the balloon force, the parameters σ and K in equation (8.15), the number of time steps before reinitialization of the distance function, the placement of the initial contour, and whether we should use the doublet term or not.

Both the level sets and the traditional edge detectors require the parameters to be tuned to the image and the type of structure that is to be segmented. As the level sets have several parameters, it may require several attempts of an expensive algorithm before we find an appropriate combination. If there are several images which must be segmented in an automated way, level sets might not be the best first choice of a segmentation algorithm. The active contour approaches also require a good knowledge of the computational solution of PDEs, making them difficult to implement. However, they do have a very appealing quality of generating closed contours, thus eliminating the tedious postprocessing required to convert the edges from an edge detector into closed boundaries of an object.

8.1.3 The USAN approach

The USAN principle [553] was introduced in Chapter 7 for the structure preserving denoising of images. As observed earlier, one of the main applications of the principle is in the calculations of edges and corners in the image.

The basic idea behind USAN, or the Univalue Segment Assimilating Nucleus, is to consider the area in a circular mask where the pixel intensities are similar to the intensity of the central pixel, which is referred to as the nucleus. Referring to Figure 7.2, when the nucleus of a circular mask is near a straight edge of an object of uniform intensity, the region it covers will be half in the object and half in the background. Thus, the area of pixels which are similar to the nucleus will be half the area of the mask. If the nucleus is located such that the entire mask is within the object, the USAN is the area of the mask. And, if the nucleus is located at a corner, the USAN will be less than half the area of the mask. This observation led to the SUSAN principle, or the Smallest Univalue Segment Assimilating Nucleus principle, which states that an image processed to give the inverted USAN area at each pixel will have the edges and corners enhanced, with the corners (whose USAN is smaller than the USAN at the edges), enhanced more strongly.

Originally, the SUSAN method was implemented by considering a pixel to contribute a fixed value of unity to the area if its intensity was within a threshold of the intensity of the pixel at the location of the nucleus. However, instead of contributing all or nothing to the area, the similarity function was changed so that the contribution of each pixel was a function of the difference between the two pixel values. This allowed the contribution to be unity when the difference was close to zero, and allowed a gradual fall in the contribution as the difference became closer to the threshold.

The edge response is then obtained by comparing the USAN with a fixed threshold, usually set to 0.75 times the maximum value of the USAN. The smaller the USAN relative to this threshold, the larger the edge response. Additional details on the SUSAN approach for edge detection, including the similarity function, are described in [553].

It should be noted that the SUSAN technique is somewhat unusual in its approach to edge detection as it uses no derivatives in its implementation.

8.2 Region-based methods

Unlike the edge-based techniques, the region-based approaches focus on the "inside" of an object instead of its boundary. They exploit the fact that the pixels which constitute an object are more similar to each other than to the background. This similarity can be implemented via a homogeneity metric which is often based on the pixel intensities. However, depending on the problem, the metric can be more complex or include combinations of simpler metrics. For example, the metric may require that the variance of pixel intensities in a region be below a particular threshold or that the texture features calculated using a small window around each pixel be similar for all pixels in a region. In addition, problem-specific constraints based on the shape or size of the regions could be imposed to discard regions which do not meet these requirements.

A very simple approach to region-based segmentation is thresholding, where pixels with intensities below or above a certain value are considered as the objects. This approach can work surprisingly well in problems where the objects of interest all have pixels with similar intensities which is different from the intensities of nonobject pixels. A key challenge in this approach is the determination of the threshold, which is done either experimentally by trying out different threshold values, by considering the shape of the histogram of the image [559], or by exploiting known information about the intensities of the objects in an image.

More complex region-based methods, which are described next, often use the concept of connectivity or neighbors of a pixel. This general topic is discussed further in Section 8.6.

8.2.1 Region splitting

In this approach to region-based segmentation [227, 559], a region is split into subregions if it does not meet the homogeneity criterion. In the simplest version, the process starts with the entire image being considered as a region. As there are likely to be several objects in the image, as well as a background, it is likely that the full image will not meet the homogeneity criteria. So, the image is split into four quadrants. Next, each of these quadrants is considered individually, and if the corresponding region does not meet the homogeneity criteria, it is split further into four. The process continues until all regions meet the criteria or are small enough that further splitting is deemed unnecessary. The latter condition occurs naturally when each region is comprised of a single pixel, but a threshold on the smallest size of a region can also be imposed.

Note that this approach does not require the image to be square or the length of each side to be a power of two. If the length of a side of an image or region is odd, the region is split into unequal quadrants. As a result, if a cropped version of the image were segmented, the results may not be identical to those obtained by applying the technique to the full image.

8.2.2 Region merging

This approach is the opposite of the region-splitting approach. Here the original image is first split into many small regions, which are then merged as long as the newly created region continues to satisfy the homogeneity metric. This requires the identification of seed regions, which are the initial regions chosen as the ones which will grow as a result of the merging process. The neighboring regions around each seed pixel are considered for merging with the seed pixel. If the newly formed region satisfies the homogeneity criteria, the regions are merged. The process is stopped when a region cannot grow anymore and each pixel in the image has been assigned to a region.

The image is first divided into small regions each composed of either a single pixel or a small number (say 4 or 16) of pixels. In the absence of seed regions which can be used to start the region growing, the process considers each region in turn. The results usually depend on the order in which the regions are considered in the merging process. This can lead to undesirable results, for example, when the homogeneity metric allows one region to grow at the expense of others simply because the subregions of that region were considered for merging first. This problem can occur even if the seeds are known and care must be taken to grow regions around each seed in turn, instead of growing the region around the first seed to completion before starting on the next seed.

Several approaches have been proposed to suitably merge the regions. In seed competition [175], regions are allowed to compete for pixels, and a pixel originally assigned to one region may be reassigned to another if, in the process of region growing, it is found to be more similar to the new region. In boundary melting [559], the "edge" between two neighboring pixels from two regions is first assigned a value based on the similarity of the intensity values between the pixels. Then, two regions are merged if a significant part of the edge between them is "weak"; that is, pixels on either side of the edge are very similar to each other.

The identification of the seed regions is usually dependent on the problem and the segmentation results can be sensitive to the choice of these seed pixels. In some problems, one can use the brightest pixels as the seeds, specially if it is known that the brighter pixels form the objects of interest. Alternatively, we can use regional maxima (minima), where we consider connected pixels with the same intensity which are surrounded by pixels with lower (higher) intensity.

Instead of starting with seed pixels, we can also combine the processes of splitting and merging, and start the merging with the results from splitting, as discussed next.

8.2.3 Region splitting and merging

The process of region splitting may not always give the desired segmentation. Consider as an example, the simple 4×4 pixel image in Figure 8.3(a), where the object of interest is in grey on a white background. Assume a homogeneity metric which requires all pixels in a region to have the same pixel intensity. The first step of region splitting will create the four regions shown in Figure 8.3(b). Of these, only region IV satisfies the metric and therefore, is not split further. The other three regions are split, resulting in the image in Figure 8.3(c). Note that this split creates 13 regions in the image, though only two should result if the segmentation is correct.

However, the image suggests a solution to the problem as these 13 regions can be considered the starting point for a region-merging approach. If we start with region I_a, we can grow it to form the grey object. Next, starting with region I_d, we can grow it to form the white background, thus separating the object of interest from the background.

In reality, however, a metric such as the one we have used here is too stringent as an object rarely consists of pixels all of which have the same intensity value. Suppose we use a more realistic homogeneity metric, where we allow the pixel intensity in a region to vary a little so that 25% of the pixels in a region can have a different intensity value from the rest. If we start the merging process with region I_a and consider the four neighbors of each pixel (see Figure 8.3), then we would consider in turn, the pixels I_b, I_c, II_a, I_d, III_a, II_b, II_c, and so on. So, when the region I_d is considered for the merge, it could merge into the grey region, which, after the merge, would have four grey pixels and one white pixel. If however, the order of regions considered for the merge is based on the eight neighbors of each boundary pixel, then we would consider, in order, the pixels I_b, I_d, I_c, II_a, II_c, IV, and so on. Then, based on the metric, region I_d will not be included in the grey object as at the time of its merge, the grey object would have two grey pixels (I_a and I_b) and adding I_d would not satisfy the constraint. The results of the merging process are thus sensitive to the order of regions considered for the merge.

Another issue with the split and merge technique is one of data structures. For the splitting process, the simplest data structure to use is the quad-tree data structure [530, 531]

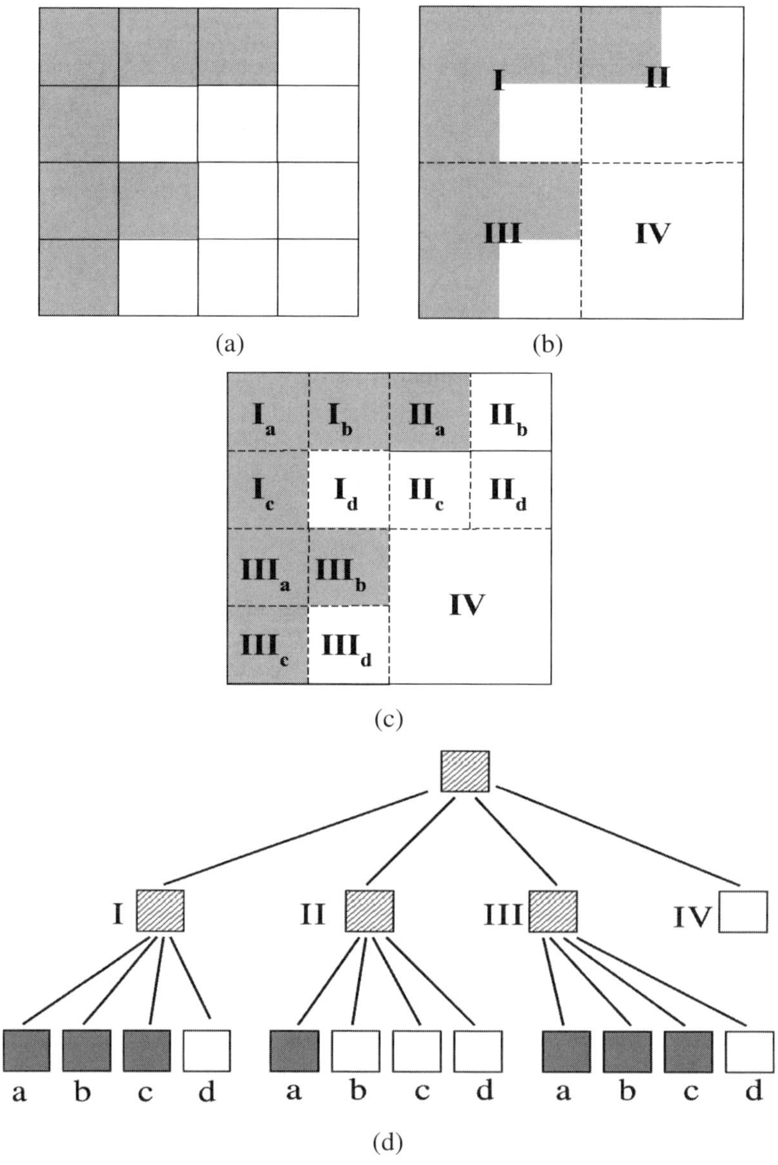

Figure 8.3. *Segmentation based on region splitting and merging. Consider the grey object in the* 4×4 *image in panel* (a). *Let the splitting of the regions continue until each subregion has a uniform intensity. Then, panel* (b) *illustrates the first split into the four regions, and panel* (c) *illustrates the second split, where only three of the four subregions are split further. Panel* (d) *shows the quad-tree data structure which can be used to store the split. Cross-hatched regions indicate a mix of pixels while white or grey regions have uniform intensity. Note that if a quad-tree is used in the splitting process, it is difficult to identify neighboring regions for the merging process.*

as it supports the split of a region into four. However, with such a data structure, it is difficult to determine neighboring regions for the merge and we may need to store adjacency information about the regions as well. This issue of data structures is discussed further in the text by Sonka [559].

Note that region-merging and splitting techniques can be easily extended to three-dimensional data on a regular grid.

8.2.4 Clustering and classification

The pattern recognition techniques of clustering and classification (see Chapter 11) can also be applied to image segmentation tasks. In clustering, we attempt to find natural groupings of the objects (in this case, pixels) in feature space such that pixels in a group are similar to each other but dissimilar to pixels in other groups. By applying clustering techniques, we can group pixels from each object together, thus separating them from the background pixels. In classification, given a sample of pixels assigned to various objects, we attempt to build a model which, given a pixel, will assign it to an object. This allows us to identify the pixels which form each object.

Pattern recognition techniques are often used in the segmentation of multispectral images, where each pixel is associated with several values from the different bands. For example, satellite imagery from the IKONOS satellite [212] is available both in grey-scale as well as in four-band multispectral, with red, green, blue, and near-infrared bands. In the latter case, each pixel can be considered as a point in a four-dimensional space represented by the four bands.

Consider, for example, the identification of human settlements in satellite imagery described in [323], which exploits both the multispectral and the panchromatic (that is, grey-scale) imagery from IKONOS. In the multispectral images, we can clearly see regions such as tarred roads and tarred parking lots, concrete buildings and concrete roads, lush vegetation, and so on. Ideally, a pixel from a tarred road should lie in the same region of feature space as a pixel from a tarred parking lot. Thus, if we identify regions of different types in an image and extract sample pixels representing the regions, we can use these pixels as a training set to build a decision tree or a neural network classifier. Then, given a pixel and the values of the four bands associated with it, we can identify its class, for example, tarred or concrete surface. Similarly, if after looking at the image, we can say that there are k different types of regions in the image, such as regions of lush vegetation or tarred surface, we can take the pixels of the image and apply a clustering algorithm, which will group similar pixels regardless of where they are in the image. Thus, pixels representing tarred surfaces will lie close to each other in feature space, regardless of where in the image they came from, or whether they came from a tarred road or a tarred parking lot.

The technique of graph partitioning for clustering (see Section 11.1.3) has also been applied in the context of image segmentation [541]. In this case, the image pixels are considered as nodes of the graph, with a key challenge being the assignment of weights to the edges of the graph, where the weight between two nodes indicates the similarity between the corresponding pixels.

When we use pattern recognition techniques, whether from classification or clustering, it is often the case that an object will have a majority of pixels from one class and a few stray pixels from another class. For example, an area of lush vegetation will likely have a few pixels which are classified as not-so-lush vegetation. These must be addressed in a postprocessing step (see Section 8.7); unless the stray pixels form a large enough region, the usual solution is to merge them into an appropriate neighboring region.

8.2.5 Watershed segmentation

The watershed approach borrows ideas from geography to generate a region-based segmentation of an image. It belongs to a class of related techniques developed somewhat independently in topography for digital elevation models, in mathematical morphology (see Section 7.4) for image segmentation [145], and more recently, in scientific visualization for the topological analysis of simulation data using the Morse–Smale decomposition [118]. Each of these three domains has provided different insights into the algorithm and created enhancements suitable for their domains.

Watershed algorithms usually operate on the gradient image and consider the greyscale gradient image as a topographic surface, where the grey levels represent altitudes. Region edges in the original image correspond to high altitudes in the gradient image, or watersheds in topography. The interior of regions in the original image are more homogeneous and correspond to lower altitudes in the gradient image, or catchment basins in topography. The watersheds separate the catchment basins from each other.

There are two ways in which watershed segmentation can be implemented. The first is based on the topographic interpretation of watersheds. It essentially considers a rain drop falling on a topographic surface and moving under the influence of gravity to the local minimum of the image surface altitude. A catchment basin is then defined as the set of pixels for which these downstream paths all end at the same altitude minimum. The alternative approach is to imagine the landscape, with holes pierced in its local minima, being immersed in a lake. The catchment basins will start filling up with water as the landscape is immersed further. At points where water coming from different basins would meet, dams are built all the way to the highest point in the landscape. The process stops when the water reaches the highest level. The dams, or the watersheds, essentially separate the basins, thus segmenting the image into regions.

A fast implementation of the immersion technique is given in [604]. However, a straightforward implementation of the watershed algorithm can lead to severe oversegmentation and the regions must be merged to correctly segment an image [392]. Techniques for addressing this problem are described in [559, 513].

8.3 Salient regions

In recent years, there has been interest in the information retrieval community in identifying objects using salient points or regions. Salient points, also referred to as interest points, are landmarks in an image which are often intuitively obvious to a human. These include corners of a building, the eyes on a human face, a high-contrast region, and so on. Of course, the edges and corners of an object are also considered salient as these features often help draw attention to the object by making them stand out from the background. I discussed the extraction of edges earlier in this chapter. I next briefly discuss the extraction of corners before describing the extraction of more sophisticated salient regions.

Recall that salient regions are also useful in registering images, as discussed in Section 6.3.2, where a transformation to align two images can be obtained by first aligning the salient regions. By using salient regions, instead of extracting the entire object of interest from the image as one would in segmentation, we are extracting only a subset of regions from that object.

8.3.1 Corners

A corner is a salient region where two edges intersect. Alternatively, it can be considered as a feature formed at the boundary between two regions in an image with different brightness levels, where the boundary curvature is sufficiently high [553]. One of the earliest "corner" detectors was developed by Moravec [438], with the intent of finding "points of interest" which could be used in image registration across consecutive frames of a video sequence in a computer vision problem. He defined a point of interest as one where there is large intensity variation in every direction. This intensity variation was obtained by considering a small square window around each pixel and obtaining the cross correlation of this window with a window obtained by shifting a pixel in the eight directions (horizontally, vertically, and along the two diagonals). The intensity variation at a point was the minimum of the variation in the eight directions. This idea is somewhat similar to the approach used in the SUSAN algorithm [553] discussed in Section 7.2.1.

However, there are some issues with the Moravec detector as it is not rotationally invariant and identifies a different set of corners if the image is rotated. Further, the square window, with its unequal distances from the center pixel, results in a noisy calculation of the variation in intensities in the different directions. A circular window, with perhaps a weight associated with each pixel, can help address this concern. The Moravec detector is also sensitive to small variations along an edge and can incorrectly identify edge pixels as corner pixels. This is an issue in problems where the images are noisy and variations in intensity along an edge are to be expected.

These concerns are addressed in a rather computationally intensive approach proposed by Harris and Stephens [257], which is also referred to as the Harris corner detector or the Harris–Plessey corner detector. It uses first-order derivatives to calculate the autocorrelation, making it isotropic, as well as a circular window with Gaussian weights to make the variation in intensity less noisy relative to a square window with equal weights. The issue of imperfections in an edge resulting in an incorrect identification of a corner is addressed by calculating a corner-ness measure for each pixel and using an appropriate threshold to minimize false positives.

Several alternative techniques have since been proposed to detect corners with varying degrees of success. These include the SUSAN approach (see Section 7.2.1), different methods to extract the curvature, as well as techniques where more than one method is used to detect the corners in an image reliably and localize them accurately. Several of these techniques are reviewed in [553, 436].

8.3.2 Scale saliency regions

The Scale Saliency algorithm, proposed by Kadir and Brady [311], is a method for measuring the saliency of image regions and selecting the optimal scales for their analyses. They were motivated by earlier work of Gilles [220] which used salient points to match and register two images. Gilles' definition of saliency was based on the local signal complexity. Specifically, he used the Shannon entropy of local attributes such as the intensity of the pixels in a neighborhood around the current pixel. Image areas with a flatter distribution of pixel intensities have a higher signal complexity and thus a higher entropy. In contrast, a flat image region has a peaked distribution and thus, lower entropy. Gilles used a fixed-size neighborhood, which led to his algorithm selecting only those salient points which were appropriate to the size of the neighborhood. As an extension, Gilles proposed the use of a global scale for the entire image, which was automatically selected by searching for peaks in the average global saliency for increasing scales.

Kadir and Brady [311] extended Gilles' algorithm by incorporating a local scale selection procedure. They defined salient regions (not points) as a function of the local signal complexity weighted by a measure of self-similarity over scale space. As a result, the Scale Saliency algorithm detects regions at multiple scales and the size of the circular salient patches is determined automatically.

The algorithm of Kadir and Brady has a simple implementation. For each pixel at location x in the image, a histogram of the intensities in a circular region of radius s (which reflects the scale) is obtained. The entropy $E(x,s)$ of each of these histograms is calculated and the local maxima are considered in further processing as candidate scales for determining the salient regions. Specifically,

$$E(x,s) = -\sum_{d \in D} p(d,x,s) \log_2 p(d,x,s), \tag{8.20}$$

where the entropy E is defined as a function of both the location x and the scale s; and $p(d,x,s)$ is the probability of the value d occurring in the region of scale s around the pixel at location x. D is the set of all values that d can take. For example, for 8-bit grey-scale images, assuming a histogram with a bin width of 1, D is $[0,\ldots,255]$. The set of scales at which the entropy peaks is defined as

$$s_p = \{s : E(x,s-1) < E(x,s) > E(x,s+1)\}. \tag{8.21}$$

These peaks of the entropy are weighted by the interscale saliency measure, which is defined as

$$W(x,s) = \frac{s^2}{2s-1} \sum_{d \in D} \left| \, p(d,x,s) - p(d,x,s-1) \, \right|, \tag{8.22}$$

resulting in the following definition of the Scale Saliency at a point x:

$$S(x,s_p) = E(x,s_p) \cdot W(x,s_p). \tag{8.23}$$

The top n scale-salient regions in an image, ordered based on their scale saliency, are used for further processing. Note that since the entropy is calculated for scales ranging from s_{min} to s_{max}, the calculation of the peaks implies that they can occur only between scales $s_{min}+1$ and $s_{max}-1$. By their definition, scale-salient regions are invariant to scaling, intensity shifts, and translations. The use of circular regions also ensures invariance to planar rotation.

The Scale Saliency algorithm can be summarized as follows.

Scale Saliency Algorithm: For each pixel location x

- For each scale, s, between s_{min} and s_{max}
  - Obtain the histogram of the pixel intensities in a circular region of radius s centered at the pixel.
  - Calculate the entropy $E(x,s)$ from the histogram using equation (8.20).
- Select the scale for which the entropy is a local maximum using equation (8.21). Note that there may be no scales that satisfy this constraint.
- Weight the peaks by the appropriate interscale saliency measure from equation (8.22).

Evaluating the scale saliency at each pixel in an image can be computationally very expensive. In addition, there is often little difference in the scale saliency at a pixel and its neighbor. As a result, if every single pixel is processed, and only a small number (say 100) of the top scale-salient regions are considered, many of these regions will be right next to each other, and many of the true salient regions will be missed. A simple way to address this problem, and reduce the computational cost of the algorithm at the same time, is to calculate the scale saliency at every nth pixel, where n can range from 2 to 10 or more, depending on the data. Further, instead of considering every scale between s_{min} and s_{max}, we can use a stride for the scale as well [320].

8.3.3 Scale-invariant feature transforms

A recent development that has attracted much attention in the content-based image retrieval community is the work of Lowe on the Scale-Invariant Feature Transform (SIFT) [395]. This uses a two-step process—the selection of keypoints or regions and the extraction of features for these regions which are invariant to scale, rotation, and translation. In this chapter, we focus on the first of these steps, namely the selection of the regions. The extraction of features will be considered in Chapter 9.

Like the scale-saliency approach, SIFT is also a scale-space approach to the detection of salient locations, also called keypoints, in an image. Lowe considered locations that are maxima or minima of a difference-of-Gaussians function. The convolution of a two-dimensional Gaussian function is first implemented as two passes of a one-dimensional Gaussian function with $\sigma = \sqrt{2}$ using 7 sample points. The resulting image, A, is again convolved with the Gaussian function with $\sigma = \sqrt{2}$, to give an image B, which is equivalent to the original image convolved with a Gaussian of $\sigma = 2$. The difference-of-Gaussians function is obtained by subtracting image B from image A.

The idea of a scale-space approach is next introduced by considering a pyramid of such difference-of-Gaussians images. The image B is resampled using bilinear interpolation with a pixel spacing of 1.5 in each direction and the process is repeated on this new image. The 1.5 spacing implies that each new sample will be a linear combination of four adjacent pixels. Next, maxima and minima of this scale-space function are determined using an efficient and stable method described in [393]. A pixel is retained if it is a maximum or minimum when compared with the 8 neighbors at its level and the nine neighbors at each of the levels above and below it.

In his later work [395, 394], Lowe discussed enhancements to the original method to improve the way in which the sampling was done in the image and the scale domains. In addition, instead of just locating the keypoints at the location and scale of the center sample point, further improvements were proposed to locate the keypoints more accurately and remove points that were poorly localized along an edge or had low contrast. These keypoints formed the salient regions in the image.

8.4 Detecting moving objects

In problems where the data are available as a video sequence, we can exploit the motion of the objects to detect them. If we can identify the moving objects accurately, we can track them over time. This allows us to model the interactions between the moving objects and use these models to detect anomalous events in a scene. This tracking can be invaluable in applications such as surveillance, traffic monitoring, and gesture recognition in human-machine interface.

Detecting moving objects in video in a large scene over a long period of time can be difficult for several reasons. We need to account for issues such as changes in illumination; occlusions; moving objects which are not really moving such as trees blowing in the wind; small motion of a stationary camera; motion of a camera which is mounted on a moving platform; objects which are moving but come to a stop in the scene; or objects which were not moving for a while, but suddenly start moving again. In addition, objects may enter and leave a scene, and then enter again, or we may have several cameras, each covering part of a scene, and we need to detect and track moving objects across regions covered by different cameras. This task of detecting and tracking moving objects in video is a difficult one and the subject of active research, especially in algorithms which are robust and accurate enough to handle all of these challenging conditions.

I next describe two rather broad categories of techniques for the detection of moving objects in video. One category is based on maintaining a background image consisting of nonmoving objects, while the other focuses on identifying where a region in one frame has moved to in the next frame.

8.4.1 Background subtraction

The process of background subtraction involves comparing each video frame against a reference or background frame. Pixels in the current frame that deviate significantly from the background are considered to form the moving objects. These "foreground" pixels are further processed for object localization and tracking. A good background subtraction algorithm must adapt quickly to changes in the environment, be able to detect objects moving at different speeds, provide good localization of the objects, and have a low computational complexity.

At the heart of any background subtraction algorithm is the construction of a statistical model that describes the background state of each pixel in the frame. There are different ways in which the background model can be built. The simplest is to use the previous frame as the background for the current frame, leading to the frame-differencing approach. This

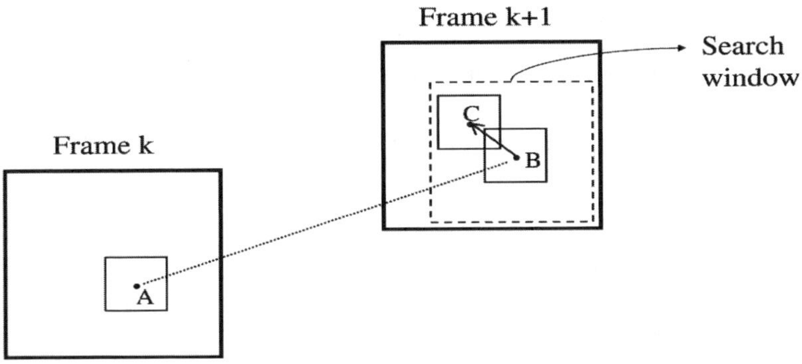

Figure 8.4. *Block matching for motion estimation. Consider the block centered at location A in frame k. The corresponding point is location B in frame* $(k+1)$, *which is used as the starting point of the search for a best match. The search window indicates the region over which the search is conducted. If the block centered at location C is the best match, then the motion vector is given by the arrow connecting B to C.*

technique suffers from the aperture problem, as a result of which only the front and back parts of a large, uniform intensity object will appear to have moved. The middle part, whose intensity does not change between the two frames (assuming the object has not moved far between the first and second frame) will appear to have not moved at all.

More complex methods obtain the median value of the previous few frames and use it to create a single-state model of the background for the current frame [222, 128, 387, 649, 127]. Others use the Weiner filter [587] or the Kalman filter [325, 631, 266, 239, 52]. A recent approach which has gained popularity is a full density estimation based on Gaussian mixture models [206, 565, 211, 312, 482, 366] or histograms [169].

All these algorithms have several parameters which must be tuned to the characteristics of the video and the objects being detected. These parameters range from the number of frames to include in the frame buffer for the median filter to the adaptation rate used in the Kalman gain matrix. They determine how adaptive the algorithm is to changes in the pixel values. Once a background estimate has been created, a simple way to determine the moving objects, or the foreground, is to subtract the background image from the current frame and threshold the result to keep only those pixels which deviate significantly from the background.

The background algorithms described in this section all have their pros and cons; a detailed study of the performance of the algorithms on several videos taken under different situations is presented in [111, 112].

8.4.2 Block matching

Block matching is a simple approach to motion estimation which uses a pixel domain search procedure to determine the best estimate to the motion vector of an object [577]. Given a block (a square or rectangular region of pixels) in the current frame of a video sequence, the block-matching approach determines the new location of the block in the next frame. The displacement of the block is the estimate of the motion vector for that block. Figure 8.4 shows a block at location (i, j) in the current frame (frame k). Starting with the corresponding location in the next frame (frame $(k+1)$), we search for a block of the same size which is the best match in a search window centered around (i, j). The displacement from (i, j) to the center of the block which is the best match is the estimated motion vector.

There are essentially three main components to a block-matching algorithm with several options available for each component:

- **Matching criterion:** The matching criterion is a metric used to determine how similar a block in the current frame is to a block in the next frame. These criteria include the sum of absolute difference, which considers the sum of the absolute value of the difference between pixels in the identical locations in the two blocks; and the mean squared difference, which is the mean of the square of the difference between the corresponding pixels in the two blocks. Typically, all the pixels in a block are considered for the matching criterion, though we can reduce the processing time by considering a smaller sample of the pixels, at the risk of increasing the error in the motion vector estimate.

- **Search strategy:** This determines where and how we search for the best match for a block. The simplest approach is the window search (shown in Figure 8.4), where the block at location (i, j) in the current frame is compared against blocks in a search window of a fixed size centered around (i, j). Essentially, for each pixel in the search window, we consider a block of the same size centered at the pixel, compute the matching criterion, and select the best match over all pixels in the search window. The most exhaustive case is where the search window is the same size as the video frame. Typically, the size of the search window is determined by the maximum distance the objects in a video are expected to move between frames. It is possible to speed up the search by not considering every pixel in the search window for a possible match, but by considering, for example, every other pixel. Alternatively, we can use more sophisticated searches such as the three-step search or the cross search [577], where the matching criterion is evaluated only at a certain predetermined subset of locations. These may lead to suboptimal solutions as they are not exhaustive searches.

- **Block determination:** This specifies the position and size of the blocks in the current frame, the scale of the blocks, and the starting location of the search in the next frame. Typically, a fixed-size block is used for a frame at a given scale; that is, the size of the block does not vary within the frame. The blocks can be either overlapping or disjoint and span the frame. The starting location of the search is the same as the center of the block in the current frame. The size of the blocks is chosen based on the application and the size of the objects moving in the video. If it is too small, there may not be enough variation in pixel intensities in the block to enable a good match. If the block is too large, it may be difficult to accurately detect the motion in the block especially if more than one moving object is contained in it. In more complex approaches to block determination, a multiresolution approach is used. First, a multiresolution pyramid using a Gaussian filter is created for the current frame and the next frame. Then, block matching is applied at the coarsest resolution and the best match found. The process is repeated at the next finer resolution, where the search starts at the location of the best match from the previous level. This use of multiresolution introduces the idea of a scale in block matching, where the matching is done starting with the coarsest scale and progressing to ever finer scales.

A simple way to improve the performance of block-matching algorithms is to incorporate a zero-motion bias. A block in the current frame is compared against the block at the corresponding location in the next frame. If the matching criterion indicates that the block has not moved, then the search is stopped. To allow for small changes in the pixel intensities, for example, those caused by small camera movement, a zero-motion bias threshold is used. This threshold varies based on the video. The zero-motion bias, as well as a comparison of different options in block matching are discussed in [390, 391].

Another approach to substantially reducing the cost of the block-matching algorithm is to apply it to select regions rather than an entire image. For example, we can first identify the salient regions in an image and then consider only these regions in the block-matching process. This approach is taken in the ASSETT-2 project [552], where the corners in an image are considered in detecting the moving regions in a video sequence.

Note that unlike the background subtraction methods, where we determine the objects which have moved, the block-matching approach also provides us information on the new location of the objects, and therefore, the motion vector. These motion vectors can then be processed further to track the objects over time. For example, if an object is composed of many blocks, the motion vectors corresponding to the blocks must be close to each other. The motion vectors can therefore be clustered to identify a moving object [552].

8.5 Domain-specific approaches

In some problems, it is possible to exploit the characteristics of the domain to identify the objects in the data. For example, if the objects of interest are straight lines such as airport runways or railway lines, or in some cases, roads, we can use the Hough transform [227]. Or, if the objects of interest have a particular template, for example, they are circular with a given intensity variation, we can use template-matching approaches [297, 227, 559]. Here, we look for a match between the template of an object of interest and locations in the image where the mismatch energy is a minimum or the correlation is high. This kind of a matched filter approach was used in the identification of focus of attention regions in the detection of volcanoes in the imagery from Venus [73]. Sometimes, the shape of an object, or parts of an object, can be exploited. For example, in the FIRST astronomical survey [29, 174] the locations of the galaxies are identified by finding pixels of high intensity in an image and fitting an elliptic Gaussian to it. The information on these ellipses, such as the location of the center and the lengths of the major and minor axes, are collected to form a catalog. This catalog can be considered to be a summary of the objects found in the images, and can then be mined to find the patterns among the galaxies [315]. Another example of exploiting domain-specific characteristics to detect objects of interest is face detection using the distance between the eyes or the skin color. Needless to say, such techniques are not viable in situations where a person is wearing a hat and the eyes cannot be seen clearly; or is looking sideways so both eyes cannot be seen; or the illumination is poor, so the skin color cannot be exploited.

In other problems, the characteristics of the data are such that new techniques must be developed for the identification of the objects of interest. This is especially true for simulation data sets. The characteristic of mesh data, where the values of the variables are available at discrete points in the mesh, can make it difficult to identify objects of interest using existing technology. If the underlying mesh is a regular Cartesian mesh, with equal spacing in all dimensions, then the mesh can be considered an image and image processing techniques can be applied directly, both in two- and three-dimensional problems. However, it should be noted that the mesh variables are floating-point values and not 8-bit or 16-bit data as in most images. The analysis techniques must therefore be modified so they are applied to floating-point data [318].

If the mesh is more complex, such as those resulting from the application of the Adaptive Mesh Refinement (AMR) technique or from an unstructured mesh, then, a simple application of traditional image processing techniques is not sufficient to identify the objects of interest in such data. In this case, one option would be to sample the mesh at regular intervals, thus converting the data to be similar to an image. Alternatively, one could use the underlying mesh as is, and use the techniques based on PDEs (see Section 8.1) to detect the edges in the data. A positive aspect of such data is that they are multivariate, with several variables associated with each grid point. This additional information can be invaluable in improving the quality of segmentation or edge detection as we can use more than one variable in the processing. Further, as simulation data are available over time, we can exploit the temporal information as well. For example, we can obtain the boundary of an object using active contours for one time step and use the boundary as the initial contour in the following time step. This topic of identifying the objects in mesh data from simulations is a particularly challenging one, especially when the objects are distributed among the files output by different processors used in running the simulation.

8.6 Identifying unique objects

Once we have identified the pixels in an image which belong to the objects of interest, there are still several additional steps which must be completed before we can extract characteristics of these objects as discussed in Chapter 9. One of these steps is identifying the pixels which form an object, or, in other words, assigning a unique identifier to each pixel which associates it with either a specific object, all of whose pixels have the same identifier, or the background, which is usually assigned an identifier of zero. This process is called connected component analysis or labeling [227, 559].

Essential to the task of connected components labeling is the concept of connectivity, which, given a pixel, defines which of the surrounding pixels are considered to be its neighbors. The two definitions often used are 4-neighbors or 4-connected pixels and 8-neighbors or 8-connected pixels. Figure 8.5 illustrates the four and eight neighbors of a given pixel. The object indicated by the grey pixels in panel (c) can be considered a single object if we use 8-connectivity, or two objects if we consider 4-connectivity. In the latter case, the two squares are decoupled as the corner pixels of each square are not 4-neighbors of the other.

Note that this idea of connectivity is also used in region growing, when we evaluate the neighbors of a pixel for merging. The idea can be easily extended to three-dimensional objects, where we can have 6-connectivity or 26-connectivity corresponding to 4- and 8-connectivity in two dimensions, respectively.

Once an image has been segmented, each of its pixels is assigned a unique object identifier using connected component labeling. Consider a binary mask image where the

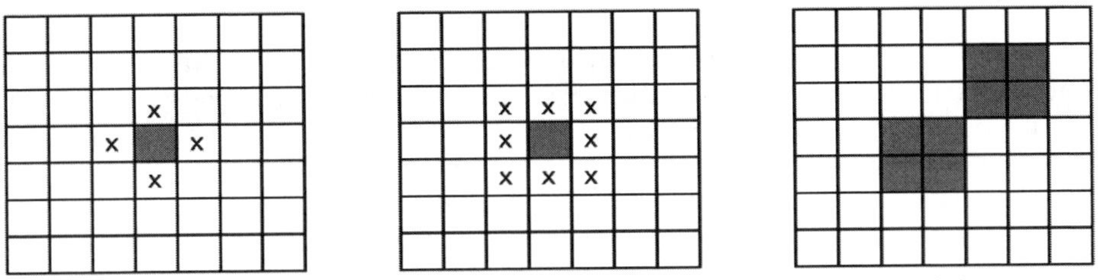

Figure 8.5. *Connectivity of the pixels in an image, illustrating the* (a) *four and* (b) *eight neighbors (shown by an* $\times$*) of a center pixel in grey.* (c) *The region indicated by the grey pixels could be either two objects if* 4*-connectivity is used or a single object if* 8*-connectivity is used.*

pixels corresponding to the objects of interest are set to 1 and the background is set to 0. If the objects of interest in an image are represented using a format other than a binary mask, such as run-length encoding or a quad-tree structure (see Section 8.8), then the labeling can be applied to these alternative data formats as well [559]. In the connected component labeling, we first select a desired level of connectivity (4 or 8 for two-dimensional data), and then proceed to assign a unique identifier or object number to each of the objects of interest.

Connected Component Labeling Algorithm:

- Scan the columns of the image, row by row, and assign a label to each pixel with value 1 (i.e., an object pixel) as follows:
  - If all neighbors of the pixel have value 0 (background) or 1, then assign the pixel a new, that is, not yet used, label.
  - If there is one neighboring pixel with a label not equal to 0 or 1, assign this label to the current pixel.
  - If there is more than one neighboring pixel, all with identical labels (excluding 0 or 1), assign this label to the current pixel. If these labels of the neighbors are not identical, we have a label collision, that is, two pixels in a single object have different labels. We keep track of the fact that the two labels are equivalent and revisit the issue in the second pass through the image. In this case, the current pixel is assigned any one of the labels of its neighbors.
- The image now has a label assigned to all nonzero pixels. However, due to label collision, some regions may have different labels assigned to the pixels in the region. In the second pass, the information on label collision is used to assign a single label to the entire region. This is done by replacing equivalent labels by a unique label.

8.7 Postprocessing for object identification

In many cases, once the objects have been identified in an image, some additional processing may be needed to clean up the results after segmentation. Depending on the type of cleanup, this postprocessing may be done before or after the component labeling. For example, some object identification methods, such as the ones based on thresholding or region growing, can result in objects which are too small, consisting of one or two pixels. In such cases, we could use connected component labeling first to identify the objects. The size of the objects could then be used either to remove them or to merge them with other nearby objects which are sufficiently close. The use of a size criterion is particularly useful if we know the approximate size of the objects of interest. Alternately, one could use morphological operations to perform postprocessing cleanup. Nearby objects could be merged by first dilating each object (thus connecting nearby ones) and then eroding them to revert back to their original size. Small objects could be removed by first eroding the objects, and then dilating them. These steps may need to be applied multiple times to achieve the desired results.

Sometimes, more sophisticated postprocessing techniques may be needed. For example, gradient-based segmentation techniques often result in gaps in the edges. If these gaps are not too long, they can be completed using various edge relaxation techniques as described in [559]. These approaches consider an edge in the context of its neighbors and use simple heuristics to complete edges or remove spurious ones. Other techniques for finding and completing salient edges are the structural saliency approach [537] and the tensor voting method for inferring structure from sparse information [425, 426].

8.8 Representation of the objects

There are many ways in which we can represent the objects found in an image. The simplest is by using a mask image, where each object is assigned a unique identifier and all pixels in the object are labeled using the identifier. The background is assumed to be assigned the identifier zero. Though this representation is the simplest in terms of its use in further processing, it makes poor use of memory as the entire image including the background pixels are stored, and these pixels could constitute a large part of the image.

An alternate approach would be to store only the pixel locations corresponding to each object or use run-length encoding [559], which stores the locations of only the beginning and end of each row of object pixels, with the locations of the intermediate pixels being implicitly defined. However, such approaches can make further processing cumbersome and it may make sense to reconstruct a mask image first for ease of processing.

It is useful to note that many of the schemes for the compact storage of images were developed at a time when computer memories were small and compression techniques were not very effective. Such schemes were therefore necessary if images of a reasonable size had to be analyzed; as a result, extensive research was done to determine efficient algorithms for the manipulation of data stored in these data structures. However, given the availability of larger memories in modern computers, as well as sophisticated compression techniques, the use of such storage schemes must be carefully evaluated to ensure that the additional complexity introduced in the implementation of the image analysis algorithms is indeed worth the savings in storage. Often times, it may make sense to store a compressed binary mask image, of the same size as the original image, and uncompress it into a full mask image just prior to use.

8.9 Summary

In this chapter, I described different ways to identify or segment objects in image data, where a part of the image may be the background and the task is to separate the object pixels from the background pixels. These techniques can be applied to both image and simulation data in two and three dimensions.

The techniques for segmentation range from simple thresholding and gradient-based approaches to more complex techniques based on PDEs. We also considered ways of identifying salient regions, where we focus on landmarks in an image instead of entire objects, as well as techniques where we use the motion of an object to identify its extent in a video sequence.

These results from the application of segmentation techniques often need to be postprocessed to complete boundaries and remove spurious objects. Then, a unique identifier is associated with the pixels which constitute an object. Next, we extract features or descriptors to characterize each object, a topic I discuss further in Chapter 9.

8.10 Suggestions for further reading

Image segmentation is a topic which has been studied extensively, and new techniques are being developed to identify objects in new media such as video, or to address long-standing problems such as edge completion, especially in the presence of noise. The early work in image segmentation is well summarized in the papers by Haralick and Shapiro [256]. A relatively recent overview of edge detection techniques is presented by Ziou and Tabbone in [651]. For those interested in comparing different edge detection techniques for suitability to their problems, several evaluation strategies are considered in the papers by Heath et al. [264], by Bowyer, Kranenberg, and Dougherty [54], and by Shin, Goldgof, and Bowyer [543].

The use of PDEs in segmentation is a topic of active research. Several books have been written covering various aspects of the topic: the use of deformable models in medical imaging is the subject of the book by Singh, Goldgof, and Terzopoulos [549] and the edited collection of papers by Malladi [408]; the fundamentals of level sets, including a chapter on shape recognition, are discussed in the text by Sethian [536]; snakes and active contours for segmentation form the focus of the text by Osher and Fedkiw [464], while active contours and stochastic methods are discussed in the book by Chan and Shen [94]; and several chapters on edge detection and boundary extraction using geometric level sets are included in the edited collection by Osher and Paragios [465].

Domain-specific texts describe how one can exploit the characteristics of the domain to improve the quality of segmentation or use different modalities of images available in the domain. For example, in the field of medical imaging, a variety of techniques are used for segmenting images; good references are the handbook by Sonka and Fitzpatrick [558] and the text by Rangayyan [502]. In contrast, in remote sensing, pattern recognition algorithms are used frequently to assign labels to pixels as discussed in the text by Lillesand and Kiefer [381] and by Schowengerdt [527].

Chapter 9

Extracting Features Describing the Objects

Next, when you are describing
A shape, or sound, or tint;
Don't state the matter plainly,
But put it in a hint;
And learn to look at all things
With a sort of mental squint.

—Lewis Carroll [88]

In Chapter 8, we considered methods to identify objects in two- and three-dimensional image and mesh data and separate them from the background. At the end of this step, we have identified the pixels or the grid points which comprise an object. The next step in the scientific data mining process is to identify and extract characteristics to represent these objects. As mentioned earlier in Section 3.2.10, we are often interested in identifying patterns among the objects in the mesh or image data. Therefore, it is important first to find the objects accurately in the data, and then to represent them by some higher-level characteristics, or descriptors, which I refer to as "features."

The word "feature" is unfortunately used in two different contexts in many scientific domains. For example, scientists may refer to "features of interest" in their data, by which they mean structures in the data such as a flame front, an eddy, a road, or a storage tank. In contrast, I use the word "feature" in a more primitive sense to mean any measurements which can be easily extracted from the data. These could range from simple characteristics, such as the aspect ratio or the area of an object, to more complex characteristics representing the shape of the object or various moments indicating the distribution of pixel intensities. These features, or descriptors, are then used to find the patterns among the objects identified in the data. For example, if we are interested in finding round storage tanks in a satellite image, we can use a feature such as the aspect ratio of the object, along with some measure of the "circularity" of the object. Or, mushroom-shaped objects in a simulation of a fluid-mix problem may be characterized by shape features.

It should be obvious that the features we extract to represent the objects in the data should be relevant to the pattern of interest. For example, if instead of round storage tanks, we are interested in finding rectangular buildings in the satellite image, we would focus on features such as the presence of sharp corners in the object, with an angle close to 90°. This observation indicates that the identification of features is likely to be very problem dependent. However, there are several general features we can extract which have worked well for several different problems. These will be the focus of this chapter.

This chapter is organized as follows. First, I discuss some general criteria which should be helpful in guiding the choice of features. Next, in Sections 9.2 through Section 9.4, I describe various features one can extract to represent the data. It should be kept in mind that, as appropriate, these features can be extracted for objects in the data, for regions in the data such as subimages or even a whole image, or at points in the data such as pixel locations or mesh points in simulation data. Further, in case of objects, we can choose to consider the pixel intensities or focus just on a binary image, ignoring the actual values of the data. The latter is often the case when we consider shape-based features where the focus is on aspects of the extent of an object. Then, in Section 9.5, I give examples of problems where it is necessary to extract specific features tailored to the problem. Section 9.6 describes various postprocessing steps which may need to be applied to the collection of features describing the objects in a data set, before we proceed to identifying the patterns among the objects. Finally, Section 9.7 summarizes the chapter and Section 9.8 provides suggestions for further reading.

9.1 General requirements for a feature

As we have observed, a key criterion a feature must satisfy is that it must be relevant to the pattern of interest. In many scientific problems, the best way to identify such features is by working with the domain scientists. They have frequently given much thought to what they are interested in finding in the data and have a good idea of what characterizes these objects of interest. Consider, for example, our experiences with the problem of classification of bent-double galaxies [315]. The astronomers had spent many hours visually analyzing the images of the galaxies and could explain why they believed a galaxy should be considered to be a bent-double galaxy. They would indicate the number of blobs in a galaxy and the spatial arrangement of the blobs as factors behind their assignment of a class label. Or, they would refer to the relative intensities of the blobs and indicate patterns such as a bright round blob with two less-bright, elongated blobs around it. This led us to initially focus on features such as distances and angles between blobs. As we continued our interaction with the astronomers, we refined the set of features, considering more than one approach to defining the angles, or adding other features if we found we were consistently misclassifying galaxies with a certain structure.

This might lead one to believe that a good approach would be to extract as many features as possible during the first iteration of the feature extraction step of the data mining process. However, this is not only time consuming, it can also cause problems with the identification of patterns in the data. I explore this issue further in the next chapter on dimension reduction (Chapter 10). Instead, a more prudent approach is to identify an initial set of features in collaboration with the domain scientists and use them to identify the patterns in the data. Then, based on the results, either add new features or refine the current set by perhaps using a more robust method for extraction, weighting the features appropriately, or using dimension reduction techniques to identify the key features. Several iterations of this process are usually needed to identify and extract the features which will result in a successful solution to the problem. However, since the quality of the end result after pattern recognition is only as good as the features which are input to the pattern recognition algorithm, it is worth spending the time to extract high-quality features relevant to the problem.

Care must also be taken to ensure that the features used do not bias the results. For example, when we first used decision trees in the classification of bent-double galaxies, the tree indicated that the right ascension and declination (which are analogous to longitude and latitude, respectively, and can be considered the coordinates of the galaxy in the sky), were important features in identifying if a galaxy was a bent-double or not. As this seemed improbable, we investigated the issue further. We found that the astronomers had generated the training set of bent-double galaxies during the early days of the survey, when only a small part of the sky had been surveyed. The non–bent-doubles, in contrast, were extracted later on, somewhat randomly from different locations in the sky. As a result, the coordinates of the galaxy appeared to play an important role in determining if the galaxy was a bent-double or not. Needless to say, we removed the right ascension and declination features in future iterations of the data mining process.

It goes without saying that the features of interest should be easily extractable from the data. Recall that by features, I refer to low-level characteristics which represent the objects in the data. Science data sets are often massive being measured in terabytes or more. Features which may be very complicated or time consuming may not be the best choice, at least in the initial iterations. It makes sense to start with the simpler features and move to more complex ones only if the results are unsatisfactory. Further, if complex features are required, then it might make sense to work with a smaller sample of the data to ensure that the features are effective before processing the whole data set.

Another factor driving the choice of features is that in many cases, the patterns of interest are invariant to rotation, translation, and scale. For example, a bent-double galaxy remains a bent-double even if it rotated, moved to another location in the sky, or scaled to be larger or smaller. This would imply that, to the extent possible, the features we extract should also be scale, rotation, and translation invariant. This may involve additional postprocessing of the features after extraction. For example, if we are analyzing satellite images at different resolutions, instead of measuring the area of an object in pixels, we may need to convert the area into square centimeters to bring the size of all objects to the same scale. Features which are sensitive to rotation may require that the objects be aligned along a certain axis before the extraction of the feature.

Some features may require the data to be preprocessed prior to feature extraction. For example, many image features require the images to be scaled such that the pixel intensities lie between 0 and 255. Or, if we are extracting features for images, we may need to scale all images to have the same size. Else, features such as the histogram of pixel intensities will be adversely affected by the size of the images. For other features, we may need to quantize the data into a few levels to make the feature extraction tractable.

A different aspect of feature extraction is the robustness of the features. Ideally, we want to extract features which are insensitive to small changes in the data. Science data are often noisy and if the values of a feature change dramatically with minor perturbations in the data, then it is unlikely that the feature will represent the object correctly. An example of such a situation is illustrated in the calculation of angles between the blobs which constitute a galaxy in the problem of classification of bent-double galaxies [315] (see also Section 9.5).

In addition to features which represent the objects, it is advisable to associate with each object or item other features which help to identify its provenance. These "housekeeping" features could include the object identification number, the name of the image file containing the object, the location of the image file (for example, a directory structure which can be used to locate the file), the location of a subimage within a larger image, the date a data set was created, the name of the person creating the data set, a file name containing information on the instrument used to obtain the data as well as the settings of the instrument, and so on. While such features are not used directly in the process of pattern recognition, they are invaluable in tracking the objects back to their original source. This is useful for displaying the results of the pattern recognition step, where a user might want to see the images from which the objects were extracted, or relate the patterns in simulation data with the values of the parameters used to run a simulation.

I next describe several general features which have been used in a variety of different problems. As mentioned earlier, these features can be extracted for an image, for objects in an image or the output of a simulation, or for pixels in the image or mesh points in a simulation output. For convenience, I refer to all of these as "objects." Further, in cases where more than one variable is available at each pixel or mesh point, as is usually the case in multispectral images or in simulation data, the features can be extracted for each variable separately, or the variables can be fused prior to feature extraction.

In some problems, the features describing an object may be relatively simple and obvious such as the values of different sensor measurements in an experiment (where each experiment is an object of interest) or the history of a patient's blood pressure in a clinical trial. In such cases, the feature extraction methods described in this chapter may not be relevant, though postprocessing of these features, as described in Section 9.6, may be necessary.

9.2 Simple features

There are several simple features which can be extracted easily from the data, regardless of whether the feature is used to represent an image, a subimage, or an object in the image. As appropriate, these may require the data values to be scaled to lie within a certain range so that when we try to identify patterns among objects in different images or different time steps of a simulation, the results will not be adversely affected by the range of values. Such simple features may include:

- **Simple statistics:** Simple statistics include the mean, the standard deviation, the maximum, and the minimum of all pixel values or mesh variables in an object. These features often work surprising well, not necessarily in identifying patterns of interest, but in rejecting objects not of interest. For example, areas of simulation where nothing interesting is happening, and the variable values are almost constant can be rejected easily from further consideration.

- **Histograms:** A histogram indicating the distribution of pixel values or variables at mesh points is another feature which works surprisingly well, given its simplicity. We can also use histograms of more complex structures, such as edges, as in the MPEG-7 edge descriptor [418].

  To ensure that the comparison of histograms across objects is done correctly, care must be taken to scale the pixel intensities appropriately and to choose a suitable metric for comparing the histograms. There are several commonly used metrics. Bin-to-bin comparisons using standard metrics, such as χ^2 distance, Minkowski distance, or the Kullback–Leibler divergence, work well when the histograms are aligned. However, they can be sensitive to quantization errors and techniques such as the earth-mover's distance [518, 384], which considers cross-bin distances, have been proposed to alleviate these errors. Other metrics include the diffusion distance [383] and a scale-space decomposition using scale trees for multidimensional histograms [218]. The number of bins used to obtain the histogram should also be chosen appropriately, perhaps by trying different values on a subset of the data.

Note that histogram comparison arises in many tasks, including retrieval and feature selection. Some additional details are provided in Section 10.3.1.

- **Location of the centroid:** In many applications, the location of the centroid of an object may serve as a "housekeeping" feature to enable one to identify the object in an image. However, it can also serve as a feature in problems such as tracking, either to display the tracks of a moving object or to enable the implementation of constraints which require that an object, represented by its centroid, can move only within a certain neighborhood of its location in the previous frame as determined by its speed.

- **Scalar object descriptors:** There are several basic features which one can extract to describe an object. The size of an object, that is, its area in two dimensions or volume in three dimensions, is a simple feature which can work well in rejecting patterns which are not of interest. Scaling is of course required if we are considering data at different resolutions. Other related features which might be useful are the perimeter in two dimensions or surface area in three dimensions and features derived from a bounding rectangle around the object such as elongatedness, aspect ratio, and direction [559].

- **Geometric moments:** Geometric moments are among the simplest moment functions and can be considered as simple representations of the shape of an object. The $(p+q)$th order, two-dimensional, geometric moment m_{pq} is defined as

$$m_{pq} = \sum_{x=-\infty}^{x=+\infty} \sum_{y=-\infty}^{y=+\infty} x^p y^q f(x,y), \tag{9.1}$$

where $f(x,y)$ is the pixel value at location (x,y). To make this definition invariant to translation, we consider the central moment μ_{pq} defined as

$$\mu_{pq} = \sum_{x=-\infty}^{x=+\infty} \sum_{y=-\infty}^{y=+\infty} (x-x_c)^p (y-y_c)^q f(x,y), \tag{9.2}$$

where x_c and y_c are the x and y coordinates of the centroid of the object and can be written as

$$x_c = m_{10}/m_{00} \quad \text{and} \quad y_c = m_{01}/m_{00}. \tag{9.3}$$

If we consider an additional scale factor, we obtain the scaled central moments which are invariant under scale and translation:

$$\eta_{pq} = \mu_{pq}/(\mu_{00})^{(p+q+2)/2}. \tag{9.4}$$

From the central moments, Hu [279] obtained seven polynomials of m_{pq} which remain invariant under rotation [297]. The corresponding moment characteristics which are invariant to scale, rotation, and translation can be obtained using scaled central moments [559]. These invariant moments are

$$I_1 = \eta_{20} + \eta_{02}, \tag{9.5}$$

$$I_2 = (\eta_{20} - \eta_{02})^2 + 4\eta_{11}^2, \tag{9.6}$$

$$I_3 = (\eta_{30} - 3\eta_{12})^2 + (3\eta_{21} - \eta_{03})^2, \tag{9.7}$$

$$I_4 = (\eta_{30} + \eta_{12})^2 + (\eta_{21} + \eta_{03})^2, \tag{9.8}$$

$$\begin{aligned} I_5 = {} & (\eta_{30} - 3\eta_{12})(\eta_{30} + \eta_{12})[(\eta_{30} + \eta_{12})^2 - 3(\eta_{21} + \eta_{03})^2] \\ & + (3\eta_{21} - \eta_{03})(\eta_{21} + \eta_{0}3)[3(\eta_{30} + \eta_{12})^2 - (\eta_{21} + \eta_{03})^2], \end{aligned} \tag{9.9}$$

$$\begin{aligned} I_6 = {} & (\eta_{20} - \eta_{02})[(\eta_{30} + \eta_{12})^2 - (\eta_{21} + \eta_{03})^2 \\ & + 4\eta_{11}(\eta_{30} + \eta_{12})(\eta_{21} + \eta_{03})], \end{aligned} \tag{9.10}$$

$$\begin{aligned} I_7 = {} & (3\eta_{21} - \eta_{03})(\eta_{30} + \eta_{12})[(\eta_{30} + \eta_{12})^2 - 3(\eta_{21} + \eta_{03})^2] \\ & - (\eta_{30} - 3\eta_{12})(\eta_{21} + \eta_{03})[3(\eta_{30} + \eta_{12})^2 - (\eta_{21} + \eta_{03})^2]. \end{aligned} \tag{9.11}$$

The Hu moments are very commonly used features in image analysis, especially in the identification of objects with unique shapes regardless of their location, size, and orientation. Other more complex moments such as Legendre and Zernike moments are discussed in [440] and the MPEG-7 moment-based shape analysis feature is discussed further in Section 9.3.

Other simple features which can be extracted for objects include the mean value of a temporal variable in a time window for use in time series analysis, the locations and number of maxima and minima of a variable, and the locations of salient regions such as corners in an image (see Section 8.3).

9.3 Shape features

Shape analysis, which involves both shape representation and shape-matching techniques, is an important problem in many domains, including image analysis and information retrieval. An excellent description of representations for shape is given in Chapter 6 of the book by Sonka, Hlavac, and Boyle [559]. The text by Dryden and Mardia [162] addresses the problem from a statistical viewpoint, focusing on the case where objects are summarized by keypoints called landmarks. They also devote a chapter to discussing the problem in the context of image analysis. The information retrieval perspective is summarized in [359, 602], while general topics from an image analysis viewpoint are discussed in the book by Costa and Cesar [121] and the survey paper by Loncaric [389].

There are several ways in which shape features can be defined and extracted from the data. One can focus on either the boundary of an object or the interior region. We have already seen one set of features based on moment invariants which can be used to characterize the shape. This feature is region based as it focuses on the region of an object rather than its boundary. I next briefly discuss some of the more commonly used shape features.

- **Fourier descriptors:** The boundary of an object can be considered a closed curve in the complex plane (in the case of two-dimensional objects). By tracing out the boundary using 4-connectivity (see Section 8.6), which ensures a constant sampling interval, we obtain a function which is periodic in the number of pixels along the boundary. The Fourier descriptors are the coefficients of the discrete Fourier transform of this function. The descriptors can be transformed to be scale, rotation, and translation invariant [559]. It is also possible to use the Fourier descriptors of two curves to obtain a measure of the difference between the shapes represented by the curves, even if they have different sizes and orientations [297].

- **Chord distribution:** A chord is a line joining any two points on the boundary of a region. The distribution of the lengths and angles of all chords can be used to describe the shape of an object [559]. A similar idea is used in features called shape distributions in the context of three-dimensional objects [463].

- **Skeleton-based representation of a region:** Given an object, its boundary pixels may be iteratively removed by a process called "thinning," resulting in the "skeleton" of the object. There are several well-known algorithms to generate this skeleton, including Hilditch's thinning algorithm [274], the medial axis transform [48, 559], and morphological operations [557]. Skeleton construction is often sensitive to noise in the boundary of a region, resulting in additional "edges" in the skeleton. Smoothing the boundary prior to the extraction of the skeleton can alleviate this problem. Alternately, we can postprocess the skeleton to remove the extra edges, but it may be difficult to identify which edges are caused by noise in the boundary and which are real edges in the skeleton. Also, some algorithms, such as the Hilditch's thinning algorithm, may result in a skeleton which is 2 pixels wide in places. A single-pixel-wide skeleton can be obtained by further processing to ensure that there is only one path from one pixel to the next, with diagonal pixels favored over 4-connected ones as this results in a shorter edge in terms of number of pixels.

  Once a skeleton of a region is obtained, the shapes of two regions may be compared by considering each skeleton as a graph (with nodes and edges) and using graph-matching algorithms [559].

- **Angular Radial Transform (ART):** This is a region-based shape descriptor proposed in the MPEG-7 standard [418]. It can be applied to objects composed of a single region or multiple regions, possibly with holes. The shape feature is obtained by decomposing the shape into a number of orthogonal two-dimensional basis functions. The normalized and quantized magnitudes of the coefficients for each basis function are then used as the feature vector.

  ART belongs to a broad class of shape analysis tools based on moments [440]. It is a unitary transform defined on the unit circle that consists of the complete orthonormal sinusoidal basis functions in polar coordinates. The ART basis function of angular order m and radial order n in polar coordinates is given by

  $$V_{nm}(\rho,\theta) = \begin{cases} \exp(jm\theta)/2, & n = 0, \\ \exp(jm\theta)\cos(\pi n\rho), & n \neq 0. \end{cases} \tag{9.12}$$

  The ART coefficient F_{mn} of a two-dimensional signal $f(\rho,\theta)$ is defined by

  $$F_{mn} = \frac{1}{\pi}\int_0^{2\pi}\int_0^1 V_{nm}^*(\rho,\theta) f(\rho,\theta)\rho\, d\rho\, d\theta, \tag{9.13}$$

where V^*_{nm} is the complex conjugate of V_{nm}. The ART feature is implemented by discretizing (9.13). The MPEG-7 standard uses twelve angular bases and three radial bases. Rotational invariance is achieved by using only the magnitude of F_{mn}, and scale invariance is achieved by normalizing all the F_{mn} by the area of the image, i.e., F_{00}. This results in a 35-dimensional feature vector as the normalized F_{00} is always one and thus dropped from the representation.

In addition to the traditional application of image retrieval, ART features have also been used in the context of retrieval of "objects" in simulation data [314].

- **Shapelets:** Another set of shape features derived using coefficients of two-dimensional basis functions are shapelets, which were developed in the context of astronomy for describing shapes of galaxies [506]. Shapelets are a complete, orthonormal set of basis functions constructed from Laguerre or Hermite polynomials, weighted by a Gaussian. Unlike wavelets, where the basis functions have the same shape but different sizes, the shapelets model an image as a collection of discrete objects of arbitrary shapes and sizes. Thus, they are suitable for astronomy data to represent structures such as clusters of galaxies and are finding use in various applications including archival and compression of data, as well as morphological classification of galaxies.

- **The contour shape descriptor:** The MPEG-7 standard [418] also includes a shape feature based on the boundary of the object. This contour shape descriptor is derived from the Curvature Scale Shape (CSS) representation of the contour [435, 434].

  The CSS representation is based on the observation that humans tend to compare shapes by first breaking up a contour into convex and concave sections and then comparing these sections based on their position on the contour, their length relative to the length of the full contour, and so on. To mimic this, the CSS representation decomposes the contour by considering its inflection points, that is, points where the curvature is zero. The contour is then iteratively smoothed using a Gaussian. As the smoothing continues, the inflection points disappear, until the contour becomes convex, with no inflection points. The CSS image is obtained by plotting the amount of smoothing on the vertical axis against the location of the inflection points on the horizontal axis (for example, see Figure 15.13 in [418]). A contour, corresponding to a particular level of smoothing, is then defined by a horizontal line in the CSS image.

  The MPEG-7 contour shape descriptor is a set of characteristics obtained from this CSS image, the original contour, as well as the contours obtained during the process of smoothing. For example, the descriptor includes the number of peaks in the CSS image, the height of the largest peak, and the x and y positions of the remaining peaks. In addition, for each contour, the circularity

$$\frac{\text{perimeter}^2}{\text{area}} \tag{9.14}$$

  and the eccentricity

$$\sqrt{\frac{i_{20} + i02 + \sqrt{i_{20}^2 + i_{02}^2 - 2i_{20}i_{02} + 4i_{11}^2}}{i_{20} + i02 - \sqrt{i_{20}^2 + i_{02}^2 - 2i_{20}i_{02} + 4i_{11}^2}}}, \tag{9.15}$$

where

$$i_{02} = \sum_{k=1}^{M} (y_k - y_c)^2, \tag{9.16}$$

$$i_{20} = \sum_{k=1}^{M} (x_k - x_c)^2, \text{ and} \tag{9.17}$$

$$i_{11} = \sum_{k=1}^{M} (y_k - y_c)(x_k - x_c) \tag{9.18}$$

are obtained. Here M is the number of pixels inside the contour shape, (x_k, y_k) are their coordinates, and (x_c, y_c) is the center of mass of the shape.

Additional details, including the extraction of the CSS image features, without the explicit creation of the CSS image, are described in [435, 434].

This MPEG-7 shape descriptor is invariant to scale, rotation, translation, and mirroring of the object contour. It has also been shown to be robust to noise in the contour [435] and performs well in similarity-based retrieval.

- **Shape context:** An approach proposed for measuring similarity between shapes is that of shape context [32]. Like CSS, it is also a descriptor derived from the contour of a shape. First, the contour is represented by a set of points; these are not required to have any special characteristics such as being landmark points or points with high curvature. The shape context is then the descriptor, which, for a given point on the shape, describes the coarse distribution of the rest of the points on the shape with respect to that point. The distribution is in the form of a histogram of the coordinates of the remaining points, where the histogram bins are chosen to be uniform in log-polar space. This makes the descriptor more sensitive to positions of nearby sample points than points which are further away. The cost of matching two points, one each on two objects, is thus the distance between the histograms of the two points. A bipartite graph-matching algorithm can then be used to match the points from one object to another.

Note that, as appropriate, shape descriptors may be applied to either the original image with varying pixel intensities, or an object with varying pixel intensities after extracting it from the background, or a binary object after extracting it from the background. Shape descriptors for three-dimensional objects are also possible, as discussed in Section 15.6 of the description of the MPEG-7 standard [418] as well as in the article by Osada et al. [463]. In addition, texture features, which are described next, can, in some situations, be good representations for shapes as well [452].

9.4 Texture features

Image texture can be considered as the spatial dependence of pixel values. Over the years, several features have been proposed which provide a quantitative description of the characteristics commonly associated with texture, such as coarseness and directionality. Specifically, texture features have been discussed at length in Chapter 14 of the book by Sonka,

Hlavac, and Boyle [559], with additional details presented in the book chapters by Tuceryan and Jain on texture analysis [590] and those by Manjunath and Ma [417] and by Sebe and Lew [529] on texture features in image retrieval. I next describe four texture features commonly used in applications.

- **Grey-level co-occurrence matrices:** Texture features based on grey-level co-occurrence matrices (GLCMs) characterize the spatial co-occurrence of pixel values in an image. These features are computed in two steps [255]. First, the pairwise spatial co-occurrences of pixels separated by a particular angle and/or distance are tabulated using a GLCM. Then, a set of scalar quantities characterizing different aspects of the underlying texture are derived from the GLCM.

  Let $I(x,y)$ be an $N \times M$ image whose pixels take one of L levels, and let $\vec{r}$ be a specified spatial offset. A GLCM entry, $\text{GLCM}_{\vec{r}}(i,j)$, is then simply a count of the number of times a pixel with value $j \in 1,\ldots,L$ occurs at offset $\vec{r}$ with respect to a pixel with value $i \in 1,\ldots,L$:

$$\begin{aligned}\text{GLCM}_{\vec{r}}(i,j) = \#\Big\{&(x_1,y_1),(x_2,y_2) \in (N,M) \times (N,M)\\ &| I(x_1,y_1) = i \wedge I(x_2,y_2) = j \\ &\wedge \vec{r} = \overrightarrow{(x_2-x_1, y_2-y_1)}\Big\}.\end{aligned} \tag{9.19}$$

  The offset $\vec{r}$ can be an angle and/or distance. We can compute the GLCMs corresponding to different numbers of offsets, depending on the application. In our work in remote sensing [83, 451], we found that the following four offsets worked well: the adjacent pixel in the 0° direction, in the 45° direction, in the 90° direction, and in the 135° direction.

  Intuitively, the diagonal and near-diagonal entries of a GLCM will be larger for images composed of patches with the same or similar pixel values, at least with respect to the offset. The off-diagonal entries will be larger for images in which the pixel values vary locally. The distribution of the GLCM entries are summarized using five scalar quantities: angular second moment (ASM), contrast (CON), inverse difference moment (IDM), entropy (ENT), and correlation (COR). These quantities are computed as follows:

$$\text{ASM} = \sum_{i=1}^{L}\sum_{j=1}^{L} (\text{GLCM}(i,j))^2, \tag{9.20}$$

$$\text{CON} = \sum_{n=0}^{L-1} n^2 \left\{ \sum_{|i-j|=n} \text{GLCM}(i,j) \right\}, \tag{9.21}$$

$$\text{IDM} = \sum_{i=1}^{L}\sum_{j=1}^{L} \frac{\text{GLCM}(i,j)}{1+(i-j)^2}, \tag{9.22}$$

$$\text{ENT} = -\sum_{i=1}^{L}\sum_{j=1}^{L} \text{GLCM}(i,j) \log \text{GLCM}(i,j), \text{ and} \tag{9.23}$$

$$\text{COR} = \sum_{i=1}^{L} \sum_{j=1}^{L} \frac{(ij)\text{GLCM}(i,j) - \mu_{i'}\mu_{j'}}{\sigma_{i'}\sigma_{j'}}, \tag{9.24}$$

where

$$\mu_{i'} = \sum_{i=1}^{L} \sum_{j=1}^{L} i \ \text{GLCM}(i,j), \tag{9.25}$$

$$\mu_{j'} = \sum_{i=1}^{L} \sum_{j=1}^{L} j \ \text{GLCM}(i,j), \tag{9.26}$$

$$\sigma_{i'}^2 = \sum_{i=1}^{L} \sum_{j=1}^{L} \text{GLCM}(i,j)(i - \mu_{i'})^2, \text{ and} \tag{9.27}$$

$$\sigma_{j'}^2 = \sum_{i=1}^{L} \sum_{j=1}^{L} \text{GLCM}(i,j)(j - \mu_{j'})^2. \tag{9.28}$$

Thus, a complete GLCM texture feature vector corresponding to k offsets is

$$\begin{aligned}&[\text{ASM}_1, \text{CON}_1, \text{IDM}_1, \text{ENT}_1, \text{COR}_1, \ldots,\\ &\text{ASM}_k, \text{CON}_k, \text{IDM}_k, \text{ENT}_k, \text{COR}_k]\end{aligned} \tag{9.29}$$

resulting in $5 * k$ features. Note that the GLCM is an $L \times L$ matrix for an image whose pixels take on one of L intensity levels. Some quantization of the image may be necessary to generate a moderate-sized matrix, even for 8-bit data, but more so for 16-bit data.

- **Power spectrum:** Texture is often thought of as being related to periodic image patterns. Since Fourier analysis provides a mathematical framework for the frequency-based analysis of images, the Fourier power spectrum can be used for characterizing texture in images. Let $F(u,v)$ be the discrete Fourier transform of an $N \times M$ pixel digital image $I(x,y)$:

$$F(u,v) = \frac{1}{NM} \sum_{x=0}^{N-1} \sum_{y=0}^{M-1} I(x,y) \exp\left[-2\pi i \left(\frac{ux}{N} + \frac{vy}{M}\right)\right], \tag{9.30}$$

where $i = \sqrt{-1}$. The Fourier power spectrum (PS) is then $|F|^2 = FF^*$, where * denotes the complex conjugate. The radial distribution of the values of $|F|^2$ in the two-dimensional frequency space is related to the coarseness of the texture in I, and the angular distribution of the values is related to the direction of the texture. Texture features summarizing these distributions can be computed by dividing the frequency space into tiles and computing the average of the PS in these tiles. If the u and v dimensions are divided into T_u and T_v tiles, respectively (assuming N

and M are divisible by T_u and T_v, respectively), then the PS texture features can be computed as

$$\mathrm{PS}(i,j) = \frac{NM}{T_u T_v} \sum_{u=i\frac{N}{T_u}}^{(i+1)\frac{N}{T_u}-1} \sum_{v=j\frac{M}{T_v}}^{(j+1)\frac{M}{T_v}-1} |F(u,v)|^2 . \qquad (9.31)$$

In addition to these tile averages, there are four additional quantities which can be computed over the entire frequency space. These are the maximum (MAX), the average (AVG), the energy (ENG), and the variance (VAR) of the magnitude of $F(u,v)$ [9]. If we divide the frequency space into $k \times k$ tiles, the complete PS texture feature vector is

$$[\mathrm{PS}(0,0),\ldots,\mathrm{PS}(k-1,k-1),\mathrm{MAX},\mathrm{AVE},\mathrm{ENG},\mathrm{VAR}] . \qquad (9.32)$$

The PS texture feature contains $(k \times k) + 4$ components.

- **Wavelet-based features:** Wavelets (see Section 5.2) are used to characterize texture for the same reason as Fourier based techniques—they provide information about the frequency content of an image which is often associated with texture. Wavelets, unlike Fourier based approaches, are also localized in space [230]. This makes them more suitable for analyzing texture in nonstationary or nonhomogeneous images, such as in remote sensing applications.

  One approach to computing wavelet texture features is to exploit the multiresolution analysis (MRA) framework using a separable discrete wavelet transform (DWT) [410]. The DWT applies the wavelet to decompose the image into low-low (LL), low-high (LH), high-low (HL), and high-high (HH) components (see Section 5.2) . The MRA framework successively reapplies the DWT to the decimated LL band, resulting in a multilevel decomposition. The wavelet texture feature vector is then composed of the energies and standard deviations of the four components at the different resolutions. For example, the feature vector resulting from a k-level decomposition is

  $$[\mathrm{energy}(\mathrm{LL}_1),\mathrm{stdev}(\mathrm{LL}_1),\ldots,\mathrm{energy}(\mathrm{HH}_k),\mathrm{stdev}(\mathrm{HH}_k)] . \qquad (9.33)$$

  The wavelet texture feature contains $8 \times k$ components. Any appropriate wavelet can be used; we have found the Daubechies-4 wavelet to be a good compromise in terms of effectiveness and computational efficiency.

- **Gabor-filter–based features:** Texture analysis using filters based on Gabor functions is a frequency-based approach which, like wavelets, seeks to localize the analysis in both the spatial and frequency domains. Gabor functions are particularly appealing as they can be mathematically shown to achieve the optimal trade-off between localizing the analysis in the two domains [135].

  The texture analysis is accomplished by applying a bank of scale- and orientation-selective Gabor filters to an image [416]. If we use filters tuned to n scales and m orientations, the Gabor texture feature vector is then formed from the mean and the standard deviation of the outputs of each of these images:

  $$[\mu_{11},\sigma_{11},\mu_{12},\sigma_{12},\ldots,\mu_{1m},\sigma_{1m},\ldots,\mu_{nm},\sigma_{nm}] , \qquad (9.34)$$

where μ_{rs} and σ_{rs} are the mean and standard deviation, respectively, of the output of the filter tuned to scale r and orientation s. The Gabor texture feature contains $2 \times nm$ components.

Some of our work in texture features is described in [451, 452] for the problems of detection of human settlements in satellite images and shape characterization in simulation data. The choices we made for the parameters used in the different texture features were based on the computational cost of feature extraction and the effectiveness of the feature. As these parameters are application dependent, some experimentation might be necessary to determine the best parameters for each application.

9.5 Problem-specific features

The extraction of appropriate features describing the objects of interest is an important and critical step in scientific data mining. It is also a very problem-dependent step—for a given data set, the features extracted may change if the problem were to change, for example, searching for round objects instead of rectangular objects in a satellite image. As a result, one of the best sources of good features are the domain scientists themselves as they have the best understanding of the data and the problem. They can not only assist in identifying good features, but also provide guidance on robust ways of extracting these features.

In Sections 9.2 through 9.4, I described some commonly used features which have found broad applicability in a variety of problems in several application domains. However, in many problems, we may need to consider problem- or domain-specific features to either complement the generic ones, or use instead of the generic ones. I next discuss some specific examples to illustrate the different types of problem-specific features.

- **Classification of bent-double galaxies:** In this problem, our goal was to classify radio-emitting galaxies with a bent-double morphology in the FIRST (Faint Images of the Radio Sky at Twenty-cm) astronomical survey [29]. The data from the FIRST survey (sundog.stsci.edu) are available both as image maps and a catalog. The catalog [623] is obtained by processing an image map to fit two-dimensional elliptic Gaussians to each galaxy. Each entry in the catalog corresponds to the information on a single Gaussian, including the coordinates for the center of the Gaussian, the lengths of the major and minor axes, the peak flux, and the position angle of the major axis in degrees counterclockwise from North.

  We decided that, initially, we would identify the radio sources and extract the features using only the catalog. The astronomers expected that the catalog was a good approximation to all but the most complex of radio sources, and several of the features they thought were important in identifying bent-doubles could be easily calculated from the catalog. The astronomers also suggested that we first group the entries in the catalog to identify those which formed a galaxy. Then, we could extract the features using the catalog entries corresponding to each galaxy.

  We identified candidate features for this problem through extensive conversations with FIRST astronomers. Our initial expectation was that shape features might be a good option as our focus was on the "morphology" of a galaxy. However, when the astronomers justified their decision to identify a radio source as a bent-double, they placed great importance on spatial features such as distances and angles. Frequently, the astronomers would characterize a bent-double as a radio-emitting "core" with one or more additional components at various angles, which were usually wakes left by the core as it moved relative to the earth.

For galaxies composed of three catalog entries, this led us to extract several features such as the distances between the centers of the three ellipses, the angles of the triangle formed by the three centers, and so on. To order these features appropriately, we considered several methods to identify the core of the galaxy. We finally selected the one which gave the least error using a decision tree classifier and also resulted in a tree which, from an astronomy viewpoint, selected relevant features [196].

We also had to try several different ways of representing the angle feature. An obvious choice was to use the position angle which is the angle made by the major axis of the ellipse counterclockwise from North. However, the astronomers pointed out that this feature was sensitive to small changes in the data. An ellipse tilted slightly to the left of North had a small angle, while one tilted slightly to the right had a large angle. We thus had to derive more robust features to represent the angle.

Our approach to mining the FIRST data for bent-doubles has been described in detail in [196, 315, 200, 316], including the process of refinement of the features to obtain the desired accuracy of classification.

- **Classification of orbits in Poincaré plots:** Another problem where the general features described in this chapter are not appropriate is our work on the classification of orbits in Poincaré plots (see Section 2.6). An orbit is a set of points in two dimensions traced out by the intersections of a particle with a poloidal plane, that is, a plane perpendicular to the magnetic axis of a tokamak. These orbits have different shapes, such as a "closed" contour in the form of a slightly deformed circle, or a set of "islands" arranged in a circle, where each island is crescent shaped and the islands do not touch each other.

  These data are rather unique as they are very different from either mesh or image data, being represented by the floating-point coordinates of points in two-dimensional space. This makes it very challenging to identify the features for the data. Visually, given just the points, it is often easy to identify an orbit as belonging to a certain class based on its shape. This is because the human visual system is very good at filling in the space between the points to generate a closed curve. For example, a circle formed by a series of dots instead of a continuous curve can still be identified as a circle. Of course, if the dots are more concentrated in certain parts of the circle than in other parts (which is the case in some orbits), it may become difficult to correctly identify the circle.

  One approach to extracting features for an orbit defined by the set of points is to try and "fill in" the curve on which the points lie. This approach was used by Yip [638] to solve a similar problem arising in dynamical systems. His approach was to consider the graph corresponding to the minimal spanning tree of the points and extract relevant features which could then be used in a simple rule-based system to classify the orbit. These features were identified by looking at the structure of this graph for various types of orbits. For example, island orbits would have edges in the graph with a large edge length as the crescent-shaped structures did not "touch" each other.

  Unfortunately, when we considered the points in an orbit generated using a computer simulation [14], we found that certain characteristics of the orbits generated by the dynamical systems considered by Yip were no longer valid. For example, in some island orbits, the crescent-shaped structures were almost touching each other, making it difficult to set a threshold to identify long edges in the graph. Further, the crescents were sometimes very thin (considered radially). This made it difficult to use the graph-based approach to distinguish between such island orbits and incomplete quasi-periodic orbits which appear like arcs of a circle. Adding to the problem was that the data from the simulation were noisy, unlike the data shown in Figure 2.5 which were generated from the equations governing dynamical systems. This made it difficult to select the thresholds to derive the graph-based features.

The extraction of robust features representing the points in an orbit is still an active area of research. A particular challenge is the wide variation in orbits from a single class. For example, a quasi-periodic orbit could trace out a complete circle with just a few points, or, even with many points, it could trace out small arcs which never completely form a circle. Also, some island chain orbits have wide islands, while others have very thin islands, with the width of each island clearly visible only after magnification. Any features used to represent these seemingly different orbits from one class must identify what is common across the orbits, and at the same time must be scale-, rotation-, and translation-invariant as well as robust to noise in the data.

- **Identification of human settlements in satellite images:** Another problem which required domain-specific features was the identification of human settlements in satellite images [323]. A challenge in this problem was the size of the data, which, given the resolution of current satellite imagery (less than 1 meter per pixel), implied that we had to come up with an ingenious solution to generate accurate results in a computationally efficient manner.

  Our approach was first to use multispectral imagery at slightly lower resolution of 4 meters per pixel to identify regions likely to contain human settlements. Here, the choice of features was straightforward—we used the spectral bands as features for each pixel and classified the pixel as belonging to one of several classes. Regions where there was a mix of different types of pixels were considered further in the identification of human settlements.

  The next step was to analyze these regions using the higher-resolution imagery at 1 meter per pixel. This was a single band, grey-scale image. Here, we could exploit the fact that the resolution allowed us to identify straight lines and corners which are typically associated with buildings. Instead of using these features to explicitly identify each building (which would have been time consuming and possibly inaccurate), we divided the region into square tiles and considered each tile to be "inhabited" if it had a sufficient number of edge and corner pixels.

  Our use of the edge- and corner-density features in this problem was motivated by the need to extract the regions of interest at low computational cost. Further, the accuracy required was such that a coarse set of features was sufficient.

- **SIFT features for image retrieval:** The Scale Invariant Feature Transform (SIFT) [393, 395] has gained wide acceptance in the image retrieval community. It has two parts to it—the scale-based detection of keypoints, discussed in Section 8.3.3, and the extraction of features, or descriptors, for the keypoints. Once a keypoint has been identified, it is assigned a consistent orientation based on local image properties; the descriptors are then represented relative to this orientation, making them rotation invariant. These descriptors are computed by first calculating the magnitude of the gradient and its orientation at each image pixel in a region around the keypoint location. These are weighted by a Gaussian derived from the scale of the keypoint and accumulated into an orientation histogram for subregions around the keypoint. The vector containing these histograms for each of the subregions, after modifications to reduce the effects of changing illumination, forms the scale- and rotation-invariant feature vector at that keypoint.

- **Determination of velocity for moving objects:** The tracking of moving objects is often done using their velocity vectors as features to discriminate between two similar objects near each other, but moving with different velocities. Sometimes, the velocity of a moving object may be of interest by itself, for example, in problems where the velocity sheds light on some physical phenomena or is useful in other ways, such as the velocity of a hurricane moving towards land.

  There are different ways in which the velocity of an object can be determined. We have already considered one such approach, namely block-matching techniques, in Section 8.4.2. Another approach to calculating two-dimensional image motion is optical flow [28, 551, 559]. The optical flow field represents the three-dimensional motion of objects when projected onto the two-dimensional image plane. However, what is calculated is the perceived motion; for example, a rotating sphere, with no markings, will appear to be stationary under constant illumination. Also, a large object of uniform intensity moving a short distance will appear as if some of its pixels have not moved at all due to the aperture effect (see Section 8.4.1).

  The early work of Horn and Schunk [277] used spatiotemporal derivatives to calculate the optical flow. They start with the assumption that the observed intensity of any point on an object changes slowly with time. Thus,

$$I(x,y,t) \approx I(x+\delta x, y+\delta y, t+\delta t), \tag{9.35}$$

  where δx and δy are the displacements in the x and y directions over time δt. Representing the right-hand side by its Taylor series expansion, we have

$$I(x,y,t) = I(x,y,t) + \frac{\partial I}{\partial x}\,\delta x + \frac{\partial I}{\partial y}\,\delta y + \frac{\partial I}{\partial t}\,\delta t + O(\partial^2), \tag{9.36}$$

  where the last quantity represents the higher-order terms. These can be ignored if δx, δy, and δt are very small, leading to the equation

$$\frac{\partial I}{\partial x}\delta x + \frac{\partial I}{\partial y}\delta y + \frac{\partial I}{\partial t}\delta t = 0 \tag{9.37}$$

  which can be written as

$$\frac{\partial I}{\partial x}\frac{\delta x}{\delta t} + \frac{\partial I}{\partial y}\frac{\delta y}{\delta t} = -\frac{\partial I}{\partial t}. \tag{9.38}$$

  Since the optical flow is

$$\vec{v} = \left(\frac{\delta x}{\delta t}, \frac{\delta y}{\delta t}\right)^T, \tag{9.39}$$

we have

$$-\frac{\partial I}{\partial t} = \nabla I \cdot \vec{v}. \tag{9.40}$$

This single equation only provides the component of flow in the direction of the brightness gradient. To determine the flow completely, a constraint is imposed on the smoothness of the velocity which requires nearby points in the image plane to move in a similar manner. Combining this constraint with equation (9.40) results in the minimization of a squared-error quantity, which can be solved using iterative approaches.

Several different techniques [24] have been proposed to calculate the optical flow, with two of the more popular approaches being the ones proposed by Horn and Schunk [277] and by Lucas and Kanade [397], the latter in the context of image registration. These techniques have different computational complexities and vary in how they discretize the different equations and the constraints used to generate a complete velocity vector. Often, preprocessing by temporal smoothing is necessary to calculate the derivative estimates and some postprocessing may be required to assign a single velocity to the pixels comprising an object. Multiresolution approaches are also an alternative when objects move a large distance between frames and the assumptions made for the application of the traditional optical flow algorithms are no longer valid. Motion estimation has also been discussed extensively in the text by Tekalp [577].

Feature extraction is a very problem-dependent task and it is not always easy to determine which features should be extracted and how. In some problems, we can start with simple features such as the values of the different spectral bands in a multispectral image; the values of various variables such as pressure and velocity at grid points of a mesh; the input parameters used in a series of simulations; or the values of different sensors over time as an experiment progresses. In others, the features have to be extracted, often in an iterative manner, by evaluating the results using one set of features and refining them until the desired results are obtained. If, in the mean time, more data are collected and the data set becomes larger, with new types of objects or new variants of existing objects, then the process may have to be repeated all over again, with the new data.

Once the features for all the objects or items in a data set have been extracted, we are ready to consider the next step in scientific data mining, namely, dimension reduction. However, it is important to first evaluate this collection of objects, and their associated features, as a whole, and check that the quality of this reduced representation of the data set is appropriate. I next discuss various operations which may be necessary to postprocess this data item by feature matrix representing the data set being analyzed.

9.6 Postprocessing the features

Once we have extracted the features for the objects of interest in the data, some postprocessing may be required. Consider, for example, the case where we are analyzing the data from a series of experiments and the features for each experiment are the outputs measured by different sensors. If one of the sensors is not working, or working incorrectly and reporting erroneous readings, then we need to determine if we should include that experiment in our data set and if so, how we should handle the erroneous readings. Or, perhaps we are working with image data, and the parameters for the algorithms used to extract the objects are not applicable across all images, resulting in objects which do not correspond to anything physical. In such cases, the features we extract may be meaningless and we may need to determine if we should remove these objects from the data set or rerun the algorithms using a more appropriate set of parameters.

A simple way to evaluate the quality of the reduced data is to extract statistics for the features including the minimum and maximum values, the mean, and the standard deviation. We can also use visual tools, such as those described in Chapter 12, to understand the data item by feature matrix and to ensure that its quality is suitable for further processing. These tools can help us determine if the quality of the data needs to be improved, through techniques such as

- **Normalization of the data:** The normalization, or scaling, of data is an issue which should be addressed in the context of the data and the problem. Though the question appears simple enough, it is nontrivial to figure out if the features should be scaled and what is the best way to scale them. It is often recommended that the features be standardized by subtracting the mean and dividing by the standard deviation so that each feature has zero mean and unit variance. This is to prevent features with large numerical values from dominating any analysis. However, as Duda, Hart, and Stork (see [164, page 539]) point out, such normalization is valid when the spread of values is due to normal random variation. However, when the spread is due to the presence of two subclasses in the data, such normalization may be less than helpful. Scaling can have untoward effects, such as completely obscuring clusters already present in the data [338].

  Sometimes, scaling is necessary as, for example, in the use of parallel plots (see Section 12.1.3) to view high-dimensional data, that is, data items with many features. All the features are typically viewed at the same scale, else features with small values can be completely lost in the presence of features with large values. Normalizing the features to lie between, say, zero and one, can be accomplished by subtracting the minimum and dividing by the range between the maximum and the minimum values for each feature. Alternatively, we can standardize each feature to make its mean equal to zero and standard deviation equal to one.

  Alternative approaches to scaling involve simple transformations of variables, such as taking the logarithm, or using weighting to emphasize one feature over another. Care mut be taken in the application of these techniques, lest they do more harm than good.

- **Handling missing features:** As we have observed, one or more features might be missing for an object. This can be the result of a sensor malfunction, or an oversight in the generation of the feature, for example, certain tests were not done in a clinical trial or certain variables were not recorded in an experiment. There may also be situations where a feature was not included in early versions of an experiment, but was added later on as it was found to be useful. Addressing these missing features is very problem dependent and should be done in consultation with the domain scientists. The solutions may range from ignoring the feature for all objects to estimating the missing features using a suitable method, or ignoring objects with several missing features.

- **Identification of outliers:** A sensor malfunction or a bug in a computer simulation may also result in objects which are outliers. An object may be an outlier for several reasons, including having a feature with a value very different from the mean of the values of the feature for other objects or having a class label shared by few other objects. Again, a solution must include input from the domain scientists and may result in removing the object from further consideration or adding more objects to the minority class.

Detection of outliers is also an important topic by itself. In some problems, such as credit card fraud detection, network intrusion detection, or identification of unusual particles in high-energy physics experiments, the outliers may be the items of interest. A closely related topic is that of anomaly detection, with some domains using the two terms interchangeably. The topic of outlier detection is discussed further in Section 11.6.

- **Extraction of derived features:** In some situations, the features extracted are too primitive for the problem being addressed. We may need to generate derived features by considering, for example, ratios of existing features as in the circularity feature (equation (9.14)) in the MPEG-7 contour shape descriptor which is a function of the perimeter and the area. Or, instead of the length and width of an object, we could consider its aspect ratio, which is scale independent.

- **Using appropriate units:** Another issue which may make one feature appear more or less important than the others is that of the units used to represent the features. For example, measuring the length of an object in meters may be appropriate while measuring it in kilometers may make it appear very small relative to other variables.

All these postprocessing approaches are very problem specific and must be addressed in collaboration with the domain scientists who may be able to advise on the best way to improve the quality of the reduced representation of the data in the form of objects and their associated features. This evaluation is particularly important when the data provided by the scientists can be directly represented as a data item by feature matrix, rather than mesh data or images. In the latter case, we have had ample opportunity to become familiar with the data during the process of extracting the objects and the corresponding features. However, this is not true in the former case, and a quality check is essential.

And finally, the importance of care in postprocessing the features cannot be overemphasized, especially as it can strongly affect the analysis as well as any conclusions drawn from the data.

9.7 Summary

In this chapter, I discussed the extraction of features, also referred to as descriptors or characteristics, to represent the objects of interest in the data. I use the word "feature" in the primitive sense to imply a measurement which can be extracted from the data. Starting with some desired properties of features such as scale, rotation, and translation invariance, I discussed simple features such as object sizes and moments as well as more complex features such as shape and texture. Through the use of examples, I illustrated the need to exploit problem-specific features in collaboration with the domain scientists. I also mentioned the need to include "housekeeping" features such as file names and locations of objects to enable one to trace the object back to its origin. Finally, I described various ways of postprocessing the object-feature data to improve their quality.

Often, in scientific data mining, not all the features we extract may be useful or relevant to the problem. Or, the features may not yield results to the desired accuracy, or they may be sensitive to small changes in the data. In such situations, we will need to revisit the step of feature extraction, discuss it further with the domain scientists, and extract additional features until the desired results are obtained. This iterative aspect of feature extraction can lead to the creation of a large number of features, resulting in the problem of high dimensionality of the feature space. The problem is aggravated if we are working, for example, with multispectral data and we choose to extract the features separately for each band in the data. In Chapter 10, I describe ways in which we can reduce this dimensionality and identify key features for a problem.

9.8 Suggestions for further reading

The extraction of features relevant to a problem is a task which is strongly dependent on the problem. Two different problems, based on the same data set, are likely to require the extraction of different features. The best source of ideas on features to extract for a problem are likely to be the domain scientists themselves as well as features extracted for other similar problems. Thus, I would recommend domain-specific journals or articles where similar problems are addressed. For example, the articles in the *IEEE Transactions on Medical Imaging* are a good source of ideas for features in medical image analysis, while the *IEEE Transactions on Geoscience and Remote Sensing* or the *SPIE Journal of Applied Remote Sensing* would be a help for remote sensing problems.

Chapter 10

Reducing the Dimension of the Data

It is of the highest importance in the art of detection to be able to recognise out of a number of facts which are incidental and which are vital. Otherwise your energy and attention must be dissipated instead of being concentrated.

—Sherlock Holmes [159, p. 254]

In the previous chapters, we have seen how we can identify objects in scientific data sets and extract features or descriptors representing these objects. This results in a matrix where each row represents an object or data item and the columns are the values of the features for that object. These features include both "housekeeping" features, or metadata, such as the name of the image file and the location of the object in the image; as well as features which describe the object such as its size, shape, and texture. The latter may number in the hundreds, and in some cases, such as gene expression data, even in the thousands. The number of features is referred to as the "dimension of the problem" as each data item can now be represented by a point in feature space, of dimension equal to the number of features, with each coordinate representing a feature. In this chapter, I discuss why we need to reduce the number of features and the ways in which we can accomplish this task. The focus is on the features describing the data, not on the housekeeping features. I consider the latter to represent the provenance of the data; they must be retained as the data are processed to enable us to track the origin of an object, but they are not used in the analysis algorithms.

A brief note on the terminology used in this section may be helpful. The idea of reducing the dimension of a problem has been explored in many fields, including statistics and machine learning, as well as several application domains such as meteorology, fluid dynamics, bioinformatics, and various social sciences. As is often the case, the terminology used is not consistent. For example, what is referred to as a "feature" in one domain, may be called an "attribute" in other, and a "variable" in a third domain. Some authors may use more than one term, distinguishing them based on subtle differences. Further, similar techniques may be rediscovered in different domains. Each time, new insights into the approach are provided, with possible enhancements to the technique, along with a new name, just to add to the confusion. These terminology issues must be kept in mind when borrowing ideas from other fields.

This chapter is organized as follows. First, in Section 10.1, I describe the need for reducing the number of features (excluding the housekeeping features). As mentioned earlier, this number is the dimension of the problem. Then, in Section 10.2, I describe transform-based methods which transform the original set of features into a set in a reduced-dimensional space. Next, in Section 10.3, I discuss feature subset selection methods which select a subset of the original features. Section 10.4 describes some domain-specific methods for reducing the dimensions, followed by Section 10.5 on ways of representing high-dimensional data. Finally, Section 10.6 summarizes the chapter and Section 10.7 provides suggestions for further reading.

10.1 The need for dimension reduction

The manner in which features are obtained for an object in the data can result in many features which are irrelevant to the task at hand. Sometimes, scientists may not have any preconceived notion of all the analyses they plan to do when they start generating or collecting the data. So, they collect as much information as possible as it may be prohibitively expensive, or even impossible, to collect the data again. For example, in a simulation, scientists may choose to output several variables and derived quantities, not knowing which ones will be important. Running these simulations may require several hours of computer time on a massively parallel system, making it expensive to run the simulation again if it is later discovered that a key variable has not been included in the output. Or, if astronomers are observing a rarely occurring event, they would tend to err on the side of collecting more data. In other cases, scientists may not know which feature is important until they do the analysis. For example, they may be interested in identifying objects with a certain shape but may not be able to identify which shape features are the most appropriate for the task at hand. All this can result in each object being described by several features.

There are several reasons why we may want to reduce the dimension of the problem and represent each object by the fewest number of features necessary for the task of analysis. The simplest reason is that if fewer features are used to describe an object, it means that fewer features must be extracted and stored, resulting in lower computational and memory requirements for data mining. Further, when the patterns identified in the data are described using fewer features, they are usually easier to understand and interpret, a quality which is important in many scientific applications.

Another key reason why we want to reduce the number of features is that it can adversely affect the task of pattern recognition in several different ways. For example, if the problem being addressed is one of classification, irrelevant or redundant features can often hurt the accuracy of the classifier induced on the data [307]. As Witten and Frank [630] explain, experiments with a decision tree algorithm have shown that adding a random binary feature can reduce the accuracy by 5% to 10% for the problems tested. This is because at some point in the creation of the tree, likely at deeper depths of the tree, the number of data items at the node is so small that the random feature gets selected for the decision at that node. Therefore, it is important that such irrelevant features be removed prior to classification.

A large number of features in a problem also leads to the phenomenon of the curse of dimensionality [31]. One manifestation of this relevant to the task of pattern recognition is that in high-dimensional spaces, the number of samples (that is, data items) are sparsely distributed. So, in classification, where we are interested in the hyperplane which separates, say, the positive examples from the negative ones, we would need far more training samples in a high-dimensional space to get the same accuracy as in a low-dimensional space. As the complexity of functions of many variables can grow exponentially with the number of variables, the size of the training set must grow exponentially as well. This can be particularly problematic in scientific data sets where training data are usually generated by the tedious process of manual inspection of the data.

Another manifestation of the curse of dimensionality is that some algorithms may lose their meaning. For example, the concept of nearest neighbor in feature space is used in the k-nearest neighbor classification algorithm, in locally weighted regression, and in similarity searches. However, recent work has questioned the meaning of nearest neighbor in problems where the dimensionality is high. This is based on the observation that, under certain assumptions, in high-dimensional spaces, the nearest neighbor may be just as far as the farthest neighbor [38]. This observation can also cause problems in clustering algorithms as it becomes difficult to interpret similarity based on the distance between two points in feature space.

The need to reduce the number of features may also be driven by the need for fast indexing in retrieval. While the raw data in scientific applications are rarely stored in databases, the features which have been extracted for each object can be, and sometimes are, stored in a database. Retrieval is a key part of content-based information retrieval applications, where the scientist is interested in retrieving objects similar to a query object, with the similarity being measured in feature space. Retrieval also plays an important role in retrieving objects which satisfy certain simple queries on the features, for example, all objects in a specific image, or all objects whose nth feature satisfies some constraint. These retrieved objects can then be analyzed further. While these retrieval tasks can be accomplished by a sequential and time-consuming search through the data, it is much faster to use indexing techniques to speed up the search [209, 177]. However, typical indexing schemes become inefficient as the dimension increases, and at around 8–12 features, their performance can be worse than a sequential scan through the data. As a result, dimension reduction is an integral part of all retrieval applications.

Another concern with a large number of features is the time for pattern recognition. Some algorithms, such as the Naïve Bayes and decision tree classifiers scale linearly in the number of attributes. However, in the absence of any of the previous considerations, one must balance the reduction in pattern recognition time resulting from the use of fewer features with the additional time necessary to reduce the number of features.

With all these reasons, it is clear that a reduction in the number of features should be considered before pattern recognition techniques are applied to the data. The simplest approach is to work with the domain scientists to identify features likely to be irrelevant or redundant. Another is to understand the relationships between the features through information visualization tools such as parallel plots (see Chapter 12). More sophisticated approaches are also possible as discussed in this chapter. However, dimension reduction techniques must be used with caution. While there are benefits to identifying the key features in a data set, we may have a data set where different features are important when different subsets of the data are considered. For example, in a decision tree, the subset of data at one of the lower-level nodes may be such that a feature which is relatively unimportant when the full data set is considered, becomes important when only the subset of the data at that node is considered. If we had removed the features considered unimportant at the root node, where the entire data set was considered, then the decision-tree model would be less accurate at the lower level as an important feature at that level would be missing from the data.

I next describe two categories of techniques for dimension reduction—one where the original features are transformed into features in lower-dimensional space and the other where a subset of the original features is selected.

10.2 Feature transform methods

In feature transform methods, the original features extracted from the data are transformed into a new, reduced-dimension space and the important features are identified in this space. As a result of the transformation, we often lose the interpretation in terms of the original features. For example, a variable considered important in the reduced-dimension space may be a function of several, or all, of the original variables.

A note on terminology—feature transform methods are also referred to as "feature extraction" methods in some domains [385]. As I use this term to refer to the extraction of the original features from the data (as in Chapter 9), I will refer to techniques which generate new features through a functional mapping as feature transform methods.

10.2.1 Principal component analysis

One of the most commonly used transform methods for dimension reduction is the Principal Component Analysis, abbreviated as PCA. If the number of different names for a method are an indication of its popularity in solving practical problems and of its simplicity in having been reinvented many times, then PCA can be considered to be an incredibly popular and simple method. The basic idea behind PCA is known as Karhunen–Loève transform (KLT) in signal processing [581], Empirical Orthogonal Functions (EOFs) in meteorology [483], Hotelling Transform in Psychometry [278], Proper Orthogonal Decomposition (PODs) in dynamics [103], Latent Semantic Indexing in text mining [137], and the underlying algebraic formulation is called Singular Value Decomposition (SVD) in linear algebra [33]. The term Principal Component Analysis itself appears to have arisen in statistics in multivariate analysis [471, 503]. The developments in each of these fields has further enriched the insights one can obtain from the application of this method and are therefore worthy of further investigation even if the problem one is trying to address is not in one of these domains.

As expected, the idea behind principal component analysis (PCA) is very simple. Let $\mathbf{x}$ be a vector of d input features. These are the features we have extracted using either the techniques in Chapter 9 or other methods. Ideally, we want these features to be uncorrelated. Our goal in PCA is to use linear combinations of the input features to transform them into a set which is uncorrelated and where only a few features are necessary to adequately describe the data. The first step is to consider the linear combination of features which results in the maximum variance:

$$\mathbf{a}_1^T \mathbf{x} = \sum_{j=1}^{d} a_{1j} x_j. \tag{10.1}$$

Next, we look for another linear combination, $\mathbf{a}_2^T \mathbf{x}$ which is uncorrelated to $\mathbf{a}_1^T \mathbf{x}$ and has maximum variance. Proceeding in this manner, at each step p, we look for a linear function

$\mathbf{a}_p^T\mathbf{x}$ which is uncorrelated to all the previous functions, $\mathbf{a}_k^T\mathbf{x}$, where $k = 1, \ldots, p-1$. The kth derived feature, $\mathbf{a}_k^T\mathbf{x}$, is referred to as the kth principal component or PC, for short.

These PCs are obtained by solving an optimization problem

$$\text{Maximize} \quad \text{Var}[\mathbf{a}_k^T\mathbf{x}] = \mathbf{a}_k^T \text{Cov}(\mathbf{x})\mathbf{a}_k, \tag{10.2}$$

where Var represents the variance of the term in the square brackets, $\text{Cov}(\mathbf{x})$ is the covariance matrix of $\mathbf{x}$, and the maximum is obtained subject to the constraint that $\mathbf{a}_k^T\mathbf{a}_k = 1$. The solution to this problem is given by selecting the vectors $\mathbf{a}_k$ to be the eigenvectors of the matrix $\text{Cov}(\mathbf{x})$, with $\mathbf{a}_1$ corresponding to the largest eigenvalue λ_1, $\mathbf{a}_2$ corresponding to the second largest eigenvalue λ_2, and so on [310]. For the vector $\mathbf{x}$ of length d, we have

$$\text{Var}[\mathbf{a}_k^T\mathbf{x}] = \lambda_k \quad \text{for} \quad k = 1, \ldots, d. \tag{10.3}$$

Note that sometimes there is confusion about what is meant by the term "principal components." Some texts refer to the vectors $\mathbf{a}_k$ as the PCs, while others refer to the derived variables $\mathbf{a}_k^T\mathbf{x}$ as the PCs. In the latter case, the $\mathbf{a}_k$ may be referred to as the vector of coefficients or loadings for the kth PC, the principal directions, or the principal vectors. These principal directions are orthogonal and of unit length, thus forming a set of basis functions for the data. The PCs are then the original data projected onto the principal directions.

This derivation of the PCs using the eigenvectors allows us to exploit several resources from linear algebra, both theoretical and computational, for dimension reduction. Let the data item by feature matrix be denoted by $\mathbf{X}$ with n rows (corresponding to the n data items) and d columns (corresponding to the d features). Typically, $\mathbf{X}$ will be rectangular, with the number of rows and columns being problem dependent. One approach to calculating the PCs is to obtain the eigenvalues and vectors of the covariance matrix corresponding to $\mathbf{X}$. However, this involves forming the covariance matrix.

Alternatively, we can use the SVD of $\mathbf{X}$, which is given as

$$\mathbf{X} = \mathbf{U} \begin{pmatrix} \mathbf{D} & 0 \\ 0 & 0 \end{pmatrix} \mathbf{V}^T, \tag{10.4}$$

where $\mathbf{U}$ is an $n \times n$ orthogonal matrix, $\mathbf{V}$ is a $d \times d$ orthogonal matrix, and $\mathbf{D}$ is a diagonal matrix of size $r \times r$ with $\mathbf{D} = \text{diag}(\sigma_1, \sigma_2, \ldots, \sigma_r)$ and $\sigma_1 \geq \sigma_2 \geq \cdots \geq \sigma_r > 0$. The σ_i are the singular values of the matrix $\mathbf{X}$ which is of rank r (hence the r nonzero singular values). The columns of the matrices $\mathbf{U}$ and $\mathbf{V}$ are called the left- and the right-singular vectors, respectively. The σ_i^2 are the eigenvalues of the matrix $\mathbf{X}^T\mathbf{X}$, the columns of $\mathbf{U}$ are the eigenvectors of $\mathbf{X}\mathbf{X}^T$, and the columns of $\mathbf{V}$ are the eigenvectors of $\mathbf{X}^T\mathbf{X}$. The SVD thus diagonalizes the matrix $\mathbf{X}$, making the features uncorrelated.

Reducing the dimension of the problem using PCA essentially exploits the idea that if we consider only the PCs corresponding to the largest eigenvalues, we have captured most of the variance in the data. When we consider all d eigenvectors, we get a complete representation (to within roundoff error). However, since the smaller eigenvalues are often quite small, they can be considered to be noise in the data, and therefore, ignored.

PCA, as defined here, is the orthogonal projection of the data onto a lower-dimensional space, sometimes called the principal subspace, such that the variance of the projected data is maximized. It can be shown that this is equivalent to the linear projection which minimizes the average projection cost which is defined as the mean-squared distance between the data points and their projections [42].

This description barely touches upon the rather broad field of PCA. There are several questions which must be addressed to obtain the representation in the reduced-dimension space. For example, how many PCs should we use and how do we determine this number? Is there a way of determining which of the original features are more important? Should the features be standardized (by subtracting the mean and dividing by the standard deviation) prior to the use of PCA? There are several texts which delve into these topics in greater detail, including the very practical book by Jackson [296], the excellent statistical text by Joliffe [310], and the posthumously published book of Preisendorfer [483] which focuses on the subject from the viewpoint of meteorology and oceanography.

There are also some excellent software packages available for obtaining the SVD in a numerically accurate and robust manner. In particular, the Linear Algebra Package, LAPACK [5], provides much of the functionality needed for dense matrices, while the functionality for the sparse version using the Lanczos method is available in SVDPACK [36] and using the Arnoldi method is available in ARPACK [370]. A parallel version of the LAPACK functionality is available in the SCALAPACK library [43].

10.2.2 Extensions of principal component analysis

PCA is extensively used as, among all the linear techniques, it gives the minimum mean-squared-error approximation in the reduced-dimension space. This is good for tasks such as the compression of the data. However, it may not be the best option for all problems. For example, it is well known that it does not necessarily lead to maximum class separability in the lower-dimension space (see, for example, Figure 6.1 in [581]). This is to be expected as the dimension reduction is not being optimized for class separability. Further, PCA finds a linear subspace and therefore cannot handle data lying on nonlinear manifolds.

Several techniques have been proposed to address these deficiencies [86, 109]. For example, projection pursuit selects interesting low-dimensional linear orthogonal projections of a high-dimensional point cloud by optimizing a certain objective function called projection index [205, 282]. Methods such as principal curves and surfaces, which are nonlinear generalizations of PCA, have been proposed for the case where the underlying distribution is unknown [259, 260]. As PCA can be represented using dot products, a nonlinear version, called the kernel PCA, can be obtained by replacing the product with a nonlinear kernel [525]. Another nonlinear variant is Locally Linear Embedding (LLE) [517], an eigenvector method which computes low-dimensional, neighborhood-preserving embeddings of high-dimensional data.

Another class of methods worth mentioning are those used in blind source separation applications [113], where independent source signals are estimated from observed signals which are linear mixtures of the source signals. One of the methods used in this context is the Independent Component Analysis (ICA) which determines a transformation which results in mutually independent features, not just uncorrelated features [285, 286].

10.2.3 Random projections

An interesting technique which has gained popularity recently is random projections [330]. Here the original high-dimensional data are projected onto a lower-dimensional subspace using a random matrix whose columns have unit length. The basic idea behind random projections is based on the Johnson–Lindenstrauss lemma, which states that any set of points of dimension n in a Euclidean space can be embedded in $O(\log n/\epsilon^2)$ dimensions without distorting the distances between any pair of points by more than a factor of $(1 \pm \epsilon)$ for any $0 < \epsilon < 1$ [133]. As the method is computationally very efficient, it has been used to first project data in a very high-dimensional space to a lower-dimensional space before using a more computationally expensive technique such as PCA [39].

There are two topics related to random projections which are relevant to the analysis of scientific data. The first is the concept of a "sketch" which is used in streaming data to build lower-dimensional models [290]. The second is "compressed sensing" where a sparse signal is reconstructed from a small number of linear measurements [22, 119]. Both these topics are relatively recent developments and therefore the subject of active research.

10.2.4 Multidimensional scaling

Given a set of objects in a d-dimensional space, along with a matrix of similarities (or dissimilarities) between them, the idea behind multidimensional scaling (MDS) is to find a k-dimensional space, $k < d$, with each object being represented by a point in this space such that the distances between the points match the original similarities. Often, MDS is used to project the data into two or three dimensions so that it can be displayed graphically to reveal the underlying spatial structure in the data.

The basic implementation of MDS is relatively simple. Given k, and the distances d_{ij} between objects i and j, the algorithm tries to map each object to a point in the k-dimensional space to minimize the stress function

$$\text{stress} = \sqrt{\frac{\sum_{i,j}(\hat{d}_{ij} - d_{ij})^2}{\sum_{i,j} d_{ij}^2}}, \tag{10.5}$$

where $\hat{d}_{ij}$ is the Euclidean distance between the lower-dimensional representations of objects i and j. The stress can be considered as the relative error in the distances in the lower-dimensional space. MDS essentially starts with a guess, assigning each object to a point in the k-dimensional space, either randomly, or using some heuristic. It then iteratively moves the points to minimize the stress, until no further improvement is possible.

MDS as a statistical technique has been described at length in the books by Borg and Groenen [51] and Cox and Cox [123]. However, as it requires $O(N^2)$ time, where N is the number of objects, it can be impractical for very large data sets. A nonlinear variant, which builds on the classical MDS, is the recently introduced Isomap algorithm by Tenenbaum et al. [578].

10.2.5 FastMap

An efficient solution to the problem of mapping objects into a lower-dimensional space while preserving the distances was proposed by Faloutsos and Lin [179] in the context of information retrieval applications. Given only the matrix of dissimilarities or distances between the N objects, they assume that the objects are points in a d-dimensional space and then project them into a smaller number of orthogonal directions. Their method, called FastMap, starts by selecting "pivot" objects, which are two objects such that the projections of the other objects onto the line joining the pivots are as far apart from each other as possible. A linear heuristic algorithm to select the pivot points starts by arbitrarily selecting an object, and identifying it as the second pivot. The first pivot is then selected as the object farthest away from this second pivot, according to the matrix of distances. The second pivot is then replaced by the object farthest away from the first pivot.

The objects are then projected onto the $(d-1)$-dimensional hyperplane which is perpendicular to the line joining the two pivot objects. The distance between the objects in this projected space can be obtained from the original distances and the projections of the objects on the line joining the pivot objects. The problem now reduces to one of N objects in $(d-1)$-dimensional space, along with the dissimilarities in this new space. The process then iterates k times until the projections onto a $(d-k)$-dimensional space are obtained. Note that in this approach, the dimension of the original space d could be unknown.

This approach requires that a matrix of dissimilarities or distances between the objects be provided. The distance function should be nonnegative, symmetric and obey the triangle inequality. When the objects are described by a feature vector, a common choice is to use the Euclidean distance between the objects.

10.2.6 Self-organizing maps

A technique commonly used to project higher-dimensional data to two dimensions for the purpose of visualization is the Self-Organizing Map, or SOM, [346]. In its basic form, it produces a similarity graph of the input data, converting the relationships between high-dimensional data into simpler geometric relationship of their representations in a regular two-dimensional grid of nodes. The SOM essentially preserves the most important topological and metric relationships between the objects.

The SOM is a specific example of what are referred to as competitive learning algorithms [581]. These start with a set of representatives in the lower-dimensional space. As each vector in the higher-dimensional space is considered in turn, the representatives compete with each other, with the winner chosen as the representative which is closest in some distance metric to the vector. The winner is updated to move toward the vector, while the losers either are not updated or are updated at a much slower rate. In the case of the SOM, the representatives are "nodes" in two-dimensional space, which are initially arranged in a regular pattern (often rectangular or hexagonal). In addition to the winner, the nodes in a neighborhood around the winner are also updated. This neighborhood is defined with respect to the indices of the winner node. As the original vectors are used in turn to move the nodes around, this neighborhood shrinks in size to ensure convergence. Eventually, the nodes are representative of the distribution of the data, and neighboring nodes lie close to each other, reflecting their distance in the original space.

Though SOMs are similar to clustering algorithms considered in Chapter 11, they do not optimize a cost function, making it difficult to set parameters and assess convergence [42]. They are also implemented in "batch" or "incremental" mode, with the nodes being updated after each vector is presented instead of at the end of an iteration.

10.3 Feature subset selection methods

In contrast to the transform-based methods, many of which have their roots in statistics, information retrieval, and signal processing, the feature subset selection methods have their beginnings in the machine learning and data mining literature. As the reduced-dimension features are just a subset of the original features, the use of such methods makes it easier for a scientist to understand the features found by the method and to interpret the results of the analysis.

Feature subset selection methods have typically been considered when the analysis task is one of classification (see Chapter 11), where a class label has been assigned to each item. The very early work in this area focused on the selection of features by searching the space of all possible combinations using optimization techniques such as branch and bound, simulated annealing, or genetic algorithms [546]. More recent approaches to the problem [345] consider two categories of methods—wrappers and filters. Wrappers use the classification algorithm to evaluate the effectiveness of the feature selection while filters are independent of the classifier and select features based on other properties such as their ability to distinguish between different classes. As they do not use the classifier to evaluate the feature subset, filter methods are computationally more efficient, while wrappers may give more accurate classification results as they select the features based on their effectiveness in the pattern recognition task which follows the feature selection step.

Feature subset selection methods are not restricted to classification problems. They can also be used in the context of regression as discussed in Section 10.3.3 and techniques such as the PCA filter, which is described in the next section, can be used in the context of tasks such as clustering.

10.3.1 Filters for feature selection

Filter methods select important features based on how well they discriminate among the different classes. One approach to this is to consider each feature in turn, and determine the distance between the histograms or distributions of the feature values for each class. Ideally, if the histograms of feature values for two different classes have little overlap, the feature can be used to discriminate between those classes. If a problem has several classes, we want this condition to hold for each pair of classes.

There are several ways in which we can compare the histograms. We assume that the histograms for each feature for each class has already been suitably aligned so that a bin in one corresponds to a bin in another. Further, the histograms have been normalized by dividing each bin count by the total number of data items to estimate the probability, $p_j(d = i|c = n)$, that the jth feature takes a value in the ith bin of the histogram for a given class n. The Kullback–Leibler class separability filter [83] uses the Kullback–Leibler divergence $\delta_j(m,n)$ between the normalized histograms corresponding to classes m and n defined as

$$\delta_j(m,n) = \sum_{i=1}^{b} p_j(d = i|c = m) \log \left(\frac{p_j(d = i|c = m)}{p_j(d = i|c = n)} \right), \tag{10.6}$$

where b is the number of bins in the histograms. Summing these distances over all possible pairs of classes gives the class separability for feature j as

$$\Delta_j = \sum_{m=1}^{c} \sum_{n=1}^{c} \delta_j(m,n), \tag{10.7}$$

where c is the number of classes. The features can then be ranked using Δ_j, as larger values indicate better separability using the jth feature. In addition, very small values of

Δ_j may indicate noninformative features while large gaps in consecutive values when the Δ_j are ranked in descending order may be exploited to determine the number of features to keep. Class separability filters using other distance metrics such as the χ^2 distance, the L_2 distance, or the Bhattacharya distance are also possible.

Some classification techniques (see Chapter 11) also provide an indication of which features are important. For example, decision trees identify the decision to be made at each node by considering each feature in turn and determining a split that minimizes (or maximizes) an impurity (or purity) measure over all possible split points for that feature. The feature which optimizes this measure across all features is chosen to split the data at that node. A stump filter [83] is obtained when we apply this process only at the root node of the tree (hence the name "stump"), where we consider all the data items and rank the features by using one of the many measures used in creating decision trees (see Section 11.2.3).

Other filter approaches perform a broader search through the feature space. For example, Almuallim and Dietterich [3] consider minimal combinations of features which perfectly discriminate among the classes. They start with a single feature in isolation, then consider pairs of features, and then triples, and so on, until they find a combination which partitions the data such that each partition has items of only one class.

Alternately, one can identify key features by evaluating how correlated they are to the target variable, for example, by using the linear Pearson correlation coefficient. Given two vectors $\mathbf{x}$ and $\mathbf{y}$, with values $x_i, i = 1, \ldots, n$, and $y_i, i = 1, \ldots, n$, respectively, the Pearson's coefficient is defined as

$$\frac{\sum_i (x_i - \bar{x})(y_i - \bar{y})}{\sqrt{\sum_i (x_i - \bar{x})^2 \sum_i (y_i - \bar{y})^2}}, \tag{10.8}$$

where $\bar{x}$ and $\bar{y}$ are the mean values of $\mathbf{x}$ and $\mathbf{y}$, respectively. If one of the variables is discrete, as would be the case for the class variable, then k correlations are calculated corresponding to the k discrete classes [246]. Feature selection is then done by calculating the coefficient for each feature and keeping only the ones most correlated with the target.

Another filter method, called the PCA filter, is derived from the PCA [83]. Usually, PCA results in a linear combination of the input features and it is difficult to relate the transformed features back to the original ones to identify which of the original features is the most important. Several ways of eliminating unimportant variables using PCA were suggested by Joliffe [308, 309] and evaluated on both synthetic and real data. One approach starts with the PC corresponding to the smallest eigenvalue of the covariance matrix and discards the variable with the largest coefficient (in absolute value) in that vector. This variable is considered the least important. Next, the variable with the largest coefficient in the PC corresponding to the second smallest eigenvalue is discarded, assuming it has not been discarded before. The process continues until all the variables have been ranked. This process can also be done iteratively, where a prespecified number of variables are first deleted, the PCA is repeated on the remaining variables, and then, a further set of variables is deleted. The process continues until no further deletions are considered necessary. A complementary approach is to start with the largest PC and keep the variable associated with the highest coefficient in absolute value. These and other techniques are also discussed in Section 6.3 of the book on PCA by Joliffe [310]. Note that these techniques do not require the use of a class label and can therefore be applied in analysis problems other than classification, such as clustering.

Another approach to feature selection, called Relief, is based on the rationale that a useful feature should differentiate between instances from different classes and have the same (or similar) values for instances from the same class [340]. The basic idea behind Relief is very simple. The algorithm considers a random instance from the data and determines its closest neighbor of the same class and the opposite class (assuming there are only two classes). The values of the features of these nearest neighbors are then compared to those of the selected instance and used to update relevance scores for each feature. By considering several instances from the data, these scores are refined and the features with scores exceeding a user-defined threshold are retained to form the final subset. Kononenko [349] extended Relief to ReliefF, which is more robust to noise in the data as it considers k nearest neighbors in the calculation of the relevance scores instead of the single nearest neighbor. The technique was also extended to handle multiclass data by considering the nearest neighbors of each class different from the class of the sampled instance and then weighting their contributions by the prior probability of each class.

10.3.2 Wrapper methods

Wrapper methods consider candidate feature subsets, evaluate them using the classification algorithm, and select the subset which yields the most accurate results. As a result of this evaluation, the wrapper methods are usually more computationally expensive than the filter methods.

The different wrapper methods differ in the way the feature subsets are selected. The most exhaustive method is to consider all possible k subsets of features, with $k = 1, \ldots, d$, where d is the total number of features, and select the subset which gives the highest accuracy. Alternatively, the process can be made computationally more efficient through the use of a greedy algorithm, which makes the best decision at each step. There are two ways in which one can implement such algorithms. In the sequential forward selection method, the algorithm first considers all subsets with a single feature and picks the one with the best accuracy on the classification algorithm. Next, it considers each feature not previously selected and retains the one which gives the best accuracy using the subset of the two features selected so far. The process continues by selecting the best new feature to add at each step and terminates when the accuracy cannot be increased by the addition of any new features. In the backward elimination approach, we start with all the features and tentatively delete each feature in turn from the set of features which have not been deleted previously. The accuracy of the feature subset selected is determined by considering the error of the classification algorithm using cross validation (see Chapter 11).

Techniques which combine the filter and wrapper approaches are also possible, as described in the work of Das [131]. This hybrid technique exploits an idea called boosting (see Section 11.2.6) from ensemble learning in the feature selection [202]. In boosting, the instances of a training set are weighted based on how accurately the class is predicted for each instance. Instances which have often been misclassified in previous iterations are given higher weights. In the hybrid filter-wrapper approach, the next feature is selected using boosted decision tree stumps. This ensures that if there is a part of the feature space that is often misclassified by the features selected thus far, then the next feature selected will be one which is a better predictor in this part of the feature space. As in the case of wrappers, the features selected are evaluated using the classification algorithm which would normally be used for the learning task.

10.3.3 Feature selection for regression

Many of the techniques proposed for feature selection tend to be for cases where the classes are discrete. However, when the problem is one of regression rather than classification, one option is to use a wrapper-based feature selection technique in conjunction with the regression algorithm. Alternatively, one can use a filter based on a correlation coefficient as in equation (10.8).

Another approach, also based on correlation, is CFS, or Correlation-based Feature Selection by Hall [245, 246]. This ranks feature subsets instead of individual features and is based on the heuristic that good feature subsets have features which are highly correlated with the class, yet uncorrelated with each other. Using a heuristic from test theory, Hall calculates a figure of merit for a feature subset of k features as

$$\frac{k\,\overline{r_{cf}}}{\sqrt{k+k(k-1)\,\overline{r_{ff}}}}, \tag{10.9}$$

where $\overline{r_{cf}}$ is the average class-feature correlation and $\overline{r_{ff}}$ is the average feature-feature intercorrelation. The numerator can be considered to give an indication of the predictive power of a group of features, while the denominator indicates the degree of redundancy among them. The CFS method first creates a matrix of class-feature and feature-feature correlation and then searches the feature subset space using a best first search. It first selects the best single feature, expands the subset by one by selecting the next feature such that the two taken together are the best, and so on. If expanding a subset does not result in any improvement, the search drops back to the next best unexpanded subset and continues from there. This feature selection method can be applied to both discrete and continuous class labels.

The feature selection algorithm, Relief, described in Section 10.3.1, has also been extended to regression problems. When the class is continuous, it does not make sense to consider the nearest instances of the same and the opposite class. Instead, Robnik-Šikonja and Kononenko [512] introduce the concept of a kind of probability that the predicted values of two instances are different based on the relative distance between these predicted values.

10.4 Domain-specific methods

The methods discussed thus far to reduce the number of features are domain independent. However, as I have mentioned earlier, the domain scientists can often provide useful input in identifying important features. In addition, several domains have specific characteristics which can be exploited in dimension reduction. Or, they may have special requirements which must be met by any technique used in dimension reduction. I next discuss some of these domain-specific methods and requirements in further detail.

An interesting problem arises in the case of information retrieval applications [91]. Here, the goal is to retrieve items which are similar to a query item, where the item can be an image, an object in an image, or a document. As the number of features used to represent an item is often very large, in the hundreds or more, it is necessary to reduce the dimension of the search space. The dynamic nature of the data makes this application rather challenging from a dimension-reduction viewpoint as the characteristics of the data, such as the distributions, could change when new data are added. As many of the techniques are computationally expensive, attempts have been made to create incremental versions of various algorithms. The numerical analysis literature has several references to work in the area of updating the SVD [69, 234, 415, 644]. In the area of image information retrieval, for example, Ravikanth, Agrawal, and Singh [505] describe an efficient method for updating the SVD of a data set as well as strategies to determine when the update should occur. An efficient solution to a more challenging problem is provided by Brand in [55], where he considers the case when the data arrives as fragments of rows or columns of the data item by feature matrix. As a result, in his method, rows, columns, or fragments thereof can be added, changed, or retracted.

It must also be pointed out that there are other problems where the data are dynamic. For example, in function estimation, where we are interested in relating the output variables from, say, a computer simulation to the input variables, we may have a dynamic data set as the input space is explored based on the results of the function estimation. This may result in new data items being added in parts of the input space not explored thus far. Some data items may even be removed if they are considered as outliers.

A different approach to addressing the high computational cost of techniques such as PCA is to exploit the characteristics of the data and create lower-cost alternatives. For example, Karypis and Han [326] consider the problem of document retrieval using a query term. They observe that the retrieval performance can be poor as a term can have many meanings and a concept of interest can be described by several different terms, not just a single term. To address this problem, they use an approach called concept indexing for dimension reduction. This can be applied both in a supervised setting, where the data items have an associated class label, as well as in an unsupervised setting, where there is no class label. The algorithm first computes a lower-dimensional space by finding groups of similar documents and then uses the centroids of the groups as the axes of the lower-dimensional space. If the data are labeled, then the number of clusters is the number of classes, while in the unsupervised setting, the number of clusters is a prespecified number. The clustering essentially groups similar documents together, with the centroid containing terms which act as synonyms for the topic of the cluster. While this application is not of direct relevance in the context of science data, except perhaps in the mining of biology documents, the idea behind it could be applied in scientific problems as well. This approach is particularly interesting as it provides an efficient way of reducing the dimension of the problem while, at the same time, improving retrieval performance by exploiting the characteristics of the data, namely, that similar text documents have a common underlying concept.

Another problem in information retrieval is similarity searches in time series data, such as sensor data from experiments, variations in a physical quantity at a grid point in a simulation over time, and medical data such as electrocardiograms. The common approach to this problem is first to reduce the dimensionality of the data and then use a fast indexing scheme (see Section 10.5) for the retrieval [178]. Typically, the dimensionality is reduced by using the SVD, the discrete Fourier transform, or the discrete wavelet transform [336]. However, even simple approaches, where a time series is approximated by dividing it into equal length segments and using the mean in each segment, have provided surprisingly good results in comparison with the more sophisticated dimension reduction schemes. Keogh et al. [335] have further enhanced this approach by representing the time series by varying length segments of constant value, allowing a more accurate representation of the data. Their results indicate that the method is superior to the more complex dimension reduction techniques for time series data.

The field of bioinformatics has some rather special requirements for feature selection. A microarray experiment typically has between 10 and 100 tissue samples (the data items), each represented by thousands of genes (the features). Therefore, any analysis of such data must include dimension reduction techniques [632]. As the data items are few, there are few samples of each class, and any statistics derived from such small populations can be unreliable. Further, the data are often noisy, resulting in different values of a gene in two microarray measurements taken from the same sample. As Dong, Li, and Wong [154] point out, approaches based on the entropy measure could be considered for such applications. However, they also observe that many of the feature selection methods provide a ranking of the features, with the top k features used in further processing. This may result in the genes in the primary pathways being selected, while those in the secondary pathways are ignored. To address this issue, they suggest the correlation-based approach of Hall (see Section 10.3.3) which considers groups of features correlated with separating the different classes, but not correlated with each other. However, this may result in selecting the more important gene in each pathway, but neglecting the secondary genes. The topic of determining all the relevant pathways, as well as all the genes relevant in each pathway, is an open research problem which must be addressed to obtain the best understanding of gene expression profiles.

And, finally, in the context of domain-specific methods for reducing the dimension of a problem, there is the excellent book by Preisendorfer [483] on the use of PCA in meteorology and oceanography.

10.5 Representation of high-dimensional data

A practical problem which must be addressed in the analysis of data is the way the data item by feature matrix is represented. Many tasks require identifying objects whose feature values fall within a specific range, or objects which are close to a query object in feature space. For example, the Relief feature selection method described in Section 10.3.1 requires the nearest neighbors of an instance with the same and the opposite class. The k-nearest neighbor classification algorithm described in Chapter 11 also requires the nearest neighbors of an instance. In such problems, if we were to represent the data item by feature matrix in a simple and straightforward way using a two-dimensional array, such similarity search queries would become prohibitively expensive, especially when the data are too large to fit into memory.

A more efficient way to access the data is to use more sophisticated data structures for storage and indexing to allow fast access to the data. For example, if we just had a single feature, we could sort the data on the feature, allowing us to easily identify the nearest neighbors of a given data item without having to search through the entire data set. Further, by using traditional indexing methods, we could directly access the information we need.

The field of relational databases is replete with indexing techniques which support not only efficient access, but also insertions and deletions. The B-tree and its variants are the most popular methods in database management systems, though they are not suitable for higher-dimensional data, where the R-tree and its variants are more appropriate [379]. However, the performance of even these data structures degrades as the dimension of the problem increases beyond 20. Dimension reduction techniques have to be applied first to reduce the number of features before an indexing scheme can be used.

A description of various data structures, indexing schemes, and the implementations of various search algorithms on these data structures is beyond the scope of this book. An excellent reference for this topic is the detailed text on multidimensional data structures by Samet [521] and the extensive reference list included therein.

Choosing an appropriate indexing structure for multidimensional data for a particular application is very difficult. Some guidance is provided in Section 5 of the book chapter by Castelli [91] as well as Chapter 4 of the text by Samet [521]. However, any benefits provided by indexing should be considered in light of the additional time spent in identifying the appropriate indexing scheme, the building of the index itself, and the implementation of the algorithms using these data structures.

10.6 Summary

This chapter has focused on the task on dimension reduction, namely, the identification of key features used to represent the objects or data items in a data set. I first discussed the many reasons why dimension reduction is useful and necessary. I then described two broad categories of methods, one where the original features are transformed into a lower-dimensional space, and the other where we consider a subset of the original features. A section on domain-specific approaches to dimension reduction was followed by a brief note on indexing structures, a topic which may be appropriate for some problems.

It should also be mentioned that we should be careful when ignoring the features not identified as important in the dimension reduction process. Sometimes, features identified as unimportant when all the data items are considered may become important when a subset of the data items is considered. In addition, if the data set grows with time, the dimension reduction must be repeated to ensure that the new data has not made previously unimportant features, important.

While the focus in this chapter has been on reducing the number of features, I would be remiss in not mentioning the "blessings of dimensionality" [60]. In this viewpoint, the number of features is increased, not reduced. The idea is to exploit the fact that each feature potentially carries a little piece of information about the problem and we may stand to gain by extracting and integrating these pieces. For example, a problem in classification of handwritten numerals was solved to high accuracy by extracting a large number of features and creating forests of decision trees (see Chapter 11), where each tree is built using a random subset of the features [4]. Also, in support vector machines (see Chapter 11), the dimensionality of the problem is increased by adding new features which are functions of existing features, such as monomials of various degrees.

It is also interesting to note that the features which are not selected can be put to good use. Instead of just discarding these input variables, Caruana and de Sa [85] show that using some of these as output variables can increase the performance of the main output.

10.7 Suggestions for further reading

Dimension reduction techniques have seen much use in several different application domains. In addition to techniques which have been tailored to specific domains, there are also several general techniques which have been widely investigated and enhanced over the years. The more popular of these, the PCA, is described in detail in several books [483, 296, 310], which also provide practical guidance in the application of the method. The books by Liu and Motoda [385, 386] provide the data mining viewpoint on the subject of feature selection.

There are also several surveys on dimension reduction both from the machine learning perspective [47, 236] as well as the pattern recognition perspective [546]. A nice summary of filter methods is provided by Duch [163], including techniques based on information theory, while the use of feature selection for unlabeled data is discussed by Dy and Brodley in [166].

The interested reader is referred to the literature in their own problem domain for problem-specific methods. Often the subtleties associated with each problem may influence how a method is applied or its results interpreted.

Finally, I recommend the excellent book by Samet [521] for indexing schemes used to improve access to the data both before and after the dimension reduction techniques have been applied.

Chapter 11

Finding Patterns in the Data

The first principle is that you must not fool yourself—and you are the easiest person to fool.

—Richard P. Feynmann [189, p. 212]

In the previous chapter, I discussed techniques to identify key features which are relevant to the problem being addressed. Some of these techniques, such as wrappers (Section 10.3.2), incorporate the pattern recognition algorithm, while others, such as filters (Section 10.3.1) or the transform-based methods (Section 10.2), are independent of the pattern recognition algorithm.

In this chapter, I describe the different types of analyses we can do once we have the data item by feature matrix where the key features have been identified. In some problems, the scientist may be interested in the statistics of the features for the different objects in the data. For example, they may want the distribution of sizes of the objects in an image or the variation in the number of objects in a simulation data set over time. These simple summaries may provide insights on the physical phenomenon being studied or provide a "model" which can be used in other, more complex, simulations. In these cases, the information of interest can be directly obtained from the data item by feature matrix.

However, more sophisticated analyses are also possible, where the scientist is interested in finding patterns in the data. For example, they may want to group the data items so the items in one group are similar to each other, but different from the items in another group. Or, they may know that some items belong to one group and others to a different group, and they would like to build a "model," which, given an item, would assign it to one of these groups. In other cases, instead of a discrete group, there may be an output value or values associated with each data item, and the scientist may be interested in being able to predict these values. Or, given the structures of interest in a simulation, the scientist may be interested in tracking them to understand how they evolve over time, the processes associated with their birth and death, and any correlations between the evolution of the structures and other events which may shed light on how they may be able to control the behavior of the structures.

The analysis algorithms described in this chapter are referred to by names such as "data mining," "machine learning," "pattern recognition," "artificial intelligence," or "statistical learning" algorithms in the literature. Given the diversity of domains which have contributed to this area, the algorithms have been extensively described in several textbooks. They include both general texts covering a broad range of techniques [430, 252, 164, 261, 581, 630, 574, 42], as well as specialized texts on a specific technique, which I will refer to in the appropriate sections in this chapter. This diversity of names is the result of techniques being created in different domains at roughly the same time, or being independently reinvented in a different domain and given a new name, with the similarity between techniques known by different names becoming clearer over time. Each domain provides its own interpretation of a technique and it can often be helpful to exploit the wealth of information available in the literature and look at the results of an analysis from different perspectives.

As I have observed earlier, much of the analysis using "data mining" algorithms is done on commercial or nonscientific data, such as market basket analysis, customer relationship management, and text mining, where the data are presumed to have been preprocessed and the key features extracted using relatively simple techniques. In contrast, for scientific data, much of the time spent in the analysis is in the steps prior to the actual identification of patterns in the data. In these preprocessing steps, we need to identify the objects in the data correctly and extract features that accurately represent them; else, we run the risk of using garbage as input to the pattern recognition techniques. In my experience, if the features are of high quality, the actual algorithm we use for detecting the patterns is less important. On the other hand, if the features are inadequate, even the best algorithm will fail. In light of this consideration, and the extensive literature already existing on the topic of identifying the patterns in the data, I will describe the pattern recognition algorithms relatively briefly and provide references to texts where further details can be found. I will also mention some of the practical issues which must be considered when these techniques are applied to science data. These issues are often the result of characteristics unique to science data and can make the identification of patterns rather difficult.

This chapter is organized as follows. Sections 11.1 through 11.6 describe various algorithms for clustering, classification, regression, tracking, association rules, and anomaly detection. In each section, I describe the commonly used algorithms and provide extensive references for further reading. Next, in Section 11.7, I discuss some general topics such as optimization and distance metrics which are useful in the implementation of the algorithms described in this chapter. Finally, a summary of the chapter in Section 11.8 is followed by Section 11.9 which provides suggestions for further reading.

11.1 Clustering

The task of clustering is essentially one of grouping data items, each represented by a set of features, into similar clusters or groups. Typically, items within a group are similar to each other, but items from two different groups are less similar. In clustering, we do not have a priori information on the number of groups in the data or examples of items from each group. Therefore, clustering is also referred to as unsupervised learning to distinguish it from supervised learning, or classification, which is discussed in Section 11.2. In classification, we start with a collection of items, each of which has already been assigned to a group or class, and thus, has a class label. The goal is to build a "model" to distinguish among the different groups. The model can then be used to assign a class label to an item which is described by a set of features, but which does not have a class label. Since the learning of the model is supervised through a set of examples with known labels, classification techniques allow us to exploit a priori information, while clustering algorithms have to learn how to group similar items based solely on the features used to represent the items.

There are two key issues which must be addressed before we apply a clustering algorithm. The first is the choice of a similarity metric. This essentially indicates how similar (or dissimilar) two objects are in feature space (assuming that each object is described using a feature vector). The most obvious choice of a similarity metric is the distance between the two data items. The Euclidean distance is often chosen, though other distance metrics (see Section 11.7.1) are also possible. Alternatively, one can use a similarity matrix which directly provides the value of the "distance" between two objects. This matrix can be obtained using a domain-specific function whose value is large (small) when the two objects are similar (dissimilar). Sometimes, a dissimilarity matrix is used as it is more appropriate.

The second issue is the clustering criterion, which is used to evaluate a partitioning of the data items into groups. The task of clustering then becomes one of optimizing this function. A commonly used criterion is the sum of squared error. Suppose we have n objects described by the features vectors $\mathbf{x}_1, \ldots, \mathbf{x}_n$ which we want to group into c clusters, C_i, $i = 1, \ldots, c$. If n_i is the number of objects in cluster i, then the mean of these samples is

$$\mathbf{m}_i = \frac{1}{n_i} \sum_{\mathbf{x} \in C_i} \mathbf{x}. \tag{11.1}$$

Then the sum of squared error is given by

$$\sum_{i=1}^{c} \sum_{\mathbf{x} \in C_i} \|\mathbf{x} - \mathbf{m}_i\|^2. \tag{11.2}$$

By minimizing this error, we ensure that the sum of squared error in representing each object by its cluster mean is minimized. This is an appropriate metric to use when the objects form clusters which are compact and well separated from each other. However, the presence of a few outliers can result in an unnatural split of a cluster using this criterion [164]; care must be taken to evaluate the clustering results and incorporate additional constraints as needed. Other clustering criteria are also possible, such as optimizing the average, the median, or the maximum distance between objects in a cluster, or using criteria derived from scatter matrices which minimize the within-cluster scatter, while maximizing the between-cluster scatter [252, 164]. There are several different categories of clustering algorithms which I describe next.

11.1.1 Partitional algorithms

Partitional algorithms create the clusters by partitioning the data into a predetermined, or an automatically derived, number of clusters. The k-means algorithm [17, 18, 164] is the most popular and perhaps the simplest method for clustering. Here, k is the number of clusters and is assumed to be given. The algorithm essentially starts with k cluster centers, often assigned randomly. It then assigns each data item to its closest cluster center. The centers are then recomputed as the mean of the data items assigned to the cluster, and the process repeats until the cluster means do not change significantly.

The k-means algorithm, while popular, has certain restrictions [261, 574]. It is appropriate when the dissimilarity metric is the Euclidean distance, but is inapplicable when the features are such that there is no notion of a mean or center. It can also be sensitive to outliers.

One solution to these restrictions, albeit a computationally intensive one, is to require the point which represents the "center" of a cluster to be one of the items in the cluster. This is the case for the k-mediods clustering algorithm which uses as the center of a cluster, the item in the cluster which minimizes the total distance (or dissimilarity) to other items in the cluster.

The k-means technique has several modifications which vary based on how the initial cluster centers are obtained, the stopping criterion, as well as ways of splitting and merging clusters. It is also related to techniques in several other domains, including vector quantization [214], Voronoi diagrams [10, 458] and the Expectation Maximization (EM) algorithm [252].

11.1.2 Hierarchical clustering

Hierarchical clustering algorithms produce a hierarchy of clusterings and often exploit the fact that such a hierarchy may occur naturally in the data, as in biological taxonomy. Hierarchical clusterings are conveniently represented by a tree-like structure called a dendrogram [252, 164, 581]. At the lowest level, each item in the data set, which can be considered as a cluster consisting of a single item, is represented by a leaf of the dendrogram. When two clusters merge, the two branches of the tree merge. At the highest level, there is a single branch representing the cluster with all the data items. As we traverse the tree from the lowest leaf level to the highest root level, the number of clusters reduces by 1 at each level.

This indicates that there are two ways in which we can build the tree. Algorithms which start at the lowest level and build the dendrogram by merging clusters are called agglomerative algorithms, while algorithms which start at the top level and work down to the leaf level are called divisive algorithms as they divide the original cluster which consists of all the data items.

Agglomerative, or bottom up, approaches start with each item being a cluster by itself (see Figure 11.1). Nearest clusters are merged in turn, until there is just a single cluster of all the data items. The "nearest" clusters are identified using some metric like the clustering criterion. The single linkage technique uses the intercluster dissimilarity to be that of the closest (that is, least dissimilar) pair, where one item of the pair is from one group and the other item is from the other group. The complete linkage technique uses the intercluster dissimilarity to be that of the farthest (that is, most dissimilar) pair, while the group average technique uses the average dissimilarity between the items in the two groups.

The divisive technique is the less popular of the two hierarchical methods as it is less robust and requires more computations. It starts with all items in one cluster and splits it into two. One way to accomplish this is to use an algorithm such as k-means or k-mediods recursively to perform the splits. However, the results would be dependent on the initial cluster centers used at each step. A more computationally expensive approach is to start with all items in one cluster, consider all possible pairs of clusters which partition the items and select the one which optimizes a criterion. Then, we repeat the process on each of the two subclusters and, among all the partitions from each of the two subclusters, we select the one which optimizes the criterion, and so on.

11.1.3 Graph-based clustering

There is another category of clustering algorithms which shares aspects of both partitional and hierarchical approaches. This is the category of algorithms which exploits a graph

$$\begin{matrix} \mathbf{x}_1^T \\ \mathbf{x}_2^T \\ \mathbf{x}_3^T \\ \mathbf{x}_4^T \\ \mathbf{x}_5^T \end{matrix} = \begin{bmatrix} 2 & 1 \\ 1 & 1 \\ 5 & 3 \\ 7 & 4 \\ 6 & 2 \end{bmatrix} \qquad \text{Dissimilarity matrix} = \begin{bmatrix} 0 & 1 & \sqrt{13} & \sqrt{34} & \sqrt{17} \\ 1 & 0 & \sqrt{20} & \sqrt{45} & \sqrt{17} \\ \sqrt{13} & \sqrt{20} & 0 & \sqrt{5} & \sqrt{2} \\ \sqrt{34} & \sqrt{45} & \sqrt{5} & 0 & \sqrt{5} \\ \sqrt{17} & \sqrt{17} & \sqrt{2} & \sqrt{5} & 0 \end{bmatrix}$$

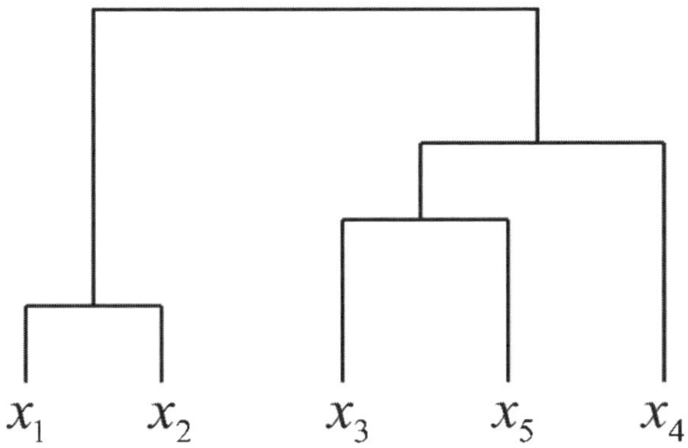

Figure 11.1. *Top: A data set with five instances,* $\mathbf{x}_1, \ldots, \mathbf{x}_5$*, each of dimension two, and the corresponding dissimilarity matrix obtained by considering the Euclidean distance between two instances. A larger distance indicates greater dissimilarity between two instances. Bottom: The dendrogram indicating how similar instances are merged to form clusters.*

representation [233] of the data. These algorithms consider the nodes of the graph to be the data items while the edges represent the similarity or the relationship between the nodes.

Simple graph-based clustering algorithms have been in use for quite a while (see, for example, Section 15.2 of [581] and the references therein and [642]). For example, some algorithms obtain the minimal spanning tree [233] of a graph and remove inconsistent edges. These are the edges with a relatively large weight, where the weight of an edge between two nodes is defined as the distance in feature space between the data items corresponding to the nodes. The idea is that edges with a large weight typically connect two different clusters.

More recently, there has been considerable interest in exploiting techniques originally developed for the solution of partial differential equations (PDEs) on large parallel machines. The clustering algorithms which arise from this field are derived from two tasks in the solution of these equations. One is domain decomposition, where the goal is to partition the grid over which the PDE is to be solved into subdomains, such that the subdomains are roughly equal in size and the connections between subdomains are minimized. This ensures that when a subdomain is assigned to a processor, the work done by each processor will be roughly the same; that is, the computation will be load balanced. Further, the communication between processors, which is time consuming, is minimized as the connections between the subdomains is minimized. The other task is the solution of a sparse set of linear equations using a parallel direct solver [213, 354]. Here, one of the goals is to reorder the equations so that the process of fill-in, which occurs during the factorization of the matrix, leads to a high degree of concurrency. Often, the other goal of re-ordering is to minimize the fill-in to reduce the computations.

This connection between the tasks in the parallel solution of PDEs and clustering has been exploited in several ways to make the graph-partitioning techniques suitable for clustering. For example, Karypis, Kumar, and coauthors have explored a multilevel approach which improved both the computational cost and the quality of partitioning obtained by directly partitioning the graph [328, 329]. This approach first coarsens the graph, partitions the smaller coarser graph, and then uncoarsens the graph to produce a partition for the original graph. The coarsening is done by collapsing the nodes and edges, and the uncoarsening incorporates refinement to produce high-quality partitions. This work resulted in the METIS software [428] which has seen extensive use for unstructured graph partitioning and sparse matrix reordering. The technique was then extended to hypergraphs, which are a generalization of graphs, where a set of edges is replaced by a set of hyperedges, with a hyperedge connecting more than two vertices at a time. Thus, the weight of a hyperedge represents the strength of the affinity among the objects represented in the hyperedge. A hypergraph-partitioning algorithm allowed clustering of the objects by identifying partitions such that the items in a partition were highly related and the weight of the hyperedges cut by the partitioning minimized [249]. This algorithm is available via the hMETIS software [428] for hypergraph partitioning.

Karypis, Han, and Kumar have also extended their hierarchical clustering algorithm to address clusters of diverse shapes, densities, and sizes. They first obtain a fine-grain clustering solution by using hMETIS. Then, they merge clusters based on measures of relative interconnectivity and relative closeness. Their algorithm, Chameleon [327], has been shown to work well on a number of data sets with points in two-dimensional space, indicating the potential of the technique for use in image segmentation, especially as it is relatively insensitive to noise.

A different approach to partitioning the graph is the use of graph Laplacians and spectral graph theory [114]. The Laplacian of a graph is a square matrix of size equal to the number of nodes in the graph. Any element in the matrix at location (i, j) represents an edge in the graph connecting the nodes i and j. The value of this element is zero if there is no corresponding edge in the graph and it is -1 if there is an edge in the graph (for $i \neq j$). For diagonal elements in the matrix ($i = j$), the value of the matrix element is the degree of the node, that is, the number of nodes adjacent to it. However, as Luxburg [400] points out, there are several definitions of graph Laplacians and one must be careful while reading the literature on the subject.

The spectral partitioning of a graph is obtained by first computing the second eigenvector of the Laplacian of the graph, which is also referred to as the Fiedler vector. The smallest eigenvalue of the Laplacian is zero, and the second smallest is thus the smallest positive eigenvalue. Then, the nodes are partitioned into two subsets using the eigenvector components. One option is to have one subset be generated using the positive values in the eigenvector, and the other subset be generated using the negative values. However, to ensure that the two subsets are balanced in size, the values in the eigenvector can be ordered and the median component be used to create the two subsets. Pothen, Simon, and Liou, in their well-known algorithm on partitioning sparse matrices [481], recursively applied the partitioning to each of the subsets. Hendrickson and Leland [271] extended this approach to create four or eight subsets at a time for the task of domain decomposition for parallel computation.

These spectral partitioning algorithms have since made their way into the clustering literature. For example, Shi and Malik [541] have used it in image segmentation. The nodes of the graph represent the pixels in an image, while the edge weights are determined by a function which incorporates both the brightness values and the spatial location of the pixels. Then, they use recursive spectral bisection to partition the graph, thus segmenting the image. Spectral partitioning is also being used in clustering protein sequences. For example, Paccanaro, Casbon, and Saqi [467] show that traditional clustering algorithms are local, as they assign a protein to a cluster by considering only the distances between that protein and the others in the set. In contrast, spectral clustering considers distances between all proteins in the set. As a result of this global perspective, it provides clusters which are of a higher quality from a biological viewpoint. Spectral clustering has also been used in the separation of speech between two speakers recorded using a single microphone, as shown in [13].

11.1.4 Observations and further reading

There are several challenges in the practical implementation and application of a clustering technique. A key problem is the number of clusters. In rare cases, we may know the number of clusters or the scientists may have an idea of what this number should be. If the number is unknown, we can try different values and compare some criterion for each clustering. A large gap in the criterion will indicate a possible value for the number of clusters (see, for example, Section 14.3.11 of the text by Hastie, Tibshirani, and Friedman [261]). For hierarchical methods, if we make a horizontal cut in a dendrogram, it will intersect the tree in a number of branches, which is indicative of the number of clusters. The issue then becomes one of determining where to make the cut. Theodoridis and Koutroumbas (Section 13.5 in [581]) suggest several techniques for choosing the number of clusters in the hierarchical method, including one based on the "lifetimes" of clusters. This is defined as the difference in the levels between when a cluster is created and when it is absorbed into another cluster. Clusters with a long lifetime are indicative of an inherent grouping in the data.

Another challenge is the choice of the initial set of cluster centers in algorithms such as k-means. Often, these are selected randomly and the clustering repeated with different random selections to ensure that the results are not sensitive to the initial cluster centers. Alternatively, if we are selecting the number of clusters as well, we can start the n-cluster solution with the centers of the clusters obtained for the $n-1$ cluster problem, plus the item which is farthest from the nearest cluster center.

It is a challenge to determine an appropriate algorithm for clustering the data, identify the different parameters, and select an implementation suitable for the size of the data. However, it should be kept in mind that a clustering algorithm will identify clusters in data regardless of whether the data are naturally clustered or random. Therefore, to avoid a misleading interpretation of structure in a data set where none exists, it makes sense to first determine if the data possess a clustering tendency. This is usually done by looking at the randomness of the data as described in Section 4.6 of the book by Jain and Dubes [298] and Section 16.6 of the text by Theodoridis and Koutroumbas [581].

Even if we assume the data have an underlying clustering structure, our choice of parameters for the algorithm and restrictions on the clusters, such as their shape, may not match the underlying structure in the data. The topic of cluster validity plays an important role in the quantitative evaluation of a clustering algorithm. This topic is discussed briefly in Section 10.10 of the text by Duda, Hart, and Stork [164] and in more detail in Chapter 4 of the book by Jain and Dubes [298], Chapter 16 of the text by Theodoridis and Koutroumbas [581], and in the article by Halkidi, Batistakis, and Vazirgiannis [240].

It is also important not to lose sight of the end goal of clustering, which is often to assign labels to the items in a cluster and then use the labeled examples in classification to build models. Different choices of clustering algorithms, clustering criteria, and distance measures might yield very different results. The domain scientist should be involved in interpreting and validating the results. It is also helpful to look at the data initially to identify any outliers—these may be the result of errors in the data or they may indicate something scientifically interesting.

Clustering is a pattern recognition technique which has been studied extensively. Several relatively recent surveys provide an excellent overview of the field, including the ones by Berkhin [34]; Han, Kamber, and Tung [250]; Jain, Murthy, and Flynn [299]; and Xu and Wunsch [635]. A survey of the literature on scalable clustering techniques is summarized in the article by Ghosh [217], while Murtagh [441] focuses on issues in the context of massive data sets. Several texts also provide good summaries and excellent practical suggestions including the early texts by Jain and Dubes [298] and Kaufman and Rousseeuw [332], as well as Chapters 11 through 16 of the text by Theodoridis and Koutroumbas [581].

Finally, the CLUTO clustering software [116] provides an extensive collection of functions for various clustering algorithms including partitional, agglomerative and graph-partitioning-based techniques. It supports various distance measures, as well as several clustering criterion functions and agglomerative merging schemes.

11.2 Classification

In classification, the goal is to build a "model" using a set of classified or labeled data items so that the model can be used to classify items without a class label. For example, we may be given a set of galaxies, each of which has been assigned a label or class indicating if the galaxy is of type bent-double or not [315]. Each galaxy is represented by a set of features. The goal is to use these data to build, say, a decision tree, which, given a new galaxy with its associated features, can be used to assign a label of bent-double or non-bent-double to that galaxy.

In this example, the decision tree is the classifier or the classification algorithm. The labeled data are referred to as the "training" data as they are used to "train" the decision tree to "learn" how to discriminate between bent-double and non-bent-double galaxies. The decision tree is a particular kind of model in that it discriminates among classes using hyperplanes parallel to the feature axes. More complex models using nonlinear discriminating functions are also possible.

Classification algorithms are usually evaluated by testing them on a part of the training data which has been set aside specifically for this purpose and not used during training. The accuracy on this "test" data is a measure of how well the classifier will perform on data it has not seen before. However, to ensure that a high accuracy is reflective of the predictive ability of a model rather than an artifact of the data, there are several issues which must be considered prior to measuring the accuracy of a classifier; I consider these in further detail in Section 11.2.7.

In addition to accuracy, another metric which is often important in scientific data is the interpretability of the model. Scientists are usually interested in understanding how a model arrives at its decision as, among other things, this can indicate which features are considered important and for what range of values. Or, the model may indicate which features are unimportant. In either case, such information can shed light on the scientific problem being investigated. Black-box models, which do not reveal their inner workings may therefore be less acceptable in some cases than a less accurate, but interpretable, model.

11.2.1 k-nearest neighbor classifier

One of the simplest classification algorithms is the k-nearest neighbor algorithm. This algorithm belongs to the category of instance-based learning, sometimes called delayed or lazy learning. In such techniques, the learning occurs only when the data item to be labeled is presented to the classifier. The classification typically involves the data items nearest to the query instead of all the data items. This allows a different model to be used for different parts of the feature space, enabling us to model a complex function using a collection of simpler functions which work well locally, but not necessarily globally.

The k-nearest neighbor algorithm assigns to an unlabeled item the most frequently occurring class label among the k most similar data items. The similar data items are obtained using Euclidean distance metric between the feature vectors. The k-nearest neighbor can also be applied using weights, where the neighbors which are closer to the query item have larger weights.

As we have observed in Chapter 10, if the feature vector is high dimensional, the most similar data items could be quite a large distance apart. This indicates a need for dimension reduction techniques prior to the use of the k-nearest neighbor classifier. Alternatively, we could weight the features in the calculation of the distance, with the more relevant features having larger weights.

Further, as we are interested in retrieving the nearest neighbors, an operation which can be quite expensive if the size of the training data is quite large, it may be beneficial to consider the use of an indexing scheme (Section 10.5) to store the data and make the access to the nearest neighbors more efficient. Alternatively, we could use the version of the algorithm which weights the data items based on distance from the query. We can then consider all the data items, with items farther away contributing a smaller amount, possibly zero, to the final result. This global version of the algorithm allows us to trade off the extra cost of building and maintaining an index with a computationally expensive sweep through the entire data set. It is also possible to determine approximate nearest neighbors as discussed by Arya et al. [7].

It is interesting to note that when $k = 1$, a query item is assigned the class of the closest item in the training data. There is a convex polyhedra around this data item which indicates the region of feature space closest to the data item; any query item in this region will be closer to this data item than any other item in the training set. The combination of these polyhedra, which form the decision surface for the problem, is essentially the Voronoi diagram of the training data (see, for example, Section 8.2 of [430]).

11.2.2 Naïve Bayes classifier

Another simple classifier is the Naïve Bayes classifier, so called as it is based on the Bayes theorem and naïvely assumes that the feature values are conditionally independent given the target class label [430, 252]. The Bayes classifier essentially assigns to a data item the most probable class label, given the values of its features.

The conditional independence of the feature values implies that, given the target class label, the probability of observing all the feature values for a data item is just the product of the probability of observing each of the individual feature values. Suppose the query data item is represented by the following d feature values:

$$q = q_1, \ldots, q_d \tag{11.3}$$

and we need to assign one of the following m class labels:

$$C = c_1, \ldots, c_m \tag{11.4}$$

to this data item. Then, for a class label c_j, the simplifying assumption states that

$$P(q_1, \ldots, q_d | c_j) = \prod_i P(q_i | c_j). \tag{11.5}$$

Using Bayes rule, this implies that the class label assigned to the data item is the one which maximizes

$$P(c_j) \prod_i P(q_i | c_j). \tag{11.6}$$

The number of distinct $P(q_i|c_j)$ terms which must be estimated from the data is just the number of distinct feature values times the number of distinct class labels. These probabilities, along with $P(c_j)$, can be estimated based on their frequencies over the training data.

However, as Witten and Frank [630, Section 4.2] point out, things can go awry with the naïve Bayes classifier if, in the training set, we do not have a particular feature value occurring in conjunction with every class label. This would make one of the probabilities $P(q_i|c_j)$ zero. As a solution, they suggest using a Laplace estimator by adding a small constant to each count. This ensures that a feature value which does not occur even once in the training data has a small, but nonzero, probability of occurring. In the case of numeric features, which is almost always the case with scientific data, instead of probabilities, we assume that each feature has values with a normal or Gaussian distribution and consider the probability density function [630]. If we know that a feature has a distribution different from a normal distribution, we can use the estimation procedures for that distribution.

The Bayes classifier often works well in practice though it makes certain assumptions on the independence of the feature values. This is especially true when the data have been preprocessed to remove redundant, that is, nonindependent, features. Given its simplicity, it is therefore worth trying the Bayes classifier before considering more sophisticated approaches.

11.2.3 Decision trees

Decision trees [61, 490, 493] are a popular technique in classification as they are accurate, relatively simple to implement, produce a model that is easy to interpret and understand, and have built-in dimension reduction. A decision tree is a structure that is either a leaf, indicating a class, or a decision node that specifies some test to be carried out on a feature (or a combination of features), with a branch and subtree for each possible outcome of the test. The decision at each node of the tree is made to reveal the structure in the data. The traditional version of a decision tree algorithm creates tests at each node that involves a single feature. These trees are referred to as axis-parallel trees because the tests can be considered as hyperplanes that are parallel to one of the axes in the attribute or feature space. As the tests at each node are very simple, it is easy for the domain expert to interpret the trees.

In a typical decision tree, the decision at each node is of the form

$$f_i < a_i, \tag{11.7}$$

where f_i is the ith feature and a_i is a value used in the decision. Given a data item with its feature values, there are two outcomes of this decision. This node of the decision tree therefore splits the data items at the node into two subsets. Starting at the root of the tree with all the data items in the training set, the building of the tree essentially consists of identifying the test to be made at the root node, using the test to split the set of items into two subsets, and repeating the process on each of the two subsets. The process of splitting the data items stops when we have a subset where all or a majority of the items have the same class label. This forms a leaf node of the tree. Then, to assign a class label to an item with its associated features, it is "dropped" in the root node of the tree and its path followed down to a leaf node, where it is assigned the class of the majority of the data items at that leaf node.

There are two key issues in the construction of a decision tree—the type of split at a node of a tree and the split criterion to be used in making this split. The simplest type of split is the axis-parallel split, which consists of a single feature on the left side of the decision. While such trees are easy to interpret, they may be complicated and inaccurate in the case where the data are best partitioned by an oblique hyperplane. In such instances, it may be appropriate to make a decision based on a linear combination of features, instead of a single feature. However, these oblique trees can be harder to interpret. They can also be more compute intensive as the problem of finding an oblique hyperplane is much harder than the problem of finding an axis-parallel one.

There are several variants of oblique decisions trees which differ in how the linear combination is obtained. In the CART-LC method [61], Brieman et al. use an approach which cycles through the features trying to find the best split on each feature, while keeping the others constant. A backward deletion process is then used to remove features that contribute little to the effectiveness of the split. This approach is fully deterministic and can get trapped in a local minima. The oblique classifier OC1 by Murthy [442] attempts to address this limitation by including randomization, where multiple random restarts are used to escape local minima. As the problem of finding the best oblique hyperplane is essentially an optimization problem, it is also possible to use evolutionary algorithms, as described by Cantú-Paz and Kamath [81]. This allows us to work with the coefficients of all the features at a time instead of the series of univariate splits considered in OC1 and CART-LC.

There are also several different split criteria one can use in the creation of the decision tree. A split criterion is a heuristic measure which evaluates the suitability of a proposed split. Depending on whether the measure evaluates the goodness or badness of a split, it is either maximized or minimized. The basic idea is to consider each feature for all data items at a node, sort the values of this feature in ascending order, propose a split at the midpoints between the sorted values, and evaluate the effectiveness of the split using a split criterion. This will give us the best split for a single feature. The split decision at the node is then just the best split over all the features.

Let T denote the set of n examples at a node, each of which belongs to one of k classes. Let T_L and T_R be the two nonoverlapping subsets that result from the split, that is, the left and right subsets. Let L_j and R_j be the number of instances of class j on the left and the right, respectively. Then, the most commonly used split criteria are [442]:

- **Gini**: This criterion is based on finding the split that most reduces the node impurity, where the impurity is defined as follows:

$$L_{Gini} = 1.0 - \sum_{i=1}^{k} (L_i/|T_L|)^2, \tag{11.8}$$

$$R_{Gini} = 1.0 - \sum_{i=1}^{k} (R_i/|T_R|)^2, \tag{11.9}$$

$$\text{Impurity} = (|T_L| * L_{Gini} + |T_R| * R_{Gini})/n, \tag{11.10}$$

  where $|T_L|$ and $|T_R|$ are the number of examples, and L_{Gini} and R_{Gini} are the Gini indices on the left and right side of the split, respectively. This criterion can have problems when there are a large number of classes.

- **Twoing rule**: In this criterion, a "goodness" measure defined as

$$\text{Twoing value} = (|T_L|/n) * (|T_R|/n) * \left(\sum_{i=1}^{k} |L_i/|T_L| - R_i/|T_R|| \right)^2 \tag{11.11}$$

  is maximized.

- **Information gain**: The information gain associated with a feature is the expected reduction in entropy caused by partitioning the examples according to the feature. Here the entropy characterizes the impurity of an arbitrary collection of examples. For example, the entropy prior to the split in our example would be

$$\text{Entropy}(T) = \sum_{i=1}^{k} -p_i \log_2 p_i, \tag{11.12}$$

$$p_i = (L_i + R_i)/n, \tag{11.13}$$

  where p_i is the proportion of T belonging to class i and $(L_i + R_i)$ is the number of examples in class i in T. The information gain of a feature F relative to T is then given by

$$\text{Gain}(T, F) = \text{Entropy}\,(T) - \sum_{v \in \text{values}\,(F)} |T_v| * \text{Entropy}\,(T_v)/|T|, \tag{11.14}$$

  where T_v is the subset of T for which the feature F has value v. Note that the second term above is the expected value of the entropy after T is partitioned using feature F. This is just the sum of the entropies of each subset T_v, weighted by the fraction of examples that belong to T_v. This criterion tends to favor features with many values over those with few values.

- **Information gain ratio**: To overcome the bias in the information gain measure, Quinlan [490] suggested the use of information gain ratio which penalizes features by incorporating a term, called the split information, that is sensitive to how broadly and uniformly the feature splits the data. This term is defined as

$$\text{Split Info}(T,F) = -\sum_{i=1}^{c}(|T_i|/n)\log_2(|T_i|/n), \tag{11.15}$$

where T_i are the subsets resulting from partitioning T on the c-valued feature F. Note that the split information is the entropy of T with respect to the values of the feature F. The Gain ratio is then defined as

$$\text{Gain Ratio}(T,F) = \text{Gain}(T,F)/\text{Split Info}(T,F). \tag{11.16}$$

- **Max minority:** This criterion is defined as

$$L_{minority} = \sum_{i=1, i \neq \max L_i}^{k} L_i, \tag{11.17}$$

$$R_{minority} = \sum_{i=1, i \neq \max R_i}^{k} R_i, \tag{11.18}$$

$$\text{Max minority} = \max(L_{minority}, R_{minority}). \tag{11.19}$$

  This has the theoretical advantage that a tree built by minimizing this measure will have depth at most $\log n$. This is not a significant advantage in practice and trees created by other measures are seldom deeper than the ones produced by Max minority.

- **Sum minority:** This criterion minimizes the sum of $L_{minority}$ and $R_{minority}$, which is just the number of misclassified instances.

In addition to the type of split at a node and a criterion for making that split, there is an additional issue which must be considered in building a tree. In an attempt to create leaf nodes which have items of only a single class, we may overfit the model to the data. This would imply that the tree would be an accurate representation of the training data but would very likely not generalize well, performing poorly on the test data. The solution to this problem is to "prune" the tree by replacing some subtrees by leaf nodes. This increases the error on the training set as the replacement leaf node will no longer be as pure as the replaced leaf nodes. However, the tree is likely to perform better on an independently chosen test set.

There are several approaches to decision tree pruning such as the cost complexity pruning technique of Breiman [61], the minimum description length approach suggested by Quinlan and Rivest [492], the pessimistic error pruning suggested by Quinlan [494], and an error-based pruning used in the C4.5 software [493]. An important issue in pruning is determining when to prune a subtree. This is typically done by considering the error rate of the nodes and leaves of the tree on a separate verification set [630].

11.2.4 Neural networks

Decision trees algorithms are well suited to problems where the decision surfaces separating the different classes can be modeled using axis-parallel or oblique hyperplanes. However, when these surfaces are more complex, it is worthwhile to consider artificial neural networks as they are often better at modeling complex data [430].

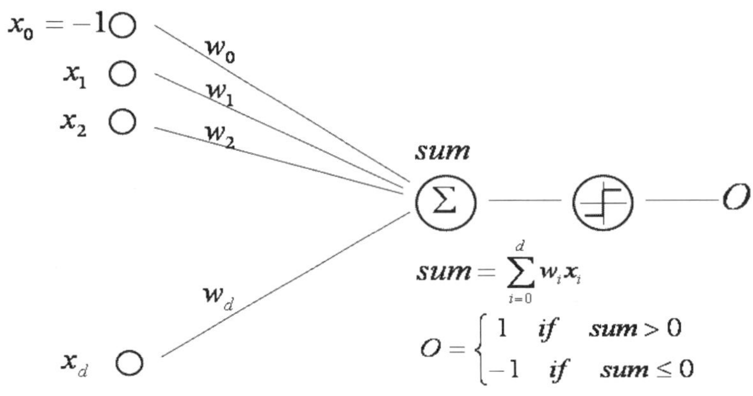

Figure 11.2. *A simple perceptron which takes as input a d-dimensional vector* $\mathbf{x}$*, obtains a linear combination using the weight vector* $\mathbf{w}$ *and outputs the value* 1 *if the result is greater than a threshold and* -1 *otherwise.*

The simplest neural network, called a perceptron, takes as input a real-valued vector of feature values, obtains a linear combination of them, and outputs a 1 if the combination is greater than a threshold and -1 otherwise (Figure 11.2). This corresponds to the linear discriminant function

$$g(\mathbf{x}) = \mathbf{w}^T\mathbf{x} - w_0, \tag{11.20}$$

where $\mathbf{x}$ is a feature vector, $\mathbf{w}$ is the vector of weights, and w_0 is the threshold. Thus $g(\mathbf{x}) = 0$ is the surface which separates items in class $C_1 : g(\mathbf{x}) > 0$ from items in class $C_2 : g(\mathbf{x}) < 0$, enabling the perceptron to act as a linear classifier for a two-class problem when the two classes are linearly separable. The perceptron must be trained, that is, the weight vector $\mathbf{w}$ is obtained, before it can be applied to classify an item without a class label. A simple approach to obtaining the weight vector is to start with random weights, apply the perceptron to classify a data item in the training set, and modify the weights whenever the item is misclassified. The process is repeated, iterating through the training examples until none are misclassified. Thus, if the perceptron misclassifies an item $\mathbf{x}$, whose true class is C_j, then the weight associated with the ith component, x_i, of $\mathbf{x}$ is updated by

$$-\eta x_i \quad \text{if} \quad C_j = C_2 \tag{11.21}$$

and

$$+\eta x_i \quad \text{if} \quad C_j = C_1. \tag{11.22}$$

Here, η is the learning rate which may be fixed for all iterations, or may vary, becoming smaller as convergence is reached.

There are several ways in which such a linear discriminant function can be extended to multiple classes, though, as Duda, Hart, and Stork [164], point out, care must be taken to ensure that we not end up with regions of feature space where the classification is undefined.

If the classes are not linearly separable, or the learning rate is not small enough, the algorithm above will not converge. When the two classes are not linearly separable, we can use a gradient descent approach to find the weights which best separate the two classes in the training data [430, 164]. This concept is perhaps best understood in the context of an unthresholded perceptron, that is, equation (11.20) with $w_0 = 0$, where we are trying to predict a real-valued output. Given the weight vector **w**, and the feature vectors corresponding to the training data, we can calculate the training error between the predicted values **o** and the desired output values **t** as

$$E(\mathbf{w}) = \frac{1}{2}\sum_{i=1}^{N}(t_i - o_i)^2, \tag{11.23}$$

where there are N items in the training set, t_i is the target output of item i and o_i is the corresponding output from the perceptron. The fraction, 1/2, is used to simplify the calculation of the gradient. The error is a function of **w** as the output **o** is a function of **w** as defined in equation (11.20). For different weight vectors, this error forms a surface (see, for example, the text by Mitchell [430] or the book by Duda, Hart, and Stork [164]). Our goal is to find the weight vector corresponding to the minimum of this error surface.

The standard approach to this problem is to start with a random set of weights, which corresponds to a point on the error surface. At each step in the iterative process, we change the weight vector in the direction which produces the steepest descent along the error surface. This direction is the negative of the gradient at that point on the error surface, where the jth component of the gradient can be obtained by differentiating (11.23) with respect to w_j, the jth component of the weight vector, as

$$\frac{\partial E}{\partial w_j} = \sum_{i=1}^{N}(t_i - o_i)(-x_{ij}). \tag{11.24}$$

Thus, given a weight vector, we can apply the perceptron to each of the training examples to determine the output o_i, and obtain the update to the jth component of the weight vector as

$$\delta w_j = \eta \sum_{i=1}^{N}(t_i - o_i)(x_{ij}), \tag{11.25}$$

where η is the learning rate and must be chosen to be small enough so that the gradient descent does not overstep the minimum in the error surface. The weight vector is then updated and the process is repeated.

A stochastic approximation to the gradient descent process [430] updates the weight vector after each data item, though it uses the gradient calculated at the end of each epoch, where an epoch is one iteration through all items in the training set. Instead of gradient descent, alternate optimization techniques such as Newton's method, conjugate gradient method, and the Levenberg–Marquardt algorithm are also possible (see, for example, Chapter 7 of [41] or any standard text on nonlinear optimization methods [454]).

A single perceptron is somewhat limited as a classifier as it can only model linear decision surfaces. To model more complex surfaces, we need several additions. First, we need multiple layers which can express a variety of complex surfaces. Second, we need units which can model a nonlinear function. Using only linear units will produce only linear functions. And finally, we need units such that the output of a unit is a differentiable function

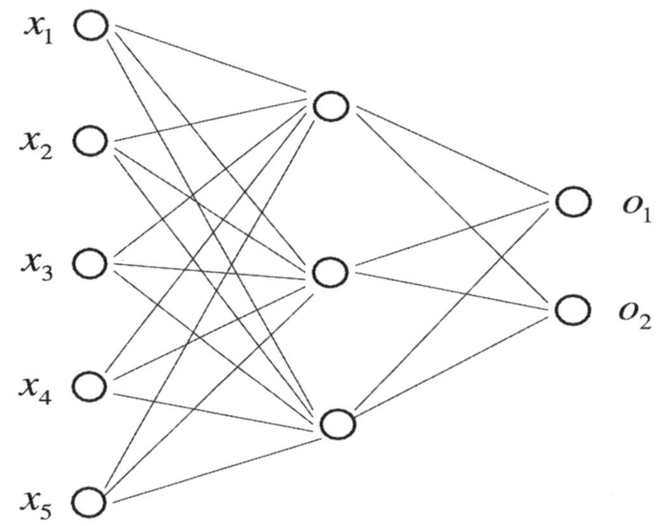

Figure 11.3. *A multilayer network with an input layer of five units, a hidden layer of three units, and an output layer of two units. The network is fully connected as each unit in a layer is connected to all units in the next layer.*

of its inputs, so we can use techniques such as steepest descent and other gradient-based optimization methods.

Figure 11.3 shows a multilayer feed-forward neural network with an input layer, a hidden layer, and an output layer, each with several units. This topology of the network, which includes the number of units and their connectivity, is typically determined first, often by trial and error, prior to the start of training. Each unit is a sigmoidal unit, where instead of a discontinuous threshold as in the case of a perceptron, we have a smoothly differentiable function (see Figure 11.4)

$$f(x) = \frac{1}{1+e^{-x}} . \tag{11.26}$$

This function is also called the logistic function or the squashing function as it shrinks a large input domain to a range between 0 and 1.

The most commonly used learning algorithm for a multilayer network using sigmoidal units is the backpropagation algorithm. As in the case of a single perceptron, the error function is defined as the sum of the errors over all outputs units. The updates of the weight vectors is now more complex as there are multiple units as well as a layer of hidden units. The derivation of the updates, as well as various practical issues related to the implementation and application of the backpropagation algorithm, can be found in several texts on neural networks, including Chapter 4 of the book by Mitchell [430], Chapter 6 of the text by Duda, Hart, and Stork [164], and Chapter 4 of the text by Bishop [41].

Once a neural network has been trained, it is applied by taking a data item characterized by its features, which are input to the network. The prediction of the network is the class associated with the output unit with the highest activation.

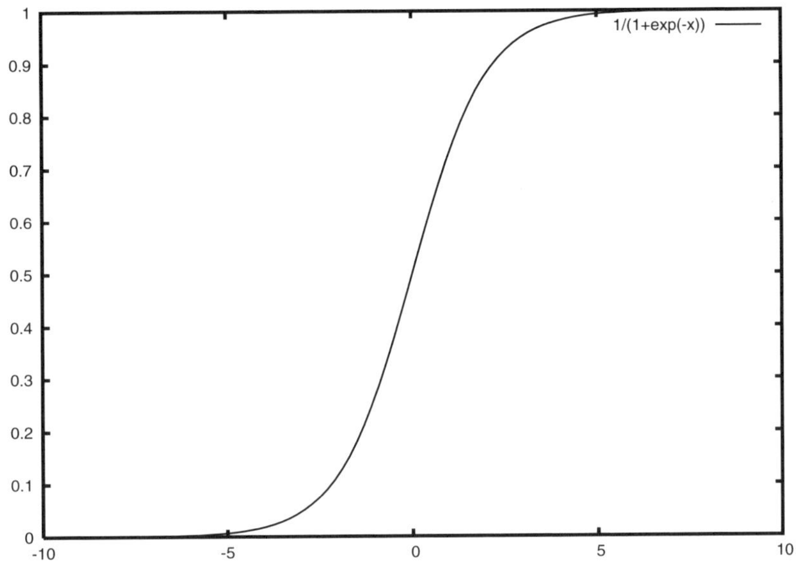

Figure 11.4. *The continuously differentiable sigmoid function used in neural networks (see equation* (11.26)*). It is also referred to as the squashing function as it shrinks values on a large domain to lie in the range* 0 *to* 1.

While the multilayer neural network can be a highly effective function approximation method, unlike the decision tree, it tends to function more as a black box, making it difficult to interpret how it arrives at a result. As with decision trees, overfitting is also an issue with neural networks. To avoid overfitting, the error on a separate test set is often used as a stopping point for the gradient descent.

Another key issue in neural networks is the determination of the topology of the network, that is, the number of layers, the number of nodes in each layer, and the connections between them. If the network has too few nodes and connections, it may not be able to learn complex patterns; if it has too many nodes and connections, it is likely to overfit the data. The topology of a network is usually determined by trail and error, though the use of sophisticated evolutionary algorithms (see Section 11.7.2) for determining the topology, as well as other aspects of neural network training, has indicated that it is often difficult to improve upon hand-designed networks with backpropagation [82].

A technique related to neural networks is projection pursuit regression from the statistics literature [204, 282]. This essentially obtains the output variable as a linear combination of a nonlinear transformation of linear combinations of the input variables. Unlike neural networks, where the transformations are fixed, in projection pursuit, they are data driven.

11.2.5 Support vector machines

A category of classification algorithms which has received much attention in recent years is the support vector machine (SVM). SVMs use linear models to implement nonlinear decision boundaries. This is done by transforming the input data using a nonlinear function into a higher-dimensional space, where the classes are linearly separable. As there are often several hyperplanes which can linearly separate the classes, SVMs choose the one which minimizes the generalization error. This is the maximum margin solution, an approach motivated through statistical learning theory. The margin is defined as the smallest distance between the decision boundary and the closest example of each class (assuming there are two classes). The data items which are closest to the maximum margin hyperplane are called the support vectors as they uniquely define the separating hyperplane. This hyperplane is obtained by solving a quadratic optimization problem subject to a set of linear inequality constraints.

There are several excellent references for SVMs, including the introductory tutorial by Burges [70], Chapter 7 in the text by Bishop [42], the text by Cristianini and Shawe-Taylor [122], and the references therein. The description above assumes that the data are linearly separable, either in the original space, or in the transformed space, though determining if the data are linearly separable is far from trivial. The technique has been generalized to the case when the data are not linearly separable [120] and a fast algorithm for training SVMs was proposed by Platt [479].

11.2.6 Ensembles of classifiers

An approach in classification which has gained much acceptance in the community is the concept of ensembles or committees of classifiers, which involves combining multiple models. The idea behind this is very simple. The training set is used to build several different models, each of which is used to assign a class label to a previously unseen instance. These class labels are then combined suitably to generate a single class label for the instance. This has been found to improve the accuracy of the resulting models, though, given that several models are used to make a decision, the process is computationally more expensive and it is now harder to understand how the decision was obtained.

Ensembles have been used extensively in the context of decision tree classification algorithms, though they can be used for other classifiers and in the case of regression (see Section 11.3) as well. There are two main issues in ensembles, how do we create the different models and how do we combine the results for classification of an unseen instance? The latter is typically done by taking a vote, or a weighted vote, of the output from each classifier and assigning the label of the class which occurs most frequently. The use of the weights allows better models to contribute more to the result than models which are not as accurate. In the case of regression, the corresponding solution is to take an average or a weighted average.

The task of building the different models is more involved. There are several ways in which this can be done [141]. They all exploit the fact that the different models are built from the same training data. However, the approaches vary in how they introduce randomization into the process of model building so that different models are generated. I next describe several commonly used techniques for introducing this randomization.

One approach used in creating ensembles is to change the instances which form the training set for each classifier in the ensemble. The most popular methods for this include:

- **Bagging:** In this approach, a new sample of the training set is obtained through bootstrapping with each instance weighted equally [56]. Given a training set with n instances, bootstrapping essentially samples the training set with replacement to obtain a new training set, also with n examples. This new training set will have some instances repeated and others which are not included at all. By creating several such training sets, we can introduce randomization into the ensemble. The results of the ensemble are then obtained by using a simple voting scheme.

This technique works very well for unstable algorithms such as decision trees and neural networks, where the classifier is sensitive to changes in the training set and significantly different classifiers are created for different training sets. In bagging, each classifier can be generated independent of the other, and randomization is introduced through the random sampling used to create each sample of the training set.

- **Boosting:** In this case, a new sample of the training set is obtained using a distribution based on previous results [202]. Unlike the bagging algorithm, which uniformly weights all the instances in the training set, boosting algorithms adjust the weights after each classifier is created to increase the weights of misclassified instances. This essentially implies that the training sets for the classifiers have to be created in sequence, instead of in parallel, as in the case of bagging. The different weights for the ensembles can either be incorporated directly into the classifier by working with weighted instances, or be applied indirectly by selecting the instances with a probability proportional to their weights. Further, in boosting, the results of the ensemble are obtained by weighting each classifier by the accuracy on the training set used to build it. As a result, better classifiers have a greater contribution to the end result than the poorer classifiers. There are several variants of boosting which differ in the way the instances are weighted, the conditions under which the algorithm stops, and the way in which the results from the ensemble are combined [202, 58, 26].

- **Pasting:** In this approach, the ensemble of trees is grown using a subsample of the entire training set [57]. This technique has been shown to be useful when the entire training set is too large to fit into main memory.

Instead of changing the instances, we can introduce randomization by focusing on the features. One approach is to create each new classifier by using a subset of the original features. For example, this technique has been used with decision trees in [275] and with neural networks in [110]. This approach tends to work only when the features are redundant, as poor classifiers could result if some important feature is left out. Alternatively, in the case of decision trees, one can introduce randomization by considering a subset of the features for making the split decision, as in the case of random forests, proposed by Breiman [59].

It is also possible to create an ensemble by changing the output targets. In [143], Dietterich and Bakiri describe a technique called error correcting output coding which can be applied to a problem with many output classes. The problem is first reduced to a two-class problem by randomly partitioning the classes into two, which are assigned labels 0 and 1. A classifier is created with this relabeled data. The process is repeated, creating a new classifier for each random partitioning of the original set of classes. To classify an unseen instance, each classifier assigns a label to the instance. If the label assigned is 0 (1), each class that was relabeled as a 0 (1) for that classifier gets a vote. The class with the maximum number of votes is assigned to the instance.

Unlike the previous techniques, where the input or output to the classifier is changed to generate the ensemble, it is possible to create the ensemble by changing the classifier itself. For example, in neural networks, the initial weights are set randomly, thus creating a new network each time. In decision trees, instead of selecting the best split at a node, one can randomly select among the best few splits to create the ensemble [142]. The random forests approach, mentioned earlier, creates different classifiers by randomly selecting the features used to determine the split at each node of the tree [59]. Alternately, one can randomize the classifier by using only a sample of the instances at a node of a decision tree in order to make the decision [313]. As the split made at a node is likely to vary with the sample selected, this technique results in different trees which can be combined in ensembles. It is also possible to use histograms to determine the split at each node, instead of sorting on the values for each feature. Then, one can introduce randomization by selecting a split value randomly in the best histogram interval [317].

Ensembles provide several benefits, including a significant improvement in accuracy [56, 202, 489, 26, 462], the potential for online classification of large databases that do not fit into memory [57], and the ease with which many of these techniques lend themselves to scalable parallelization [244]. If the techniques involve using a subsample of either the instances or the features, creating n ensembles can take less time than creating n classifiers [313].

11.2.7 Observations and further reading

There are several issues which must be considered in the practical application of a classification algorithm. The first is measuring the accuracy of a classifier. Typically, the labeled data are divided into a training set, which is used to create the classifier, and a test set, which is used to evaluate it. The test set should be selected prior to creating the classifier and should not be used in any way in the training. If the training requires two stages, one to determine a basic structure of the model and another to determine the best parameters, then the labeled data should be split into three—a training subset to determine the structure of a model, an evaluation subset to determine the parameters or to select one classifier among many, and the test subset to obtain the error of the final classifier. The actual accuracy which is required from a classifier is very problem dependent and should be set in consultation with the domain scientists. Often, the initial accuracy obtained may not be sufficient and several iterations may be required to refine the previous steps of extracting the features, postprocessing them to select key features, selecting more appropriate training data, and so on.

The selection of the training, testing, and validation subsets must be done so that each subset reflects the same distribution of the classes in the data. If the training subset is missing one of the classes, it is unreasonable to expect that the classifier built using this training set will be able to perform well on a test set which contains instances of that class. When the amount of training data are limited, which is often the case in scientific data mining as the data have to be labeled manually, a technique known as cross validation is used to divide the data and estimate the accuracy of the classification algorithms.

In cross-validation, we first decide the number of folds or partitions of the data. A typical choice is tenfold cross-validation, where the data available for training are divided into 10 approximately equal parts. Next, each part is held out in turn as the test set, and the classifier is trained on the remaining 9 folds of the data and evaluated on the subset which is held out. The 10 error estimates are then averaged to get the error estimate for the classifier. To minimize the effect of random variation in choosing the 10 folds, the process is often repeated several times. A typical number is 10, resulting in error estimates obtained from averaging ten, tenfold cross-validations.

To reduce the computational time, the leave-one-out cross validation is used, where each instance is left out in turn. This deterministic process is identical to n-fold cross validation if there are n instances. However, it does not ensure that the training and test set have the same distribution of classes. Alternatively, one can use the bootstrap approach, which is sampling with replacement. When a data set of n instances is sampled with replacement to result in another data set of n instances, it is well known that for a sufficiently large data set, the test set will contain about 36.8% of the instances and the training set will contain about 63.2% of them [630].

When we have a choice of different algorithms to use in a classification task, it is obvious to ask which method is better. In my experience, if the features selected are appropriate to the learning task and the training set accurately reflects the data, many of the classification algorithms will give similar results in term of cross-validation accuracy. In such cases, one may select the more interpretable model or the one more meaningful from the scientist's viewpoint. There are also statistical tests which can be used to determine when the mean of one set of samples, such as the accuracy resulting from each fold of cross validation, is significantly greater or less than the mean of another set of samples. Witten and Frank [630] suggest the use of the t-test, while the paper by Dietterich [144] addresses the broader problem of comparing classification algorithms in the context of the sizes of the training and test data, especially when the data set is small.

There are several approaches to increasing the often small size of the training data. It is always possible to manually label additional data items and add them to the training set, though this can be rather tedious. A more automated approach is active learning. The primary goal of active learning is to reduce the number of examples used in training, while maintaining the accuracy of the model built from the data. It can also be considered a smart way of increasing the size of the training set by judiciously adding instances which contribute more to the learning process. Active learning systems use the classifier to identify instances which are likely to be mislabeled and add them to the training data. This can be done by using a measure which indicates that the label assigned has a high uncertainty [374] or through adaptive resampling using methods such as boosting, which adaptively resample the training data biased towards the misclassified instances and then combine the predictions of several classifiers [295]. Another approach to addressing the scarcity of labeled data is semisupervised learning where the small labeled data set is extended by a larger unlabeled data set to build a better learner. Further details are available in the survey by Zhu [650] and the edited book by Chapelle, Schölkopf, and Zien [100].

Additionally, in science data, we often need to consider the nonuniform costs of misclassification. For example, the cost of misclassifying a tumor as benign is much higher than misclassifying it as cancerous. Or, predicting incorrectly that an experiment will be a success can be disastrous, if it fails causing irreparable damage to the experimental setup. Of course, predicting all tumors to be cancerous or all experiments to be failures is not an option.

This issue of misclassification cost arises when the cost of making one kind of error is much greater than the cost of other errors. Several solutions have been proposed to address this. One option is to use stratification [61] which changes the frequency of examples in the training set in proportion to their cost. However, this can adversely affect the class distribution. A more general approach to incorporating the cost information has been proposed by Domingos in his MetaCost approach [153], which wraps a metalearning stage around a classifier which minimizes cost while seeking to minimize the error rate. Alternately, one can build a classifier from the training set as is, and then make an optimal decision incorporating the cost using the probability estimates given by the classifier [170].

Another important issue which can influence the results from a classifier is the quality of the labeled data. There are several aspects to this issue, ranging from the accuracy of the labeling to the number of examples of each class in the training data. I next discuss how these problems can be addressed.

We have assumed so far that the labels assigned to the objects are consistent and that no object is mislabeled. However, in the case of classification problems in science and engineering data sets, the training data are usually labeled manually by the scientists. This is a tedious process and a scientist could assign different labels to an object at different times simply due to human fatigue during a labeling session. Further, in some problems, there is insufficient information to make an accurate decision regarding a label, especially in the borderline cases. Or, the level of understanding of the scientific phenomenon is such that the scientist is unsure of the label to assign to an object or different scientists disagree on the label. In such cases, one option is to ask several scientists to label the same data and then assign the majority label to an object. Alternatively, one can create a new class label for objects which do not belong to any of the existing classes, but are interesting, and therefore, must be flagged to separate them from uninteresting objects. Any solution to this problem must be considered in collaboration with the domain scientists to ensure that the classifier produces scientifically meaningful results.

In some problems, the data may be accurately labeled, but there may be far more objects of one class than another, resulting in an unbalanced data set. This situation arises in scientific data when the scientists are interested in one class of objects and therefore focus on objects of that class to the exclusion of others. For example, at the beginning of our work in the classification of bent-double galaxies [316], when we asked the astronomers for examples for a training set, they provided a set of bent-double galaxies they had visually identified when their survey was small. These consisted of two or more blobs. However, when we requested examples of non–bent-double galaxies to create a two-class training set, they provided a set of examples all of which were composed of a single blob. Thus, in principle, we had a set of bent and non–bent-double galaxies in our training set, but the negative examples were not very useful as they would lead to the erroneous conclusion that all single-blob galaxies were non-bent, while all multiblob ones were bent. To address this, we had to work with the astronomers to identify non–bent-doubles which were composed of two or more blobs.

Unfortunately, it is not always easy to add to the training set if the initial set is unbalanced. Consider the problem where we use a computer simulation to mimic an experiment and understand under what input conditions the experiment will be a success. The computationally expensive simulation is run many times with different inputs and the output is analyzed to determine if the simulated experiment was a success or not. Understandably, the scientists would be more interested in running the simulation to understand the range of settings over which the experiment is a success. This would result in far more examples of successful experiments in the training data and far fewer examples of experiments which failed. This problem of unbalanced data also arises when the item of interest occurs rarely, for example in the detection of oil spills in the sea from satellite images [352], as discussed in Chapter 2.

The subject of classification when the training data are unbalanced has been studied extensively in recent years; see, for example, [302, 303, 106, 105, 161], as well as the papers in the special issue of SIGKDD Explorations [104], and the references therein. The solutions range from evaluation metrics which take into account the rarity of one class, segmenting the problem into two or more subproblems such that the rare class is no longer rare in one of the subproblems, and sampling of various forms. This includes a simple subsampling of the majority class or an oversampling of the minority class. More sophisticated sampling techniques have also been considered including removal of the majority class items only when they are redundant, or adding minority class by creating synthetic examples along the lines joining the k nearest minority class items, or by combining various techniques for reducing the items of the majority class and increasing those of the minority class. Any solution approach to this problem must be taken with care and also include input from the domain scientists to ensure the scientific validity of both the approach and the results.

When the training data are unbalanced, we may need to use a different evaluation technique rather than one based on accuracy. For example, if a training set has two classes appearing in the 99:1 ratio, we can get 99% accuracy simply by predicting all examples as the majority class. In such case, alternative metrics have been proposed such as the use of the geometric mean [353] or techniques derived from the Receiver Operating Characteristic curves [487, 486, 411].

Another issue which is becoming important in the context of classification of streaming data is concept drift, where the probability distribution of the examples is nonstationary, that is, changing over time. A category of algorithms, known as incremental or online algorithms, allows the model to be built one, or a few, examples at a time, rather than all the examples. In machine learning, concept drift is handled using one of two approaches, both of which exploit the fact that in a nonstationary environment, only the more recent examples are relevant to learning the target concept. The first approach uses time windows of fixed or adaptive size and selects the examples in these time windows for the training [360]. In the second approach, either the data or earlier models are weighted based on their utility or age [626, 343, 526]. In this context, incremental versions of algorithms, such as incremental decision trees [210], are also relevant.

The subject of classification has been extensively studied and described in various books, including general texts on data mining, machine learning, and pattern recognition, as well as texts specific to a technique. In the former category, the texts by Mitchell [430], Duda, Hart, and Stork [164], Hastie, Tibshirani, and Friedman [261], Hand, Mannila, and Smyth [252], Witten and Frank [630], and Bishop [42] provide insights from various perspectives. Specific books on decision trees would include the classics by Breiman et al. [61] which presents a statistical perspective and the one by Quinlan [493] which addresses the problem from a machine learning viewpoint. Neural networks have been studied by various authors, including Bishop [41], Haykin [262], and Rojas [514], while SVMs have been described by Burges [70] and Cristianini and Shawe-Taylor [126].

11.3 Regression

Regression can be considered as a variant of classification where the desired output consists of one or more continuous variables, sometimes called target variables. Consequently, many of the techniques used in regression are variants of the corresponding techniques used in classification.

Regression problems occur in scientific data in many contexts. For example, they can be used to build code surrogates, which, given the input parameters for a computer simulation, predict the continuous output variables. These models can be used to explore the input parameter space for the problem being studied, instead of running an expensive simulation code on a multiprocessor system for each set of input parameters. This can be quite effective when the training set samples the input space adequately. Regression problems also arise in experimental and observational data when the variable being predicted is continuous.

11.3.1 Statistical techniques

The simplest approach to regression is to build a linear regression model. Here, the model is a linear function of the parameters of the model. Thus, a target variable y may be written as a function of the model parameters a_i and the d input variables x_i as

$$y(\mathbf{a},\mathbf{x}) = a_0 + a_1 x_1 + \cdots + a_d x_d, \tag{11.27}$$

where $\mathbf{a}$ is the vector of the model parameters and $\mathbf{x}$ is the vector of the input parameters. Since the model is a linear function of the input variables, it is limited in the types of functions it can model. This linear regression model can be extended by considering linear combinations of m nonlinear functions of the input variables as follows:

$$y(\mathbf{a},\mathbf{x}) = a_0 + \sum_{i=1}^{m} a_i \phi_i(\mathbf{x}). \tag{11.28}$$

Here, $\phi(\mathbf{x})$ are nonlinear functions, such as polynomials, sigmoid functions, or sinusoidal functions. These are often referred to as the basis functions. As the model is still linear in the model parameters $\mathbf{a}$, it is called a linear model, though it can model more complex functions than equation (11.27).

Given the known target variable values y and the corresponding input parameters $\mathbf{x}$ for n instances, we can obtain the model parameters by optimizing a sum-of-squares error function

$$\sum_{j=1}^{n}\left(y_j - \sum_{i=1}^{m} \alpha_i \phi_i(\mathbf{x}_j)\right)^2, \tag{11.29}$$

where the subscript j indicates the instance number. These model parameters can then be used to predict the target variable y for any input vector $\mathbf{x}$.

Linear regression has been extensively studied in statistics. Some excellent material for additional reading can be found in Chapter 11 of [252], Chapters 3 and 5 of [261], and Chapter 3 of [42]. Another approach which is suitable for problems of moderate dimension is the Multivariate Adaptive Regression Splines, or MARS, [203]. It is a data-driven technique, producing models which are interpretable. Its computational requirements unfortunately grow rapidly with the increase in dimension.

11.3.2 Machine learning techniques

Several of the techniques described in Section 11.2 on classification easily extend to regression as well. For example, the k-nearest neighbor classifier can be adapted to regression by considering the output to be the mean or the weighted mean of the output values of the k nearest instances.

One can also combine the linear regression technique described in Section 11.3.1 with the idea of using nearest neighbors to create a local approximation to the target function. Thus, instead of using all the instances to determine the model parameters $\mathbf{a}$ in equation (11.28), and creating a model which can predict the target for any input $\mathbf{x}$, we create the model only after we know the input for which we are to predict the target. Then, we assume that the target function can be modeled locally in the neighborhood of the input by a linear regression model using either linear or nonlinear functions. There are several ways in which we can obtain this local approximation [430]. We can either minimize the sum-of-squares error over the k nearest neighbors; or we can use all training examples, but weight them by their distance to the query; or we can combine the two approaches by considering only the nearest neighbors and weighting them based on their distance from the query. This idea of locally weighted regression is discussed further in [8], while techniques for solving the minimization problem can be found in the text by Bishop [41].

Corresponding to classification trees which predict a discrete quantity, regression trees predict a continuous variable [61]. As in a decision tree, the split at each node of a regression tree is made to optimize a certain quantity. Usually, this quantity is the variance of the target variables at a node. At the leaf node, instead of a class label corresponding to the majority class of the instances at the node, we use the mean value of the target variable of the instances at the node. However, this use of a constant value will cause a discontinuity at the decision boundary corresponding to each leaf.

To address this problem, we can predict the value of the target variable using a more sophisticated model at the leaf node. Such trees are referred to as "model trees." The simplest option is to use a multivariate linear model which essentially approximates the continuous function by linear "patches." However, instead of using a model based on all the input variables, Quinlan [491] suggests using only the variables which occur in the tests leading up to the leaf node. Other more complex models have been proposed by Torgo [586] and Malerba et al. [407].

It should be pointed out that using the variance of the target variable to make the decision at a node of the regression tree, while computationally efficient, is appropriate only when the model used at the leaf node is the mean value of the instances at the node. To be consistent, if we use a more complex model at the leaf of a regression tree, we must use the squared error from the same model to make the decision at each node. For example, if we use a linear model at the leaf node and the variance at each decision node, we could have the possibility that the data are well predicted by a linear model, yet have a large variance [324]. The latter condition would suggest splitting the data, while the former would suggest otherwise. Using a complex model at each decision node can be expensive. Dobra and Gehrke [152] show how it is possible to efficiently construct regression trees with linear models in the leaves which are both accurate and easy to interpret.

Another commonly used classification algorithm, namely neural networks, is by its very construction very appropriate for regression. A network with three layers of units can approximate any function to arbitrary accuracy [430], making neural networks very useful in regression when the form of the function is unknown in advance [362, 361]. SVMs are another category of classification algorithms which can be used for regression [554]. In addition, the idea of ensembles, introduced in the context of classification, can also be used for improving the accuracy of regression.

11.4 Association rules

Association rule techniques, which originated in the analysis of commercial data for market-basket transactions, are now being used in the context of science data as well. The goal in market basket analysis is to analyze the transactions at, say, a grocery store, to determine what types of items are being bought by customers. By understanding the purchasing behavior of its customers, a store could target marketing promotions and manage its inventory better.

The idea in association rules is to identify rules of the form

$$\{X\} \rightarrow \{Y\}, \tag{11.30}$$

where X is one set of items and Y is a different set of items, disjoint from X. The rule states that there is a strong relationship between X and Y; many customers who buy X also buy Y.

There are several key concepts in association rules which are best described in the context of items bought in a grocery store. Let $I = \{i_1, \ldots, i_d\}$ be the set of items in a store, and let $T = \{T_1, \ldots, T_N\}$ be the set of transactions at the store. For simplification, we assume that an item refers to all similar items, that is, the item "pasta" is a single item which includes different varieties, different brand names, as well as different package sizes.

The term "itemset" is a collection of items, that is, a subset chosen from I. A k-itemset is an itemset with k items. Associated with each itemset is its "support," S, which is the number of transactions which contain the itemset. Thus

$$S(X) = |\{t_i | X \subseteq t_i, t_i \in T\}|. \tag{11.31}$$

Each association rule also has its support which is the number of times a rule appears in the transactions; that is,

$$S(X \rightarrow Y) = S(X \cup Y). \tag{11.32}$$

The support of a rule is used to remove rules which occur rarely and therefore are uninteresting, indicating a chance occurrence. The "confidence," C, of a rule is the fraction representing the number of times Y appears in a transaction which contains X; that is,

$$C(X \rightarrow Y) = S(X \cup Y)/S(X). \tag{11.33}$$

The confidence of a rule is the conditional probability of Y given X and it measures the reliability of the inference made by a rule.

The support and confidence of a rule are key concepts in association analysis where the goal is to find rules which have support and confidence greater than user-provided thresholds. This is done by first generating all the frequent itemsets which meet the threshold constraint on the support, and then extracting all the high-confidence rules from these frequent itemsets. While computationally expensive, brute-force method is an option for these two steps, many algorithms have been proposed to speed up the process, as discussed in the book by Tan, Steinbach, and Kumar [574].

Association rules are just beginning to be accepted in the analysis of scientific data, in fields such as bioinformatics, medicine, and climate. It is important to note that an association rule of the form $\{X\} \rightarrow \{Y\}$ does not imply causality, that is, it does not say that X causes Y, just that when X occurs, so does Y. Of course, interpretation of any rules obtained from the application of these techniques must be done with care in collaboration with the domain scientists.

11.5 Tracking

Tracking is a subject not usually associated with the task of recognizing patterns in data. However, I include it here for completeness as one of the tasks of great interest to many scientists is the tracking of "objects" in data over time. The intent in these problems is to understand how these "objects," which may refer to a region of interest in simulation data or real physical objects such as cells in a culture of micro-organisms, evolve over time. Unlike physical objects, objects in simulation data may be poorly defined, with one of the main challenges being the identification of the objects prior to their tracking.

In many problems, the objects may split or merge; they may die out or be born; or change shape and size over time. For example, an object may be a low-pressure region surrounded by a flow of a particular kind, as in a hurricane, and the goal may be to track the path of the (simulated) hurricane over time. Such a region of interest may weaken, only to strengthen suddenly under the right conditions. In such problems, scientists may be interested in understanding what causes these regions to weaken or strengthen, so that in the case of a real hurricane, they can predict the path it will take.

Tracking is key to understanding the dynamics of the objects in the data. It is usually done after the objects have been identified in the data and features extracted to represent them. There are several techniques which can be used to track objects, ranging from the very simple ones which exploit the overlap between the objects from one time step to the next, to more sophisticated techniques such as Kalman or particle filters.

The simplest case arises when there are relatively few objects being tracked, the objects are far enough from each other, and move a relatively small distance between frames. In this situation, we can easily identify where an object has moved to from one frame to the next by considering the locations of the center of mass of the objects and/or the degree of overlap between objects in two consecutive frames.

The tracking problem becomes more difficult when there is more than one object in the frame at time $(t+1)$ which could correspond to a certain object at time t. In such cases, the best option is to use the features which represent the objects to correctly track the objects from one frame to the next. These features are very problem dependent. For example, if all objects in a frame are roughly the same size or the sizes of objects change significantly between frames, then the size of an object is unlikely to be a good feature. The features may also be erroneous, especially when they are the result of incorrect mapping of objects between frames. For example, this could be the case when the feature is the velocity of an object calculated using the position of its center of mass from one frame to the next.

In some cases, the task of tracking can be stated as matching objects at time $(t+1)$ to tracks which have existed till time t. This can be considered as a bipartite graph-matching problem, with the tracks forming one set of nodes, the objects forming the other set, and the weights linking the two sets being determined by how well a given object matches a track using the features of the given object and the object associated with the track. It is very easy to incorporate domain-specific constraints, for example, by removing graph links between an object which is too far from certain tracks.

A further complication arises when objects merge or split, or in some cases, appear to touch and then separate [628, 215]. Here again, domain information must be exploited to determine when two objects can be considered to have merged or split, or how long two objects should appear to be merged before they can be considered as merged. Merging and splitting also require the use of complex data structures to keep track of how an object has evolved over time. This can be complicated if an object splits into two or more while it is merging with two or more other objects.

It is often insightful to create movies which show the dynamics of various objects over time. Further, the statistics on the objects, such as their average lifetime or the variation of their features over time, can also help in understanding the gross behavior of the physical phenomenon. Often, once the objects are tracked, their behavior may need to be correlated to other phenomena for further insights.

Simple tracking techniques often work quite well on many scientific problems, especially if thousands of objects have to be tracked [215]. However, I would be remiss in not mentioning the more commonly used tracking techniques in computer vision and surveillance. These are often probabilistic in nature such as the Kalman filter [325, 631, 63, 266, 239, 21] and particle filters [403]. Many of these techniques also exploit problem-specific features, such as the availability of radar signatures, or focus on problem-specific constraints, such as tracking in the presence of clutter, tracking of rigid objects, or tracking using data from multiple cameras. The process of tuning the parameters for tracking algorithms is still somewhat of an art rather than a science and the interested reader should explore tracking techniques used in their specific domain for methods which are most suitable for their problem. For example, tracking in the context of military applications is discussed in Section 6.3.1 on multiple target tracking.

11.6 Outlier or anomaly detection

An area of increasing interest in the context of scientific applications is the detection of outliers or anomalies. When data sets were relatively small, the typical reason for identifying outliers was to remove them, lest they skew the statistics being extracted from the data. However, as data sets have become much larger, and data from new applications are being mined, the outliers themselves have become interesting as they often indicate something unusual in the data, that is, a deviation from normal behavior. Typical examples where outliers are of interest are credit card fraud detection and network intrusion detection, where a large majority of transactions or network connections are normal, but the few which are anomalous are indicative of possible fraud or intrusion of a computer network.

Scientific applications too have problems where detection of anomalies is important; these often arise in the context of streaming data, with the anomaly indicating something unusual which requires the attention of a scientist. For example, sensors monitoring an experiment in progress or a critical piece of equipment, such as an electric generator, may show an anomaly, indicating that the experiment or equipment is about to fail. Or, data being monitored from a telescope could indicate that an unusual event is about to occur, enabling scientists to focus additional resources to observe the event.

The basic idea in anomaly detection is to build models of what is considered normal in the data, and then identify deviations from it as anomalies. The models can be built using either the supervised or the unsupervised techniques described in this chapter. It is important to note that what is an outlier is very dependent on the problem and the data being monitored. For example, if we are collecting the data from different sensors in an experiment, an anomaly may occur in one of the sensor measurements, or each sensor measurement may be normal for that sensor, but when all the sensor measurements are considered together, the measurements are anomalous.

An extensive survey by Chandola et al. [95] on outlier detection describes various formulations used for the problem as well as different solution approaches, while the book chapter by Ertoz et al. [172] focuses on the subject in the context of network intrusion detection.

11.7 Related topics

There are several topics which are related to the tasks in pattern recognition, but do not fit neatly into a topic for the actual detection of patterns. These general topics are the focus of this section.

11.7.1 Distance metrics

A common issue which arises in many algorithms is the similarity or dissimilarity between two objects (which we also refer to as data items or instances), which are represented as feature vectors. This similarity, or dissimilarity, is obtained by using a proximity measure, or a distance metric. As discussed in Chapter 11 of the text by Theodoridis and Koutroumbas [581], a proximity measure must satisfy certain mathematical conditions such as the triangle inequality and transitivity to qualify as a metric.

There are several distance measures commonly used in pattern recognition. The most popular one is the Minkowski metric

$$\text{distance}(\mathbf{x},\mathbf{y}) = \left(\sum_{k=1}^{d} |x_k - y_k|^q \right)^{1/q}, \tag{11.34}$$

where the parameter $q \geq 1$ and the vectors $\mathbf{x}$ and $\mathbf{y}$, of dimension d, are the feature vectors for the two objects. Setting $q = 2$ gives the commonly used Euclidean distance, while $q = 1$ gives the city-block or Manhattan distance.

Sometimes, instead of a formal distance metric, we can use a similarity function which is defined on two vectors and is large when the two vectors are "similar" in some way. For example, in many problems, the cosine of the angle between two vectors $\mathbf{x}$ and $\mathbf{y}$

$$\frac{\mathbf{x}^T\mathbf{y}}{\|\mathbf{x}\| \ \|\mathbf{y}\|} \tag{11.35}$$

is used to measure their similarity, especially as it is invariant to rotation and scaling. The Tanimoto measure

$$\frac{\mathbf{x}^T\mathbf{y}}{\|\mathbf{x}\|^2 + \|\mathbf{y}\|^2 - \mathbf{x}^T\mathbf{y}} \tag{11.36}$$

is another metric used for real- and discrete-valued feature vectors. A common approach to handling discrete features is to take the difference between two values to be 1 if they are not the same, and 0 otherwise.

In some problems, we may need to account for the fact that some features, such as histogram values or texture features, appear in groups. For example, if we use a 16-bin histogram for the intensity values of an image, there will be 16 entries in the feature vector, whereas using an 8-bin histogram would result in only 8 entries. To avoid having such features dominate the distance metric and overwhelm the contribution of single features such as area or aspect ratio, we may need to normalize the contributions of such features by scaling by the number of bins of the histogram, or the number of texture values used.

11.7.2 Optimization techniques

Mathematical optimization is a topic very relevant to pattern recognition as many pattern recognition algorithms can be recast in the context of maximizing or minimizing certain functions. For example, the k-means clustering algorithm is essentially minimizing the sum of squared error between the data items in a cluster and the cluster mean, taken over all clusters. In other cases, such as regression, we need to fit models to the data, with the parameters of the models determined by minimizing the error between the model prediction and the actual value. Optimization techniques also form a key component of neural networks and SVMs which are used in both classification and regression.

In some problems, the traditional mathematical optimization techniques such as gradient descent, the conjugate gradient method or the Levenberg–Marquardt method, can be used effectively [41, 454]. However, in other problems, when the model whose parameters we are trying to learn becomes more complex, or the problem is high dimensional, we may need to consider stochastic methods for solving the optimization problem.

There are two classes of stochastic methods commonly used in pattern recognition [164]. The first is Boltzmann learning, specifically simulated annealing, which is derived from the field of statistical mechanics. The second is evolutionary or genetic algorithms, derived from biology, specifically, the theory of evolution. Both these methods involve stochastic searching, tend to avoid getting stuck in a local minimum, and are often computationally expensive.

Annealing is the process where a material is heated and then slowly brought to a lower temperature. The high temperature excites the atoms or molecules. As the material is cooled, these atoms or molecules assume their optimal position in the resulting crystalline lattice, resulting in fewer irregularities in the crystal. In simulated annealing, the free variables are updated as long as the value of the function to be minimized is reduced. However, to avoid getting stuck in a local minimum, if the new values of the free variables increase the value of the function, the new values are accepted based on a certain probability which is dependent on the temperature [514]. The probability is such that at high temperature, there is a higher probability of an increased function value being accepted. As the temperature reduces, this probability also goes down to zero, corresponding to the heating and cooling phases.

In contrast to simulated annealing, where local minima are avoided by explicitly allowing parameter values which increase the value of the function to be minimized, in evolutionary algorithms, local minima are avoided by including randomization which allows a better search of the parameter space [225, 136]. First, a population is created, each "individual" of which is a set of values that the parameters of the model can take. Using this population, the next generation is created through crossover, where parts of two individuals are split and spliced to each other as in genetic mating of two chromosomes; mutation, where a part of an individual is changed randomly by flipping a bit in its representation; and replication, where an individual is copied over unchanged. Only the individuals in this new generation which meet a fitness criterion, such as how well they do in minimizing the function, are kept. The expectation is that the individuals in each generation will do better at minimizing the function than the individuals in previous generations. There are several versions of genetic algorithms which vary in how the individuals are represented, the types of operations used to go from one generation to the next, the selection of individuals for "mating," and so on.

11.8 Summary

In this chapter, I described the different techniques which can be used to identify the patterns in the data. These include clustering, classification, regression, association rules, tracking, and anomaly detection. I briefly described some of the more frequently used methods in each class of techniques as well as various issues one might need to consider in their application. The class of technique used is very dependent on the category of problem being solved. Descriptive problems, where the goal is to group similar items so we can determine what makes them similar, benefit from clustering techniques, while predictive problems, with discrete or continuous output, are addressed using classification or regression techniques, respectively.

Any pattern recognition algorithm must be applied carefully to the data. The patterns that one obtains are only as good as the data which were input to the algorithm. This is one of the main reasons why most of the time in scientific data analysis is spent in the steps preceding the pattern recognition step. Also, it can be easy to fool oneself into thinking that patterns have been detected in the data, or that one has obtained reasonable accuracy on an algorithm, when the results can be traced back to some artifact of the data. This makes it important that the patterns detected be validated, first by the data miners to ensure that they are what one would expect from an analysis viewpoint, and then by the domain scientists to ensure that they are scientifically meaningful. We next consider this topic of validation, and the related topic of visualization, in Chapter 11.

11.9 Suggestions for further reading

As mentioned earlier, there are many texts which describe pattern recognition techniques in general. These include the texts by Tan, Steinbach, and Kumar [574]; Bishop [42]; Witten and Frank [630]; Theodoridis and Koutroumbas [581]; Hand, Mannila, and Smyth [252]; Duda, Hart, and Stork [164]; Hastie, Tibshirani, and Friedman [261]; and Mitchell [430]. These books provide insights from several perspectives, including machine learning, data mining, pattern recognition, and statistical learning. It is often beneficial to look at a technique which may have been developed by one community from the perspective of a different community as it provides additional insights.

There are also several texts which focus on specific techniques. These have been mentioned in their respective sections in the chapter. Oftentimes, surveys on a particular technique, either by themselves, or in the context of an application area, also provide additional insights.

Chapter 12

Visualizing the Data and Validating the Results

The greatest value of a picture is when it forces *us to notice what we never expected to see.*

—John W. Tukey [594, p. vi]

... the highest standard of statistical integrity and statistical thinking should apply to every data representation, including visual displays.

—Edward R. Tufte [592, p. 35]

An important aspect of any data mining endeavor is the visualization of the data and the validation of the results by both the data miners and the domain scientists. These tasks can occur at any step of the data mining process. We may want to visualize the raw data not only for exploratory data analysis but also to ensure its correctness. We also need to validate the results of each step. This must be done first by the data miners to ensure that the correct algorithm has been selected and applied with the right parameters and to confirm that the results are consistent with what one would expect from the application of an algorithm. Once the results from each step have been validated by the data miners, they must also be validated by the domain scientists to ensure that they are scientifically meaningful. As the data can be in the form of images, mesh data, or just a simple data item by feature matrix, we need several tools and techniques not only for visualizing the data during the initial exploratory analysis, but also for validating the results of each step of the data mining process.

But first, a caveat: the term "visualization" in this book is used in the context in which Tukey [594] and Tufte [591, 592, 593] use it, namely, quantitative visualization which includes the fundamentals of scale, orientation, and labels. Such visualization can be invaluable in scientific data mining as it can aid our abilities to draw statistically sound and scientifically meaningful conclusions about the data. It is, however, different from the work done in the scientific visualization and graphics communities where the visualization tends to be more qualitative and often results in the dequantification of the data.

This chapter is organized as follows. First, in Section 12.1, I describe ways of visualizing table data, where the data are usually represented as a table or a data item by feature matrix. Next, in Section 12.2, I focus on image data, especially images with floating-point data, followed by a short section, (Section 12.3), on how these visual tools might be enhanced for validation. Finally, I summarize the chapter in Section 12.4 and provide suggestions for further reading in Section 12.5. As appropriate, I also include pointers to free software available from the public domain.

12.1 Visualizing table data

Some scientific data sets are in the form of a data item by feature matrix. This could be the raw data for a problem where the goal is to take different runs of a simulation (the data items), with each run described by the input parameters (the features) and the corresponding outputs, and build a model which, given a set of inputs, will predict the output. Or, it could be data obtained after processing images or mesh data, identifying objects in it, and extracting features for these objects. The resulting data item by feature matrix is typically used in pattern recognition or to extract statistics on the features of interest.

Visualizing such table data can be a very useful way to gain insight into the data. When the data are large enough that it is not possible to understand or explore the data by just looking at them in the form of a table, it often helps to obtain simple statistics on the data such as the mean and standard deviation of each feature. In addition, several simple visual tools can be used, such as box plots, scatter plots, and parallel plots. These tools are available as part of several public-domain software packages mentioned later in this section. In addition, simple plotting packages, such as Gnuplot [224], can also be used to plot the necessary information.

12.1.1 Box plots

The box plot was originally proposed by Spear [560] as a "range bar" and later by Tukey as the box-and-whisker plot [594]. There are several variants of the box plot, the simplest of which provides a five-number summary of an array of numbers. This summary includes the median, the upper and lower quartiles, and the minimum and maximum values. The box plot can be drawn vertically, as shown in Figure 12.1 or horizontally. The two ends of the box represent the lower and upper quartiles (B and D in the figure), respectively, with the sample median (C) represented by a line in the box. From the box edges representing the upper and lower quartiles, lines are drawn to the maximum and minimum values (E and A) in the sample. These form the "whiskers" in the box-and-whisker plot. The width of the box does not signify anything.

The length of the box gives an indication of the sample variability. The position of the box in its whiskers as well as the location of the median in the box are indicative of the skewness of the data. A centered box and a centered median are indicative of a symmetric distribution. Whiskers which are long relative to the size of the box indicate a long tailed distribution. For a normal distribution, that is, the bell-shaped curve, the whiskers are about the same length as the box.

The box plot is frequently used for comparing multiple variables by placing the plots for each side by side with a uniform scale along the y-axis to enable the comparison. This can be particularly useful in the context of feature selection for classification. For example, we can separate a training set into subsets, one for each class, and then for each feature, create the box plots of the distributions of the feature for the different classes. If there is a substantial overlap between the values of a feature for different class labels, the feature is

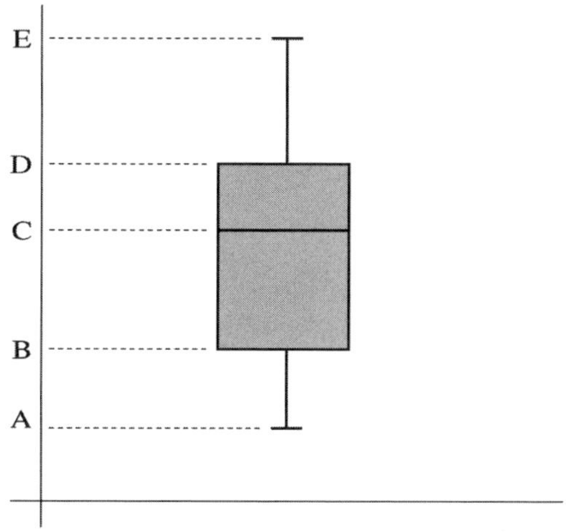

Figure 12.1. *A box plot summarizing the statistics for an array for numbers. A and E are the minimum and maximum values, respectively. B and D are the lower and upper quartiles, respectively, and C is the sample median.*

Sepal Length	Sepal Width	Petal Length	Petal Width	Class
5.1	3.5	1.4	0.2	1
4.9	3.0	1.4	0.2	1
4.7	3.2	1.3	0.2	1
7.0	3.2	4.7	1.4	2
6.4	3.2	4.5	1.5	2
6.9	3.1	4.9	1.5	2
6.3	3.3	6.0	2.5	3
5.8	2.7	5.1	1.9	3
7.1	3.0	5.9	2.1	3

Figure 12.2. *A table showing a few sample items from the Iris data set from the UCI repository* [272]. *Class* 1 *is Iris setosa, class* 2 *is Iris versicolor, and class* 3 *is Iris virginica. All values are in centimeters.*

unlikely to be a good feature for use in classification [200]. Similar information can also be obtained using parallel plots, as described later in this chapter.

12.1.2 Scatter plots

Another simple visual tool is the scatter plot. Consider the Iris data set [193] from the UCI repository [272], a few items of which are shown in Figure 12.2. This is a data set with 150 items, each of which is a set of observations of the sepal length, sepal width, petal length, and petal width of the flowers from three species of irises— setosa, versicolor, and virginica. There are 50 items for each species. All values are in centimeters.

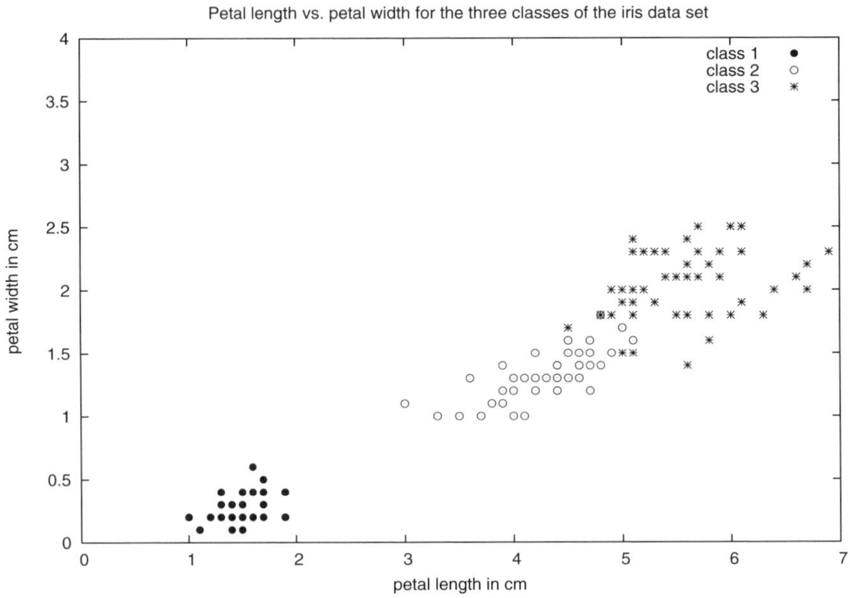

Figure 12.3. *A scatter plot showing the values of the variables petal length and petal width for the items of the three different classes in the Iris data set.*

A scatter plot essentially plots one variable against another, where each item is represented as a dot at the values of the variables for that item. These variables could be either two features or one feature and the class variable. To increase the content of the plot, the two axes could correspond to two features and instead of a dot for all items, we could use different symbols or colors for items from different classes. For example, Figure 12.3 shows that either the petal length or the petal width alone is sufficient to separate the items of class 1 from the other two classes. However, neither of these two features can distinguish between classes 2 and 3 as there is no clear demarcation between the items of the two in the scatter plot.

Figure 12.4 shows another scatter plot for the Iris data set where the derived variables petal area and sepal area are shown for each item in the three classes. As before, the items of class 1 are clearly separated from the other two classes. However, classes 2 and 3 are now better separated than before, with only three misclassifications, indicating the potential benefits of considering derived quantities as well as the effective use of scatter plots in exploratory data analysis.

12.1.3 Parallel plots

For high-dimensional data, where there are several features describing each data item, scatter plots are somewhat limiting as they reduce us to looking at three or four (if we consider three-dimensional plots) variables at a time. For such problems, parallel plots or parallel coordinates can be very helpful. These were first described by Inselberg [293, 292] and

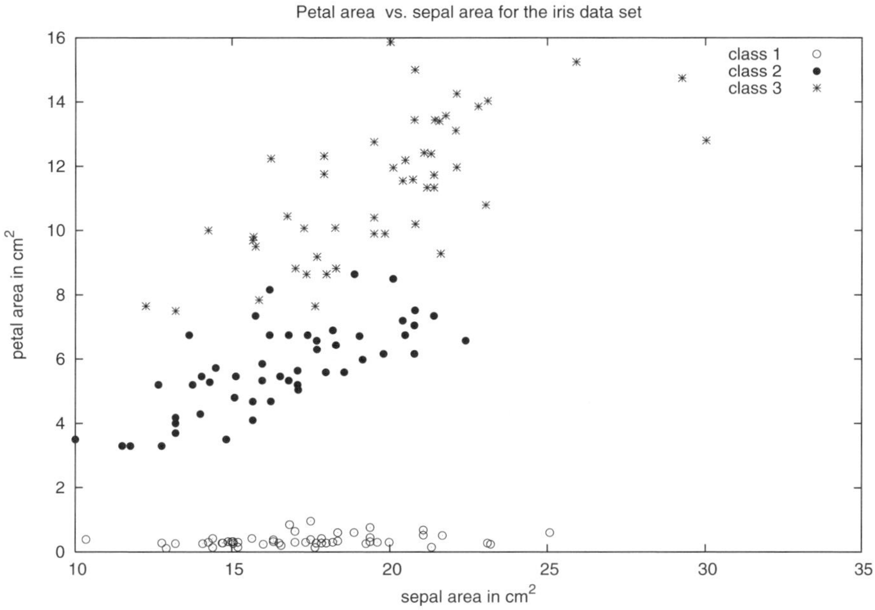

Figure 12.4. *A scatter plot showing the values of the variables petal area and sepal area for the items of the three different classes in the Iris data set. The area has been approximated as the product of the length and the width.*

have since been incorporated into several tools such as XmdvTool [610], GGobi [216], and the software developed at RoSuDa [516] .

Parallel plots provide a conceptually simple way of representing high-dimensional data. Each feature or dimension is represented along a vertical line on the x-axis. There are as many lines as there are features. Each item, with an associated value for each feature, is represented by line segments connecting the values of the features. Figure 12.5 shows the parallel plot for the Iris data. There are four features and the class represented along the x-axis. As there are three classes, the vertical line representing the class has three values, one for each class. The polyline representing an item must pass through one of these three values. This figure clearly shows that items in class 1 can be differentiated from the other two classes solely on the basis of the values of either the petal length or the petal width. This is because there is no overlap in the values of these features for items of class 1 with the values of these features for items of classes 2 and 3. The figure also indicates that it is not possible to differentiate between classes 2 and 3 using a single variable. This is because if we consider the values along any vertical line representing a feature, the values for items in classes 2 and 3 overlap.

The exploratory nature of the parallel plot can shed insight into the distribution of feature values in a data set. It can also help to identify outliers in the data, and through the judicious use of color, it is possible to identify patterns as well. For example, the parallel plot was used to check the quality of a manually created training set by considering the overlap between items from different classes [323]. It was also used to identify relevant features and outliers in our work on the classification of bent-double galaxies [200]. However, if

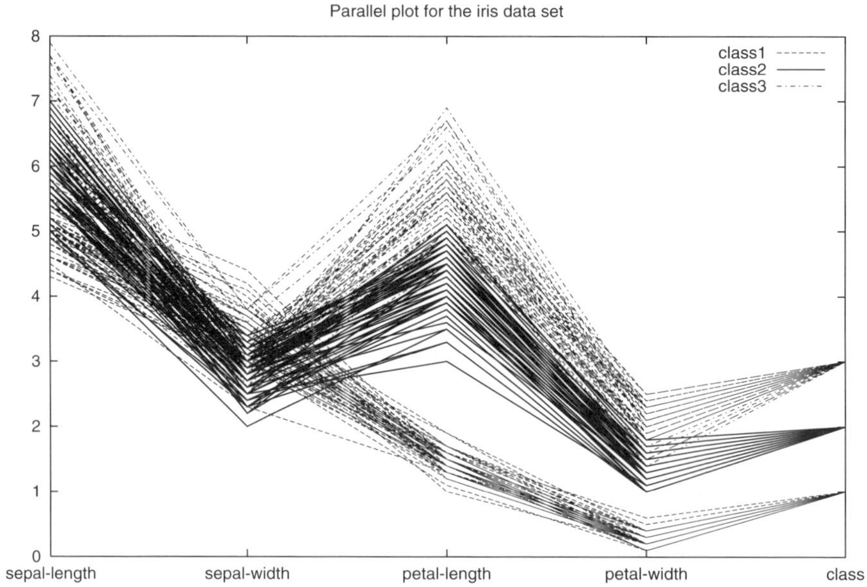

Figure 12.5. *A parallel plot for the Iris data set using the original four features.*

there are too many data items or too many features, the resulting parallel plot can be difficult to interpret.

12.2 Visualizing image and mesh data

There are several tools which are available for visualizing both image and mesh data. Some of these are quite general, while others have been developed by a specific community and are tailored to the needs of that community. However, tools developed in the context of one application area can be used in another area if it has similar needs.

In the case of mesh data, the scientific visualization community, in conjunction with the computational science and computer graphics communities, has developed software which can be used for viewing different types of meshes such as the ones described in Section 3.1.3. Some of these tools are in the public domain, such as VisIt [605] and GMV [223], and many can handle not only the smaller, two-dimensional data but also the more massive, terabyte-sized, three-dimensional data composed of several variables at a grid point. These tools provide support for several commonly used tasks such as plotting isocontours or isosurfaces, volume visualization, and the display of subsets of the data. A complete description of techniques used in scientific visualization is beyond the scope of this book; the interested reader is referred to overview texts such as the ones edited by Nielson, Hagen, and Mueller [453], Hansen and Johnson [254], and Bonneau, Ertl, and Nielson [50].

In the case of image data, there are several public domain tools which provide the basic functionality necessary to view different types of images. The more commonly used ones are ImageMagick [289] and GIMP [221]. They support different file formats and can perform several image processing tasks as well. While these tools are invaluable in scientific data mining for tasks such as conversion between file formats, visual display of images, or animation of a sequence of images, it is important to note that they typically process 8-bit data. Data which are 16-bit, or floating point, are scaled for display so the values are in the range $[0, 255]$. While this is not a major issue in viewing the data, it can have consequences in saving and processing the data as important information is lost in the conversion to 8-bits. Further, one cannot view the actual values of a floating-point image using these tools, which can be an impediment in identifying an image processing algorithm or selecting a set of appropriate parameters for an algorithm.

One tool which addresses this problem is the public-domain Fv software [208] for viewing files in the Fits format which is used in astronomy [194]. This format allows images in both two and three dimensions to be stored using floating-point values. An easy-to-use interface allows one to view an image, zoom in and out, view variation in values along a line, obtain a histogram, and create a contour plot. Various options are provided to display the intensities of the pixels in the image using different color schemes.

From a data analysis viewpoint, the most useful feature in Fv is the ability to view the floating-point value at any pixel location. This can be done in two ways—either by moving the mouse pointer over a pixel in the image or by viewing the table information which displays the values at each location of the image. For example, Figure 12.6 shows a sample image from the FIRST survey [174] using Fv. A thumbnail of the image is shown in the top right corner, indicating the zoomed-in part which is then displayed in the main pane of the window. In this example, the zoom factor is 1.0. The values at the top left are the pixel intensities at the location indicated. Note that this value is a floating-point value as the data are stored in single precision. Figure 12.7 shows the metadata corresponding to the image in Figure 12.6. While this is usually of interest to the domain scientist as it provides information on the conditions under which the image was taken, it can also provide details such as the format of the data (single precision), the size of the data (256×256), and the coordinates of the central pixel in the sky. Figure 12.8 is the tabular view of the image in Figure 12.6, where the values are displayed in floating point instead of as a scaled, 8-bit value. This table can be useful in understanding the performance of an algorithm as it shows the actual values being processed.

Fv is an example of a tool developed in a specific domain, such as astronomy, which can be used in other domains as well.

12.3 Validation of results

An important aspect of scientific data mining is the validation of the output from the different steps in the process. It essentially involves displaying the results from a particular step and evaluating them to check whether they meet our requirements. These requirements could be algorithmic, where we want to ensure that our choice of algorithms and parameters for a task perform as expected, or they could be driven by the domain scientists who want to ensure that the results are scientifically meaningful. If the results from a step match our expectations, the process of validation gives us the assurance that the analysis is proceeding correctly; if not, validation can be used to obtain feedback on ways in which the analysis can be improved to meet our requirements better.

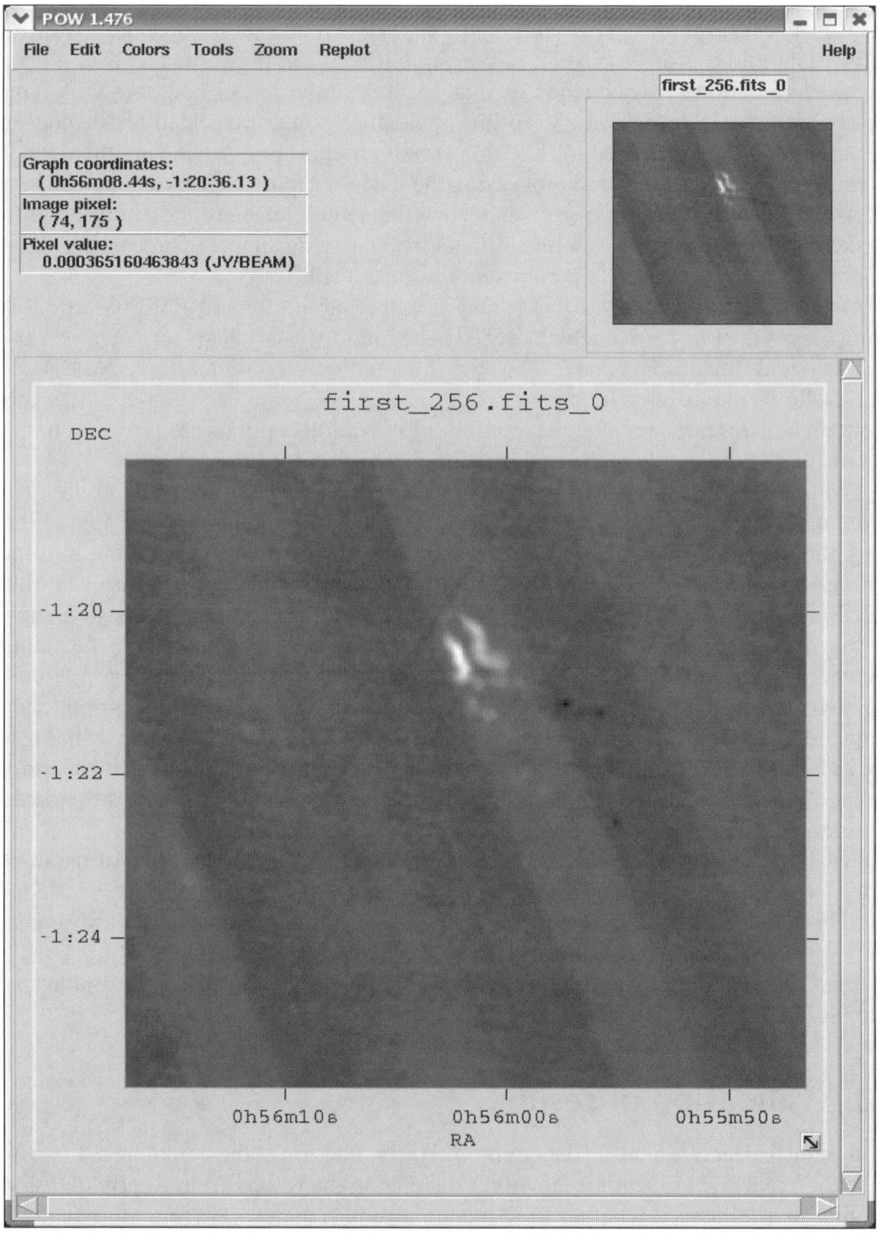

Figure 12.6. *An astronomy image from the FIRST survey* [174] *displayed using the Fv tool. The full image is shown in the top right corner, while the zoomed-in version is shown in the main pane. The intensity at a particular* (i, j) *location in the image is shown in the top left corner.*

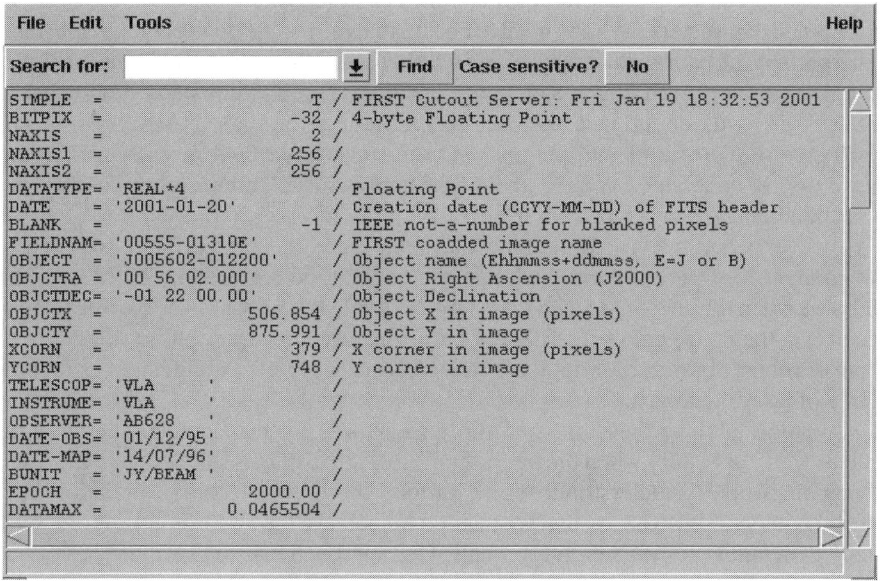

Figure 12.7. *The metadata in Fv corresponding to the image in Figure* 12.6. *This describes the instrument setting used, the units of the variable being displayed, as well as the type of data.*

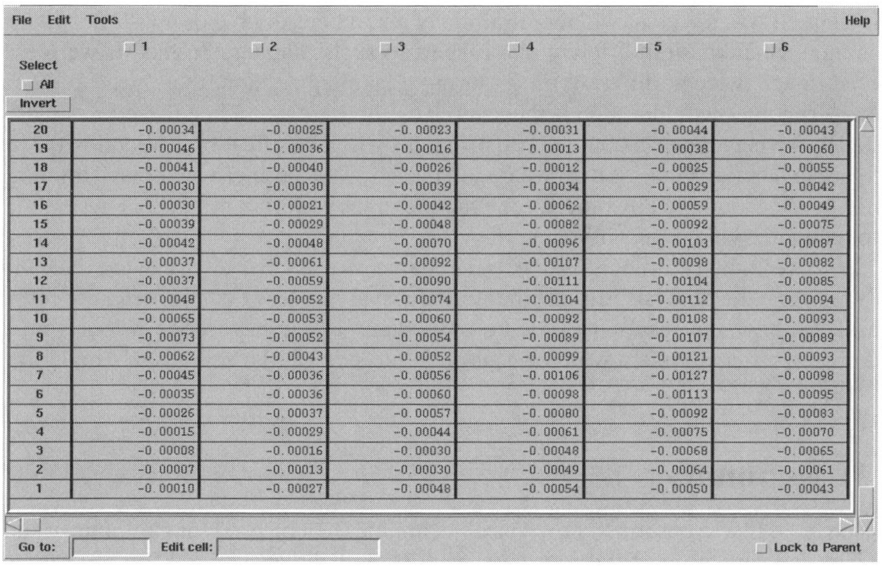

	1	2	3	4	5	6
20	-0.00034	-0.00025	-0.00023	-0.00031	-0.00044	-0.00043
19	-0.00046	-0.00036	-0.00016	-0.00013	-0.00038	-0.00060
18	-0.00041	-0.00040	-0.00026	-0.00012	-0.00025	-0.00055
17	-0.00030	-0.00030	-0.00039	-0.00034	-0.00029	-0.00042
16	-0.00030	-0.00021	-0.00042	-0.00062	-0.00059	-0.00049
15	-0.00039	-0.00029	-0.00048	-0.00082	-0.00092	-0.00075
14	-0.00042	-0.00048	-0.00070	-0.00096	-0.00103	-0.00087
13	-0.00037	-0.00061	-0.00092	-0.00107	-0.00098	-0.00082
12	-0.00037	-0.00059	-0.00090	-0.00111	-0.00104	-0.00085
11	-0.00048	-0.00052	-0.00074	-0.00104	-0.00112	-0.00094
10	-0.00065	-0.00053	-0.00060	-0.00092	-0.00108	-0.00093
9	-0.00073	-0.00052	-0.00054	-0.00089	-0.00107	-0.00089
8	-0.00062	-0.00043	-0.00052	-0.00099	-0.00121	-0.00093
7	-0.00045	-0.00036	-0.00056	-0.00106	-0.00127	-0.00098
6	-0.00035	-0.00036	-0.00060	-0.00098	-0.00113	-0.00095
5	-0.00026	-0.00037	-0.00057	-0.00080	-0.00092	-0.00083
4	-0.00015	-0.00029	-0.00044	-0.00061	-0.00075	-0.00070
3	-0.00008	-0.00016	-0.00030	-0.00048	-0.00068	-0.00065
2	-0.00007	-0.00013	-0.00030	-0.00049	-0.00064	-0.00061
1	-0.00010	-0.00027	-0.00048	-0.00054	-0.00050	-0.00043

Figure 12.8. *A part of the Fv table corresponding to the image in Figure* 12.6. *This describes the intensity values in floating point in tabular form. By scrolling the horizontal and vertical bars, it is possible to find the values for different parts of the image.*

Validation can be a challenging task for several reasons. To validate our choice of algorithms and parameters, we may want to ensure that the results do not change substantially when we perturb the parameter values or use another algorithm which performs the same task using a different approach. Such sensitivity analysis can be tedious and time consuming; however, it gives us confidence that our results reflect the data and are not an artifact of our choice of algorithms and parameters. This can be critical in problems where the science is poorly understood and the goal of the analysis is to gain insight into the scientific phenomenon being observed or simulated.

However, when the variation in the data set is large, it can be difficult to find a single algorithm and set of parameters which can accomplish the required task. The problem is exacerbated when the size of the data set is large as well, as it may not be possible to evaluate the results on the entire data set. There is unfortunately no easy solution to this problem; the topic of robust algorithms, which are relatively insensitive to small changes in the data, is an area of active research.

Validation of the results of scientific data mining by the domain scientists can be difficult as well, especially when the interpretation of the results is subjective or the scientific phenomenon is poorly understood. For example, in our work on classification of bent-double galaxies [315], we found that we needed to increase the size of our training set as the manually generated one was relatively small. Our approach was to use an iterative process, wherein we would have the astronomers validate the results from the decision tree built using the small training set, which was then enhanced using the validated examples. To do this, we created a simple tool which would show the astronomers images of the galaxies which they would classify as bent-doubles or non-bent-doubles [191]. Though the tool was very useful in capturing the feedback from the domain experts, we also found that the labeling of the examples was not very consistent. This was especially true in the hard-to-classify cases, where it was not clear whether the galaxy was a bent-double or not. Oftentimes, two astronomers would assign different class labels to such galaxies. In fact, if we presented the same image twice at different times during a labeling session, there was no guarantee that an astronomer would assign it the same label both times. Since these examples were also the ones where the decision tree had problems with classification, validating the results turned out to be more difficult than we first expected. This problem of "noisy" training data has also been encountered by others, for example in the JARTool effort [71], where the task was to recognize volcanoes on Venus.

However, when it is possible to obtain reliable user feedback, we can use the validated results to refine the analysis. In information retrieval tasks, we can incorporate relevance feedback to tailor the results to specific individuals as we have done in our similarity-based object retrieval system which was used to retrieve similar regions in simulation data sets [314].

12.4 Summary

In this chapter, I have described how visual techniques can be used for exploratory analysis as well as the validation of the results of each step of the analysis. Several visual tools, whether for images or feature data, are available in the public domain and can provide useful insights into both the raw data and the intermediate results generated during the analysis. In the case of image data, one must be aware that many tools do not preserve the floating-point format of scientific data; the conversion to 8-bit data for display can result in a significant loss of information. Therefore, appropriate tools must be used for visualization of such data sets.

12.5 Suggestions for further reading

In this chapter, I have described some simple tools, such as parallel plots, scatter plots, and Fv, which are useful in both exploratory data analysis and the validation of the results from the steps in data mining. These tools are typically used to explore and understand data sets which are small to moderate in size. As data set sizes have reached terabytes, the new field of information visualization has focused on how we can display these massive data sets, especially when the data being displayed are rather complex, such as networks or graphs, or dynamic, with the data being updated asynchronously over time.

The field of information visualization is presented in detail in several texts such as the ones by Ware [611], Chen [107], and the one by Unwin, Theus, and Hofmann [598], which focuses on very large data sets. The insights provided by Tufte in his books [591, 592, 593] are also relevant in determining better ways to present information. Finally, the work of Tukey [594] remains a classic in exploratory data analysis.

Chapter 13

Scientific Data Mining Systems

One of the basic rules of life, I have found, is that you never learn anything unless you confess your ignorance.

—Gerald Durrell [165, p. 13]

In the previous chapters, I have described the different steps in the process of scientific data mining. These steps are iterative—the validation of the results from any one step may suggest a refinement of one or more of the previous steps. The process is also very interactive, with the domain scientists closely involved in providing input and validating the results of each step. In addition, depending on the problem and the application domain, not all steps may be used in the analysis, or different algorithms may be used in a step by applying them in a cascade to get the desired results. Often, we may need to try several different algorithms and parameters before we select one that is appropriate for the task at hand. If, in addition, we are involved in the analysis of several different data sets at any one time, it seems worthwhile to identify the common threads across these problems, so that we can reuse software as appropriate.

In this chapter, I focus on the software aspect of the analysis, namely, the design and development of scientific data mining systems. There have been several scientific data mining projects where the goal was not just to solve a specific problem, but to develop software which could be used in solving other similar problems. In developing software, I have often found it helpful to understand the approaches taken by others. It enables one to anticipate unforeseen ways in which the software might be used and also provides possible solutions to common problems. This can lead to a better design of the software system architecture and the software modules, not only resulting in greater software reuse, but also enabling more time to be spent on the analysis of data, rather than the development, maintenance, and rewriting of software.

This chapter is organized as follows. In Section 13.1, I first briefly describe public domain software which is available for performing specific tasks in the data mining process. Then, in Section 13.2, I describe three projects which have taken a systems-level approach to the problem. Though the original motivation for these three projects was very different, the approaches they have chosen are very similar. I then summarize the chapter in Section 13.3.

13.1 Software for specific tasks in scientific data mining

There are several public-domain software packages which provide the functionality necessary for the tasks in the scientific data mining process. They are written in different programming languages, including Fortran, C, C++, and Java. Several have been around for many years and are relatively bug free, while others are newer, with an active community developing and maintaining the software. I divide the publicly available software into four categories—visualization and exploratory data analysis; image processing and analysis; dimension reduction; and pattern recognition.

In Chapter 12, I discussed various software tools available for visualizing data including Gnuplot [224] for simple plotting; ImageMagick [289] and GIMP [221] for visualizing 8- and 16-bit image data; Fv [208] for visualizing two- and three-dimensional floating-point data; and VisIt [605] and GMV [223] for mesh data from computer simulations. In addition, it is also possible to create tools more geared towards a specific problem using software packages such as Qt [488] which can be used for developing graphical user interfaces, OpenGL [461] for computer graphics, and Tcl/Tk [575] for development and graphical user interfaces.

The area of image processing and computer vision also has several software packages which provide useful functionality, especially for 8- and 16-bit data. In addition to ImageMagick and GIMP mentioned earlier, other public-domain packages include CVIP tools [129] for the processing of digital images, Image [287], which is a tool from the National Institute for Health for image processing and analysis, and the OpenCV package [460], which is an open-source software library for computer vision applications.

In the area of dimension reduction, the LAPACK [5] and ScaLAPACK [43] libraries provide high-performance implementations of several versions of the singular value decomposition (SVD) for dense matrices. The implementation for sparse matrices is available through SVDPACK [36] and ARPACK [370]. It is difficult to improve on these software packages, as they include extensive testing and error handling, and incorporate the latest developments in numerical analysis.

One of the better known public-domain software for pattern recognition is the WEKA toolbench [621] for machine learning and data mining. For statistical tasks, the R-project [495] provides a language and environment for statistical computing and graphics. In addition to the more general software packages such as R and WEKA, software for specific tasks such as neural networks, support-vector machines (SVMs), or clustering can also be found on the Web. Though some of these software packages are unsupported, many are supported by their developers and have an active user community.

13.2 Software systems for scientific data mining

The public-domain software libraries mentioned earlier, as well as others which are available via the Web, are often very useful in addressing specific tasks in the scientific data mining process. In some cases, it may be possible to find in the public domain all the tools one needs for analyzing a particular data set. However, in other cases, it may make sense to develop new tools, perhaps using public-domain software where appropriate. This is especially the case when publicly available software does not meet the analysis requirements. There may be several reasons for this. For example, many of the software tools operate on small- to moderate-sized data sets. If the size of the data to be analyzed is massive, one may need to develop new approaches for the analysis, or implement a parallel version of an algorithm for use on multiple processors. Alternatively, one could modify an existing algorithm so that it can be applied to the data in parallel and the results pieced back together to provide a solution. In addition, in some problems, it is possible to exploit specific characteristics of the data to simplify the task of analysis instead of using a traditional solution approach [318]. In other cases, the data may have unusual characteristics, such as noise in an image in the form of vertical streaks [321], which may require specialized algorithms. Sometimes, a problem-specific approach to handling missing values may need to be implemented, or a specific part of an algorithm, say the splitting criterion used in the induction of a decision tree, may need to be modified to be more suitable to the data being analyzed. Or, the problem may require something unique to scientific data, such as the need to quantify the uncertainty in the results.

All these requirements indicate that one may need to consider developing software specifically tuned to the characteristics of the data set being analyzed. If, in addition, we need to solve several analysis problems involving very different data sets, it may make sense to consider the development of a set of tools, or a software system. I next describe three scientific data mining projects where a systems approach has been taken in the development of the software. These are for illustrative purposes to indicate the approaches we can take to develop toolkits for the analysis of scientific data.

13.2.1 Diamond Eye

The early projects at NASA's Jet Propulsion Laboratory, such as SKICAT [184] and JARTool [71], focused on the solution of specific problems, namely, the analysis of sky objects in the Second Palomar Observatory Sky Survey and the identification of volcanoes on Venus, respectively. While both these projects were instrumental in drawing the attention of data miners to the challenges faced by scientists in the analysis of their data, it was the Diamond Eye project [72, 73] that took a more systems view to solve a broader class of problems.

The Diamond Eye project was the second generation of image mining tools from the Jet Propulsion Laboratory and incorporated many of the lessons learned from JARTool. Its goal was to mine useful information in large image collections, such as those resulting from various spacecraft missions. To support this more general goal, Diamond Eye not only incorporated several image mining algorithms, it also considered system architecture issues in its design and implementation.

From the viewpoint of image mining algorithms, the developers of Diamond Eye realized that a key challenge was to determine what mathematical processing should be applied to the low-level image data, in the form of pixels, to identify the spatial objects of interest. They chose to support several different algorithms, categorized based on how much prior knowledge in the form of a training set was required by each technique. For example, the category of *recognition* algorithms used a large number of examples to build a precise statistical model, while *discovery* algorithms used no examples, but relied on generic assumptions to look for interesting objects without a specific target. In the middle were *query* algorithms which used one or a few examples, or even a sketch. This focus on retrieving similar objects made Diamond Eye similar to other content-based retrieval projects, with the main difference being the focus on objects in the image, rather than the whole image.

From a system architecture viewpoint, Diamond Eye was designed to let users interact with the system. It used a Java-based client-server architecture to provide platform-independent access to the mining services. The core mining algorithms were written in a higher-level language (C or C++) for computational efficiency. A middleware interface tied these algorithms to the client-server system. A server program was colocated with a large image repository to enable mining of the data in place. The servers were coupled with an object-oriented database and a computational engine, such as a network of workstations. The database provided persistent storage while the computational engine enabled the parallel execution of the more compute-intensive parts of the image mining task, such as the image processing and the object recognition. The Diamond Eye infrastructure was application independent, easily allowing domain-specific algorithms to be incorporated.

The architecture of Diamond Eye resulted in several benefits, including a design which allowed trial evaluation of advanced data mining and machine learning techniques by potential new users, the reuse of a software infrastructure that was common across a range of science mining applications, and a system that facilitated closer collaborations between algorithm developers and domain experts.

13.2.2 Algorithm Development and Mining System

Another system which has been developed for the analysis of many different data sets is the Algorithm Development and Mining System (ADaM) developed by the Information Technology and Systems Center at the University of Alabama in Huntsville [1, 500]. The main application domain for ADaM is remotely sensed data, especially problems in earth sciences, though its modules can be used in the context of other scientific data sets as well.

ADaM consists of several modules, primarily in the area of data mining and image processing. For data mining operations, it provides several algorithms for clustering, classification, association rules, and feature selection. The image processing algorithms supported are mainly for analysis of satellite imagery and include various filters, texture features, boundary detection, polygon circumscription, as well as simple image processing functions such as cropping, rotation, and sampling.

A key differentiating characteristic of ADaM is its system architecture. By using autonomous components in a distributed architecture, it supports the easy integration of third-party algorithms. Each component is provided with a C, C++, or other application programming interface; an executable in support of generic scripting tools, such as Perl, Python, or shell scripts; and eventually web service interfaces to support web and grid applications. This is especially useful in the context of earth systems applications as the community has embraced the use of computational grid technologies [35]. Since ADaM components can be accessed via multiple external interfaces, it is easy to use them in different applications. They can be used either by themselves or combined with other specialized software modules. For example, general-purpose image analysis modules were combined with problem-specific modules to detect tropical cyclones and estimate their maximum sustained winds.

13.2.3 Sapphire

The Sapphire scientific data mining software was developed as part of the Sapphire project at the Lawrence Livermore National Laboratory [524]. The system architecture was designed from the beginning to meet the diverse requirements that arose in the analysis of scientific data from observations, experiments, and simulations. In particular, some of the main considerations which determined the architecture, and our approach to accommodating these considerations, were the following:

- Instead of having a single monolithic code for each problem, the tasks common to several problems were identified and implemented as separate modules, enabling us to reuse the software. Thus, there was a separate module for denoising image data, one for decision trees, another for neural networks, and so on. A data set was then analyzed by combining appropriate functions from several modules.

- As no single algorithm for a task, such as denoising, was likely to be appropriate for all problems, we supported several algorithms for each task. A uniform interface across these algorithms enabled us to easily swap one algorithm for another. It also enabled new algorithms to be added easily as necessary.

- As an algorithm often needs to be fine tuned to a data set and problem, we provided easy access to the parameters that are used in an algorithm. These parameters could be set by the user, or the default values used as a first guess.

- To make the data flow more streamlined, each module was designed so that the output from one phase of data mining could be input into the next appropriate phase. Thus, if the image segmentation algorithm produced a mask image with each object identified by a unique identifier, the follow-on step of feature extraction would take as input the mask image, along with an object identifier, and extract features for the corresponding object. This minimized any transformations necessary to convert from one data format to another between the steps in the scientific data mining process.

- It was clear that some tasks, such as those of pattern recognition, were domain independent, that is, a decision tree could be applied to problems in remote sensing, astronomy, computer simulations, and so on. Other tasks, such as denoising and feature extraction, could be problem dependent. For example, the noise characteristics could depend on the type of sensor used to collect the data, while the features extracted could depend on the problem being addressed with a data set. By clearly identifying and separating the domain-dependent and domain-independent tasks, we were able to maximize the re-use of our software as we moved from one application to another.

- As scientific data mining is essentially an iterative process, and we sometimes had to work with data as they were being collected, we provided the option to store the results of the processing at each stage. Thus, when the new data arrived, we did not need to reprocess the old data.

Figure 13.1 shows the system architecture of the Sapphire software system. Note that it closely resembles the data flow diagram in Figure 4.1. The modules in the white boxes at the beginning and end were for the reading, writing, and display of the data. To the extent possible, we leveraged existing domain software, writing additional code only to convert between the domain data format and our internal data format. By having a single internal format for, say, image data, we were able to support different input image format, such as Fits, pnm, BSQ, by first converting them to our internal format.

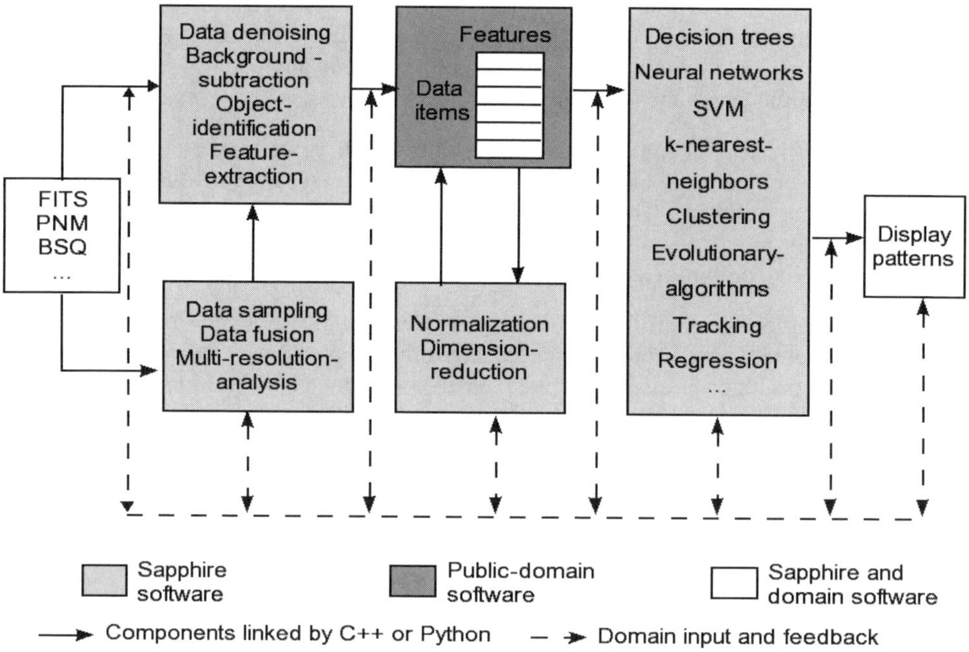

Figure 13.1. *The Sapphire system architecture.*

The module in dark grey in the middle was a data store to store the data items by features matrix. We used a public-domain software package called RDB [276], written in Perl, for this data store. This is a simple tool which stores the data in a table format and enables querying based on the row and column locations. It is very similar to the comma-separated format for storing a table, with some additional header information for the metadata as well as format information for printing the table.

The modules in lighter grey are the compute-intensive parts of the scientific data mining process and were the main focus of the Sapphire project. The entire software was written in C++, using object-oriented design principles. We used inheritance to support a single interface for different algorithms for a task and templating on the data type to write a single code for single- and double-precision data. As appropriate, we used a driver code in C++ or a Python interface to the software to create a solution for an analysis problem.

We also collected similar tasks and packaged them as a single library. For example, there was a library for basic image processing algorithms, one for wavelets and wavelet denoising, one for more complex algorithms for segmentation and feature extraction, and another for video processing. Similarly, there was a library each for decision trees, neural networks, regression, and so on. A toolkit library provided basic functions for sampling, histograms, and random number generators, while another library provided the functions to read and write files in various image formats as well as the RDB format. This breakup of the functions into separate libraries allowed us to link only to those parts of Sapphire which were necessary for analysis. If required by a problem, we would also write additional code, which was added into Sapphire if it was found to be broadly applicable.

Sapphire software was used to solve a broad range of problems in different scientific domains, including:

- classification of bent-double galaxies in observational astronomy data [316];
- separating signals, such as El Niño and volcano signals, from surface temperature data obtained from climate simulations [199];
- similarity-based object retrieval in two- and three-dimensional simulation data [314];
- detection of human settlements in satellite images [323];
- identification of key features associated with edge-harmonic oscillations in sensor data from a tokamak [83];
- detection and tracking of moving objects in video sequences [112, 320];
- estimating missing features in multimedia information retrieval [15];
- comparing simulation and experimental data for code validation [321];
- characterizing and tracking bubbles and spikes in three-dimensional simulations of the Rayleigh–Taylor instability [318, 215];
- classification of orbits in Poincaré plots [14]; and
- detection of blobs in experimental images from fusion plasma [392].

Our approach to the design and implementation of Sapphire software showed that it was possible to build a toolkit which could be used to address many different problems in scientific data mining.

13.3 Summary

In this chapter, I discussed the topic of software architectures for scientific data mining from a systems viewpoint. Using three examples, I showed that it is possible to build a software infrastructure which can support the many different needs of analysis problems in science and engineering domains. There are several common themes across these projects, including the use of modularity to group similar tasks and enabling flexibility through the easy incorporation of third-party or domain-specific software. In addition, I also mentioned various public-domain software packages which can be used in various tasks in the scientific data mining process.

In the next chapter, which is the last chapter in the book, I discuss the challenges and the opportunities that await a practitioner of scientific data mining. Having described the various tasks in the end-to-end process of scientific data analysis, I start the next chapter by returning to the beginning of the process and provide some guidelines on how one might proceed when given a data set and a problem to solve.

Chapter 14

Lessons Learned, Challenges, and Opportunities

> *The same thrill, the same awe and mystery, come again and again when we look at any problem deeply enough. With more knowledge comes deeper, more wonderful mystery, luring one on to penetrate deeper still. Never concerned that the answer may prove disappointing, but with pleasure and confidence we turn over each new stone to find unimagined strangeness leading on to more wonderful questions and mysteries—certainly a grand adventure!*
>
> —R. P. Feynmann [190, p. 144]

The purpose of this book was to bring together, in one place, many of the different techniques used in the analysis of science and engineering data sets arising from simulations, experiments, and observations. The early chapters described the many ways in which data analysis techniques are being used in various scientific domains, followed by an identification of the common tasks, which, in turn, were used to motivate an end-to-end process of scientific data mining. A large part of the book was devoted to the detailed description of the algorithms used in these tasks, which enabled us to process the raw data to find the useful information in them. Then, in the previous chapter, I described various public-domain software packages available for these tasks, as well as three projects which have taken a systems-level approach, building software infrastructures which can be used to analyze data from several different problems in different domains.

In this, the final chapter of the book, I start by going back to the beginning. Based on my experiences with mining data from several scientific disciplines, I first provide some guidelines for getting started with the analysis of a data set. This is followed in Section 14.2 by looking forward to the many open research problems which remain to be addressed, and the challenges which await data mining practitioners as they analyze data from new domains, with new characteristics, and new algorithmic requirements. Finally, I conclude the chapter, and the book, with a few thoughts on the rewards of embarking on a scientific data mining endeavor.

14.1 Guidelines for getting started

Getting started on the analysis of a scientific data set can oftentimes be rather daunting. This is especially true if the science behind the phenomena being analyzed is poorly understood, leading to a poorly defined analysis problem and the domain scientists having few preconceived notions of what they are looking for in the data. If, in addition, the data set is very large, perhaps terabytes in size, the sheer size of the data themselves can be overwhelming. Under these circumstances, it can be a challenge just to determine a place to start the analysis.

Based on my experiences in mining science and engineering data sets in different domains, I have found that no one solution approach works for all problems. However, there are some general guidelines which are helpful in getting started.

The first step is to understand the analysis goals of the scientist. These goals may be somewhat poorly defined at the beginning, in which case, it is helpful to spend more time doing exploratory analysis to understand the data better and gain some insights into what the scientists hope to achieve through the analysis. This is also true when the data set is very large and it is unlikely that anyone has ever looked at all the data. In many cases, the scientist may start with one goal in mind, and then, as the analysis progresses, may want to investigate further any interesting issues uncovered during the analysis. Of course, depending on what is learned during the analysis, the goals of the analysis may change over time.

It is often helpful to start the initial exploratory analysis of the data by viewing the data. This is easy if the data are small enough and in a commonly used format. Otherwise, we may have to write code to read in the data or extract a small sample of it. If the structure of the data is uncommon, for example, graphs or particles in three-dimensional space, we may also need to write new tools to view the data. If, however, the data are in the form of mesh data or images, existing tools may suffice.

However, a word of caution may be appropriate regarding the task of viewing, or looking at, the data, especially if one of the many scientific visualization or image analysis software packages available in the public domain are used. These packages usually process the data first to make them suitable for display; this processing may alter the data so that what is visualized is not exactly what is in the data. For example, images are often scaled to between 0 and 255 for display. If the original range of the data is much smaller, the scaling may amplify minor differences. Or, the process of interpolation and smoothing which is done in scientific visualization may hide important details. Thus, any visual tools must be used with care during exploratory analysis, and the focus must be on the actual floating-point or integer values in the data as these are the values which will be processed during analysis.

If the data set is very large and we want to work with a small sample, it is obvious to ask what is the best way to sample the data. Domain scientists can be very helpful as they may already have smaller samples they work with on a regular basis. Or, they may have examples of data which they use to discuss their analysis goals. If the data are from simulations, we can extract a smaller sample by considering a few time steps sampled uniformly. If the simulation was run on a multiprocessor, and the output from each time step is available in separate files, it might also make sense to select a sample of the files instead of all the files at a time step. If the data and the analysis do not have a time aspect, the sampling can be done either randomly or based on suggestions from the domain scientist. Of course, the sample size must be large enough to capture all the variation in the data so that we can select an analysis algorithm which can handle the variation. However, it may be unclear if all the variations present in the data are also present in the subsample. One solution approach is to start small and keep increasing the size of the subsample as the analysis proceeds. If the new data added have characteristics very different from the earlier samples, the analysis algorithms may no longer work. This may require changing the parameters used in the algorithms or selecting a different algorithm which is more suitable to the characteristics of the full data.

It is also helpful, especially at the beginning, when one is looking at the data, to identify any possible issues with the quality of the data. If the data have not been looked at extensively, it is possible that there are outliers in the data or variables with scientifically meaningless values. These could be indicative of bugs in the code which generated the data, a sensor malfunction, or a simulation being run at, or beyond, the limits of what it can simulate.

For each task in the scientific data mining process, the choice of an analysis algorithm can be difficult. Obviously, we want an algorithm which, when used with appropriate parameter settings, will transform the data as expected. We must also ensure that the data satisfy any assumptions required by the algorithm. It is often very useful to start by applying the simplest algorithm which is suitable for the data. Even if the algorithm does not work, it can provide further insights into the data and also indicate possible approaches which might work. In my experience, the simpler algorithms often work surprisingly well and may be preferable to complex algorithms as they are often faster or have fewer parameters which require tuning. If an algorithm is simple, it is easy to understand why it not working on a particular subsample of the data. Simpler algorithms are also preferable as they are easier to explain to the domain scientists, which is very helpful when the scientist has to validate the results at each step of the analysis.

If the simpler algorithms do not work, we may need to look into why they are not working and see if there is a way to address the problem. It may be that the simpler algorithm is not working because the data are not normalized or there are outliers which must be removed. If, however, the simple algorithm is not working as it is not appropriate for the data, we may need to try out progressively more complex algorithms until we find one that works. Alternatively, we can try combinations of simple algorithms and use them in the analysis by cascading them.

If the variation in the data is large, for example in the case of a simulation with small-scale structures at early time and large-scale structures at late time, we may need to use more than one algorithm for some tasks in the analysis. In such cases, care must be taken to ensure that the effects of using different algorithms do not influence any conclusions drawn from the analysis. This is also true even if the same algorithm is used for all the data, but with different parameters to account for different characteristics of different subsamples of the data.

In problems where the task is to discover the science in the data, it often helps to obtain similar results with different algorithms and to confirm that the results do not change substantially with small changes in the parameters used in the algorithms. This provides a sanity check to ensure that the conclusions drawn are driven by the data and not an artifact of the algorithm or the parameters. Of course, the scientists are more likely to believe the results if they have been verified using alternate approaches.

In addition to these guidelines, it is also worth revisiting Sections 3.2 and 3.3, which describe the characteristics of the data and the analysis and discuss how they can influence the steps one should take to analyze the data. An overview of these steps is given in Chapter 4.

14.2 Challenges and opportunities

As we have seen through the many examples in this book, mining scientific data can be challenging—the data can be noisy, with missing values; extracting objects in mesh and image data can be difficult; identifying relevant features to represent the objects is critical, but nontrivial; the number of features describing an object can be very large, making the problem high dimensional; and the training data for classification can be small and unbalanced. These are but a few of the many problems we face in scientific data mining in the context of current data sets and applications. As new application areas, such as materials science and cheminformatics, start adopting data mining techniques for the analysis of their data, the problems are bound to increase.

However, each challenge creates opportunities for new research and innovative solutions. There are several factors which will drive these opportunities in the future, including:

- **The size of scientific data sets:** As data sizes reach terabytes and petabytes, we will need to consider appropriate ways of analyzing these data sets, many of which are moved to storage systems as soon as they are collected or generated. The analysis techniques to handle massive data sets currently include traditional approaches such as parallelization of existing algorithms; the development of new methods which are computationally inexpensive, but robust and accurate enough for analysis; and ingenious ways of analyzing data so that few passes are required over the full data set, minimizing access to the data in storage. In addition, newer approaches are also needed, such as innovative sampling techniques which use only a subset of the data to draw accurate conclusions about the data; techniques for reducing the size of the data generated by identifying and storing only the "interesting" parts; and, in the context of simulations, building and using predictive models to reduce the number of simulations necessary to understand a scientific phenomenon.

- **The advent of new computer architectures:** We will also need to develop algorithms suitable for new types of computer architectures, such as multicore systems, field programmable gate arrays, and graphical processing units. As new programming models, such as cloud computing, gain broader acceptance, we will need to incorporate them into the implementation of the analysis algorithms as well.

- **The analysis of new types of data:** As new application areas look towards data mining techniques to solve their problems, new types of data are being analyzed. For example, in some problem domains such as the analysis of the power grid, the Web, the Internet, and cheminformatics, the data being analyzed are best represented as a graph. In other domains, such as bio-informatics, a rich source of data is technical documents consisting of text, images, tables, figures, and links to other documents, as well as the information in these documents. The data structures used to represent such heterogeneous data will be rather complex; we will need to develop new algorithms to successfully mine these data.

- **The need to analyze streaming data in real time:** As sensors become ubiquitous, they are being used to measure various quantities during scientific experiments and observations. The analysis of these data can be difficult for several reasons. In some problems, the data may be sampled at different rates by different sensors, or the data may be available only in an aggregated form. Missing data are also common if a sensor is not working, or the quantity it was measuring is not recorded. In addition, a key issue in streaming data is the real-time or near-real-time requirement for analysis. This is especially true if the streaming data are from sensors monitoring equipment to predict when it might fail, from telescopes observing the sky to detect unusual events, or from sensors monitoring computer networks to detect intrusions. The sheer volume of streaming data as well as the need to identify anomalous events with few false positives make this an active research area.

- **The analysis of geographically distributed data:** Until recently, the term "distributed data" referred to data at a single geographic location, but distributed among many files, for example, the output from a simulation run on multiple processors. However, the data can also be geographically distributed, as would be the case for a network of sensors, or astronomy surveys taken by different telescopes. New approaches have to be considered to avoid moving large volumes of data to one location for analysis.

- **The need to draw conclusions in the presence of uncertainty:** As we have observed, the quality of data can make scientific data mining rather difficult, whether it is due to noise in image data, missing data values, or an unbalanced training data set in classification. A related issue which is becoming increasingly important is uncertainty quantification which can give us some idea of how much we can trust the results of our analysis, given the uncertainty associated with the input data. Statistical and machine learning techniques will play an important role in addressing this problem.

- **Multiscale analysis:** In multiscale phenomena, the data collected are also at multiple scales, posing new problems for analysis. The phenomena at finer scales may influence the coarser scales, often nonlinearly, and coarse-scale behavior may feed back into the finer scales. We therefore need new ways to represent this multiscale information and new techniques to understand how the information at different scales interact to determine the macroscale behavior.

While these new challenges offer new opportunities for data miners, many of the old challenges still remain. For each step in the scientific data mining process, there continues to be a need for reliable, accurate algorithms which are robust to the parameters used in the algorithm. The interpretability of the algorithms is also important, even if it means a slightly less accurate algorithm; scientists are unlikely to accept the analysis results if they cannot understand how an algorithm processed the data to arrive at the results. And finally, algorithms which are computationally inexpensive, and have few parameters, are preferred as they allow easy experimentation and provide a fast turnaround of results.

14.3 Concluding remarks

In this book, I have described how techniques from a variety of domains including signal and image processing, pattern recognition, machine learning, statistics, high-performance computing, and data mining can be used to analyze data from scientific simulations, experiments, and observations. Though challenging, the field of scientific data mining, and the application of these techniques, can be a rewarding experience for both the scientists and the data miners. The scientists no longer have to be concerned that the sheer size and complexity of their data are resulting in a loss of serendipitous discoveries which are vital to progress in their fields. The data miners not only get an opportunity to work with a rich source of data, which is replete with new and interesting problems, they also share in the thrill of scientific discovery. For both, it is indeed a grand adventure!

Bibliography

[1] ADaM: Algorithm Development and Mining System Web page, 2008. http://datamining.itsc.uah.edu/adam/.

[2] A. Adams and A. Woolley. Hubble classification of galaxies using neural networks. *Vistas in Astronomy*, 38:273–280, 1994.

[3] H. Almuallim and T. G. Dietterich. Learning with many irrelevant features. In *Proceedings of the Ninth National Conference on Artificial Intelligence*, pages 547–552, 1991.

[4] Y. Amit and D. Geman. Shape quantization and recognition with randomized trees. *Neural Computation*, 9:1545–1588, 1997.

[5] E. Anderson et al. *LAPACK Users' Guide, Third Edition*. SIAM, Philadelphia, 1999. Software available at http://www.netlib.org.

[6] M. S. Arulampalam, S. Maskell, N. Gordon, and T. Clapp. A tutorial on particle filters for on-line nonlinear/non-Gaussian Bayesian tracking. *IEEE Transactions on Signal Processing*, 50:174–188, 2002.

[7] S. Arya et al., An optimal algorithm for approximate nearest neighbor searching. *Journal of the ACM*, 45:891–923, 1998.

[8] C. G. Atkeson, S. A. Schaal, and A. W. Moore. Locally weighted learning. *AI Review*, 11:75–133, 1997.

[9] M. F. Augusteijn, L. E. Clemens, and K. A. Shaw. Performance evaluation of texture measures for ground cover identification in satellite images by means of a neural network classifier. *IEEE Transactions on Geoscience and Remote Sensing*, 33:616–626, 1995.

[10] F. Aurenhammer. Voronoi diagrams—a survey of a fundamental geometric data structure. *ACM Computing Surveys*, 23:345–405, 1991.

[11] AVIRIS Web: NASA's Airborne Visible/Infrared Imaging Spectrometer Web page, 2008. http://aviris.jpl.nasa.gov/.

[12] S. G. Azevedo et al. HERMES: A high-speed radar imaging system for inspection of bridge decks. In *Proceedings of SPIE, Volume 2946*. SPIE Press, Bellingham, WA, 1996.

[13] F. R. Bach and M. I. Jordan. Learning spectral clustering, with application to speech separation. *Journal of Machine Learning Research*, 7:1963–2001, 2006.

[14] A. Bagherjeiran and C. Kamath. Graph-based methods for orbit classification. In *Proceedings of the Sixth SIAM International Conference on Data Mining*, pages 574–578, 2006. Available online at http://www.siam.org/proceedings/datamining/2006/dm06_064baghera.pdf

[15] A. Bagherjeiran, N. S. Love, and C. Kamath. Estimating missing features to improve multimedia retrieval. In *Proceedings of the IEEE International Conference on Image Processing, Volume II*, pages 233–236, 2007.

[16] P. Baldi and S. Brunak. *Bio-Informatics: The Machine Learning Approach*. The MIT Press, Cambridge, MA, 2000.

[17] G. H. Ball and D. J. Hall. ISODATA: A novel method of data analysis and pattern recognition. Technical Report, Stanford Research Institute, Menlo Park, CA, 1965.

[18] G. H. Ball and D. J. Hall. A clustering technique for summarizing multi-variate data. *Behavioral Science*, 12:153–156, 1967.

[19] A. Banerjee, H. Hirsh, and T. Ellman. Inductive learning of feature tracking rules for scientific visualization. In *Proceedings of the IJCAI-95 Workshop on Machine Learning in Engineering*, 1995.

[20] I. N. Bankman, editor. *Handbook of Medical Imaging: Processing and Analysis*. Academic Press, San Diego, CA, 2000.

[21] Y. Bar-Shalom, X. R. Li, and T. Kirubarajan. *Estimation with Applications to Tracking and Navigation: Theory, Algorithms, and Software*. John Wiley, New York, 2001.

[22] R. Baraniuk. Compressive sensing. *IEEE Signal Processing Magazine*, 24:118–121, 2007.

[23] D. Barash and R. Kimmel. An accurate operator splitting scheme for nonlinear diffusion filtering. Technical Report HPL-2000-48 (R.1), Hewlett-Packard Laboratories, Israel, 2000.

[24] J. L. Barron, D. J. Fleet, and S. Beauchemin. Performance of optical flow techniques. *International Journal of Computer Vision*, 12:43–77, 1994.

[25] B. G. Batchelor. *Practical Approach to Pattern Classification*. Plenum Press, New York, 1974.

[26] E. Bauer and R. Kohavi. An empirical comparison of voting classification algorithms: Bagging, boosting, and variants. *Machine Learning*, 36:105–139, 1999.

[27] A. Baxevanis and B. F. F. Ouellette. *Bio-informatics: A Practical Guide to the Analysis of Genes and Sequences*. John Wiley, New York, 2001.

[28] S. S. Beauchemin and J. L. Barron. The computation of optical flow. *ACM Computing Surveys*, 27:433–467, 1994.

[29] R. H. Becker, R. L. White, and D. J. Helfand. The FIRST survey: Faint Images of the Radio Sky at Twenty-cm. *Astrophysical Journal*, 450:559, 1995.

[30] J. Behnke and E. Dobinson. NASA workshop on issues in the application of data mining to scientific data. *SIGKDD Explorations Newsletter*, 2:70–79, 2000.

[31] R. E. Bellman. *Adaptive Control Processes*. Princeton University Press, Princeton, NJ, 1961.

[32] S. Belongie, J. Malik, and J. Puzicha. Shape matching and object recognition using shape contexts. *IEEE Transactions on Pattern Analysis and Machine Intelligence*, 24:509–522, 2002.

[33] E. Beltrami. Sulle funzioni bilineari. *Giornale di Matematiche di Battaglini*, 11: 98–106, 1873.

[34] P. Berkhin. Survey of clustering data mining techniques. Technical Report, Accrue Software, San Jose, CA, 2002.

[35] D. Bernholdt et al. The earth system grid: Supporting the next generation of climate modeling research. *Proceedings of the IEEE*, 93:485–495, 2005.

[36] M. Berry. Large scale singular value computations. *International Journal of Supercomputer Applications*, 6:13–49, 1992.

[37] J. Beutel, H. L. Kundel, and R. L. Van Metter, editors. *Handbook of Medical Imaging, Volume 1: Physics and Psychophysics*. SPIE Press, Bellingham, WA, 2000.

[38] K. Beyer, J. Goldstein, R. Ramakrishnan, and U. Shaft. When is "nearest neighbor" meaningful? *Lecture Notes in Computer Science*, 1540:217–235, 1999.

[39] E. Bingham and H. Mannila. Random projection in dimensionality reduction: Applications to image and text data. In *Proceedings of the Seventh ACM SIGKDD International Conference on Knowledge Discovery and Data Mining*, pages 245–250, 2001.

[40] H. Bischof and R. Frühwirth. Recent developments in pattern recognition with applications in high energy physics. *Nuclear Instruments and Methods in Physics Research A*, 419:259–269, 1998.

[41] C. M. Bishop. *Neural Networks for Pattern Recognition*. Oxford University Press, Oxford, UK, 1998.

[42] C. M. Bishop. *Pattern Recognition and Machine Learning*. Springer, New York, 2006.

[43] L. S. Blackford et al. *ScaLAPACK Users' Guide*. SIAM, Philadelphia, 1997. Software available at http://www.netlib.org.

[44] S. Blackman and R. Popoli. *Design and Analysis of Modern Tracking Systems*. Artech House, Norwood, MA, 1999.

[45] L. Blanc-Feraud, P. Charbonnier, G. Aubert, and M. Barlaud. Non-linear image processing: Modeling and fast algorithm for regularization with edge detection. In *Proceedings of the IEEE International Conference on Image Processing*, volume 1, pages 474–477, 1995.

[46] Blue Brain Project Web page, 2008. http://bluebrain.epfl.ch/.

[47] A. L. Blum and P. Langley. Selection of relevant features and examples in machine learning. *Artificial Intelligence*. 97:245–271, 1997.

[48] H. Blum. Biological shape and visual science. *Theoretical Biology*, 38:205–287, 1973.

[49] R. S. Blum and Z. Liu, editors. *Multi-Sensor Image Fusion and Its Applications*. CRC Press, Boca Raton, FL, 2005.

[50] G.-P. Bonneau, T. Ertl, and G. M. Nielson, editors. *Scientific Visualization: The Visual Extraction of Knowledge*. Springer, Berlin, 2005.

[51] I. Borg and P. Groenen. *Modern Multidimensional Scaling: Theory and Applications*. Springer, New York, 1997.

[52] T. E. Boult et al. Frame-rate omnidirectional surveillance and tracking of camouflaged and occluded targets. In *Proceedings of the Second IEEE Workshop on Visual Surveillance*, pages 48–55, 1999.

[53] K. Bowyer, L. Hall, N. Chawla, T. Moore, and W. Kegelmeyer. A parallel decision tree builder for mining very large visualization data sets. In *IEEE International Conference on Systems, Man, and Cybernetics*, pages 1888–1893, 2000.

[54] K. Bowyer, C. Kranenberg, and S. Dougherty. Edge detector evaluation using empirical ROC curves. *Computer Vision and Image Understanding*, 84:77–103, 2001.

[55] M. Brand. Fast online SVD revisions for lightweight recommender systems. In *Proceedings of the Third SIAM International Conference on Data Mining*, pages 37–46, 2003. Available online at http://www.siam.org/proceedings/datamining/2003/dm03_04BrandM.pdf

[56] L. Breiman. Bagging predictors. *Machine Learning*, 26:123–140, 1996.

[57] L. Breiman. Pasting bites together for prediction in large data sets and on-line. Technical Report, Statistics Department, University of California, Berkeley, 1996. Available at ftp.stat.berkeley.edu/pub/users/breiman/pastebite.ps.Z.

[58] L. Breiman. Arcing classifiers. *Annals of Statistics*, 26:801–849, 1998.

[59] L. Breiman. Random forests—random features. Technical Report 567, Statistics Department, University of California, Berkeley, 1999.

[60] L. Breiman. Statistical modeling: The two cultures. *Statistical Science*, 16:199–231, 2001.

[61] L. Breiman, J. H. Friedman, R. A. Olshen, and C. J. Stone. *Classification and Regression Trees*. CRC Press, Boca Raton, FL, 1984.

[62] Chris Brislawn. The FBI fingerprint image compression standard, 2008. http://www.c3.lanl.gov/~brislawn/FBI/FBI.html.

[63] E. Brookner. *Tracking and Kalman Filtering Made Easy*. John Wiley, New York, 1998.

[64] D. Brown and J. Brence. Discovering corrosion relationships in eddy current nondestructive test data. In *Proceedings of the Fourth Workshop on Mining Scientific Data Sets*, pages 49–55, 2001. Workshop held in conjunction with KDD2001.

[65] L. G. Brown. A survey of image registration techniques. *ACM Computing Surveys*, 24:325–376, 1992.

[66] A. Bruce and H. Gao. *S+ WAVELETS User's Manual*. StatSci Division of MathSoft, 1994.

[67] A. Bruce and H. Gao. *Applied Wavelet Analysis with S-PLUS*. Springer, New York, 1996.

[68] V. Bruni and D. Vitulano. A generalized model for scratch detection. *IEEE Transactions on Image Processing*, 13:44–50, 2004.

[69] J. R. Bunch and C. P. Nielsen. Updating the singular value decomposition. *Numerische Mathematik*, 31:111–129, 1978.

[70] C. J. C. Burges. A tutorial on support vector machines for pattern recognition. *Data Mining and Knowledge Discovery*, 2:121–167, 1998.

[71] M. Burl, L. Asker, P. Smyth, U. Fayyad, P. Perona, L. Crumpler, and J. Aubele. Learning to recognize volcanoes on Venus. *Machine Learning*, 30:165–195, 1998.

[72] M. Burl, C. Fowlkes, J. Roden, A. Stechert, and S. Mukhtar. Diamond Eye: A distributed architecture for image data mining. In *Data Mining and Knowledge Discovery, Proceedings of the SPIE, Volume 3695*, pages 197–206, 1999.

[73] M. C. Burl. Mining large image collections. In R. Grossman, C. Kamath, W. Kegelmeyer, V. Kumar, and R. Namburu, editors, *Data Mining for Scientific and Engineering Applications*, pages 62–84. Kluwer, Boston, MA, 2001.

[74] M. C. Burl, D. DeCoste, B. L. Enke, D. Mazzoni, W. J. Merline, and L. Scharenbroich. Automated knowledge discovery from simulators. In *Proceedings of the Sixth SIAM International Conference on Data Mining*, pages 82–93, 2006. Available online at http://www.siam.org/proceedings/datamining/2006/dm06_008burlm.pdf

[75] C. S. Burrus, R. A. Gopinath, and H. Guo. *Introduction to Wavelets and Wavelet Transforms: A Primer*. Prentice–Hall, Upper Saddle River, NJ, 1998.

[76] P. J. Burt and E. H. Adelson. The Laplacian pyramid as a compact image code. *IEEE Transactions on Communications*, COM-31:532–540, 1983.

[77] Canada Center for Remote Sensing. Fundamentals of remote sensing tutorial, 2008. http://www.ccrs.nrcan.gc.ca/resource/tutor/fundam/index_e.php

[78] G. Candela and R. Chellappa. Comparative performance of classification methods for fingerprints. Technical Report NISTIR 5163, National Institute of Standards and Technology, Gaithersburg, MD, 1993.

[79] M. Cannon, P. Fasel, R. Fortson, and J. Hogden. FEEMADS: Extracting Features from Hydrocode Output. Technical Report LA-UR-01-6597, Los Alamos National Laboratory, Los Alamos, NM, 2001.

[80] J. Canny. A computational approach to edge detection. *IEEE Transactions on Pattern Analysis and Machine Intelligence*, 8:679–698, 1986.

[81] E. Cantú-Paz and C. Kamath. Using evolutionary algorithms to induce oblique decision trees. In *Proceedings of the Genetic and Evolutionary Computation Conference (GECCO)*, pages 1053–1060, 2000.

[82] E. Cantú-Paz and C. Kamath. An empirical comparison of combinations of evolutionary algorithms and neural networks for classification problems. *IEEE Transactions on Systems, Man, and Cybernetics—Part B: Cybernetics*, 35:915–927, 2005.

[83] E. Cantú-Paz, S. Newsam, and C. Kamath. Feature selection for scientific applications. In *Proceedings of the SIGKDD International Conference on Knowledge Discovery and Data Mining*, pages 788–793, 2004.

[84] A. S. Carasso. The APEX method in image sharpening and the use of low exponent Lévy stable laws. *SIAM Journal on Applied Mathematics*, 63:593–618, 2002.

[85] J. W. Carl. Contrasting approaches to combine evidence. In D. L. Hall and J. Llinas, editors, *Handbook of Multisensor Data Fusion*, pages 7-1–7-32. CRC Press, Boca Raton, FL, 2001.

[86] M. Á. Carriera-Perpiñán. A review of dimension reduction techniques. Technical Report CS-96-09, Department of Computer Science, University of Sheffield, UK, 1996.

[87] L. Carroll. *Complete Works of Lewis Carroll.* Vintage Books, New York, 1936. From "Through the Looking Glass."

[88] L. Carroll. *Complete Works of Lewis Carroll.* Vintage Books, New York, 1936. From the verse: "Poeta Fit, non Nascitur."

[89] R. Caruana and V. R. de Sa. Benefitting from the variables that variable selection discards. *Journal of Machine Learning Research*, 3:1245–1264, 2003.

[90] V. Caselles, R. Kimmel, and G. Sapiro. Geodesic active contours. In *Proceedings of the IEEE Fifth International Conference on Computer Vision*, pages 694–699, 1995.

[91] V. Castelli. Multidimensional indexing structures for content-based retrieval. In V. Castelli and D. Bergman, editors, *Image Databases: Search and Retrieval of Digital Imagery*, pages 373–433. John Wiley, New York, 2002.

[92] V. Castelli and D. Bergman, editors. *Image Databases: Search and Retrieval of Digital Imagery*. John Wiley, New York, 2002.

[93] A. Chambolle, R. A. DeVore, N. Lee, and B. J. Lucier. Nonlinear wavelet image processing: Variational problems, compression, and noise removal through wavelets. *IEEE Transactions on Image Processing*, 7:319–335, 1998.

[94] T. F. Chan and J. Shen. *Image Processing and Analysis: Variational, PDE, Wavelet, and Stochastic Methods*. SIAM, Philadelphia, 2005.

[95] V. Chandola, A. Banerjee, and V. Kumar. Outlier detection—a survey. Technical Report TR 07-017, Department of Computer Science, University of Minnesota, Minneapolis, MN, 2007.

[96] Chandra X-ray Observatory Web Page, 2008. http://chandra.harvard.edu.

[97] S. G. Chang, B. Yu, and M. Vetterli. Adaptive wavelet thresholding for image denoising and compression. *IEEE Transactions on Image Processing*, 9:1532–1546, 2000.

[98] S. G. Chang, B. Yu, and M. Vetterli. Spatially adaptive wavelet thresholding based on contect modeling for image denoising. *IEEE Transactions on Image Processing*, 9:1522–1531, 2000.

[99] S. G. Chang, B. Yu, and M. Vetterli. Wavelet thresholding for multiple noisy image copies. *IEEE Transactions on Image Processing*, 9:1631–1635, 2000.

[100] O. Chapelle, B. Schölkopf, and A. Zien, editors. *Semi-Supervised Learning*. MIT Press, Cambridge, MA, 2006.

[101] P. Charbonnier, L. Blanc-Feraud, G. Aubert, and M. Barlaud. Two deterministic half-quadratic regularization algorithms for computed imaging. In *Proceedings of the IEEE International Conference on Image Processing*, volume 2, pages 168–172, 1994.

[102] C. Chatfield. *The Analysis of Time Series*. Chapman and Hall/CRC, Boca Raton, FL, 2003.

[103] A. Chatterjee. An introduction to the proper orthogonal decomposition. *Current Science*, 78:808–817, 2000.

[104] N. Chawla, N. Japkowicz, and A. Kołcz. Editorial: Special issue on learning from imbalanced data sets. *SIGKDD Explorations*, 6:1–6, 2004.

[105] N. Chawla. C4.5 and imbalanced data sets: Investigating the effect of sampling method, probabilistic estimate, and decision tree structure. In *Proceedings of the Workshop on Learning from Imbalanced Datasets* II, 2003. Available online at http://www.site.uottawa.ca/~nat/Workshop2003/schedule.html.

[106] N. V. Chawla, K. W. Bowyer, L. O. Hall, and W. P. Kegelmeyer. SMOTE: Synthetic minority over-sampling technique. *Journal of Artificial Intelligence Research*, 16:321–357, 2002.

[107] C. Chen. *Information Visualization: Beyond the Horizon*. Springer, London, 2006.

[108] C. H. Chen. Pattern recognition in nondestructive evaluation of materials. In C. H. Chen, L. F. Pau, and P. S. P. Wang, editors, *Handbook of Pattern Recognition and Computer Vision*, pages 493–509. World Scientific, Singapore, 1993.

[109] V. Cherkassky and F. Mulier. *Learning from Data: Concepts, Theory, and Methods*. John Wiley, New York, 1998.

[110] K. Cherkauer. Human expert-level performance on a scientific image analysis task by a system using combined artificial neural networks. Technical Report, Working notes of the AAAI Workshop on Integrating Multiple Learned Models, 1996. http://www.cs.fit.edu/~pkc/imlm/imlm96/papers.html.

[111] S.-C. Cheung and C. Kamath. Robust techniques for background subtraction in urban traffic video. In *Proceedings of Video Communications and Image Processing, SPIE Electronic Imaging*, volume 5308, pages 881–892, 2004.

[112] S.-C. Cheung and C. Kamath. Robust background subtraction with foreground validation for urban traffic video. *Eurasip Journal on Applied Signal Processing*, 14:2330–2340, 2005.

[113] S. Choi, A. Cichocki, H.-M. Park, and S. Y. Lee. Blind source separation and independent component analysis. *Neural Information Processing—Letters and Reviews*, 6:1–57, 2005.

[114] F. Chung. *Spectral Graph Theory*. CBMS Regional Conference Series in Mathematics, Number 92, American Mathematical Society, Providence, RI, 1997.

[115] A. Cichocki and S. Amari. *Adaptive Blind Signal and Image Processing*. John Wiley, New York, 2002.

[116] CLUTO—family of data clustering software tools, 2008. http://glaros.dtc.umn.edu/gkhome/views/cluto.

[117] R. R. Coifman and D. L. Donoho. Translation-invariant de-noising. In A. Antoniadis and G. Oppenheim, editors, *Wavelets and Statistics*, Lecture Notes in Statistics, pages 125–150. Springer, New York, 1995.

[118] L. Čomić, L. De Floriani, and L. Papaleo. Morse-Smale decomposition for modelling terrain knowledge. In *Spatial Information Theory*, Lecture Notes in Computer Science, volume 3693, pages 426–444. Springer, New York, 2005.

[119] Compressive Sensing Resources Web page, 2008. http://www.dsp.ece.rice.edu/cs/.

[120] C. Cortes and V. Vapnik. Support vector networks. *Machine Learning*, 20:273–297, 1995.

[121] L. F. Costa and R. M Cesar. *Shape Analysis and Classification: Theory and Practice*. CRC Press, Boca Raton, FL, 2000.

[122] D. D. Cox and R. L. Savoy. Functional magnetic resonance imaging (fMRI) "brain reading": Detecting and classifying distributed patterns of fMRI activity in human visual cortex. *NeuroImage*, 19:261–270, 2003.

[123] T. F. Cox and M. A. A. Cox. *Multidimensional Scaling*. CRC Press, Boca Raton, FL, 2001.

[124] J. I. Cox and S. L. Hingorani. An efficient implementation of Reid's multiple hypothesis tracking algorithm and its evaluation for the purpose of visual tracking. *IEEE Transactions on Pattern Analysis and Machine Intelligence*, 18:138–150, 1996.

[125] S. B. Crary. Design of computer experiments for metamodel generation. *Analog Integrated Circuits and Signal Processing*, 32:7–16, 2002.

[126] N. Cristianini and J. Shawe-Taylor. *An Introduction to Support Vector Machines and Other Kernel-Based Learning Methods*. Cambridge University Press, Cambridge, UK, 2000.

[127] R. Cucchiara, M. Piccardi, and A. Prati. Detecting moving objects, ghosts, and shadows in video streams. *IEEE Transactions on Pattern Analysis and Machine Intelligence*, 25:1337–1342, 2003.

[128] R. Cutler and L. Davis. View-based detection. In *Proceedings of the Fourteenth International Conference on Pattern Recognition*, volume 1, pages 495–500, Brisbane, Australia, 1998.

[129] CVIPtools Web page: A software package for the exploration of computer vision and image processing, 2008. http://www.ee.siue.edu/CVIPtools.

[130] E. B. Dam. Evaluation of diffusion schemes for multi-scale watershed segmentation. Master's thesis, University of Copenhagen, Denmark, 2000.

[131] S. Das. Filters, wrappers and a boosting-based hybrid for feature selection. In C. Brodley and A. Danyluk, editors, *Proceedings of the 18th International Conference on Machine Learning*, pages 74–81. Morgan Kaufmann, San Francisco, CA, 2001.

[132] B. V. Dasarathy. *Decision Fusion*. IEEE Computer Society Press, Los Alamitos, CA, 1994.

[133] S. Dasgupta and A. Gupta. An elementary proof of the Johnson-Lindenstrauss lemma. Technical Report TR-99-006, University of California, Berkeley, 1999.

[134] I. Daubechies. *Ten Lectures on Wavelets*. SIAM, Philadelphia, PA, 1992.

[135] J. Daugman. Complete discrete 2D Gabor transform by neural networks for image analysis and compression. *IEEE Transactions on Acoustics, Speech, and Signal Processing*, 36:1169–1179, 1988.

[136] K. Deb. *Multi-objective Optimization Using Evolutionary Algorithms*. John Wiley, Chichester, England, 2001.

[137] S. Deerwester, S. T. Dumais, T. K. Landauer, G. W. Furnas, and R. A. Harshman. Indexing by latent semantic analysis. *Journal of the Society for Information Science*, 41:392–407, 1990.

[138] M. Deshpande, M. Kuramochi, and G. Karypis. Automated approaches for classifying structures. In *Proceedings of the BIOKDD02 Workshop on Data Mining in BioInformatics*, pages 11–18, 2002.

[139] M. Deshpande, M. Kuramochi, and G. Karypis. Mining chemical compounds. In J. Wang, M. J. Zaki, H. T. T. Toivonen, and D. Sasha, editors, *Data Mining in Bioinformatics*, pages 189–215. Springer, New York, 2004.

[140] R. A. DeVore and B. J. Lucier. Fast wavelet techniques for near-optimal processing. In *Proceedings of the IEEE Military Communications Conference*, pages 48.3.1–48.3.7, New York, 1992.

[141] T. Dietterich. Ensemble methods in machine learning. In *Proceedings of the First International Workshop on Multiple Classifier Systems*, pages 1–15. Springer-Verlag, New York, 2000.

[142] T. Dietterich. An experimental comparison of three methods for constructing ensembles of decision trees: Bagging, boosting, and randomization. *Machine Learning*, 40:139–158, 2000.

[143] T. Dietterich and G. Bakiri. Solving multiclass learning problems via error-correcting output codes. *Journal of Artificial Intelligence Research*, 2:263–286, 1995.

[144] T. G. Dietterich. Approximate statistical tests for comparing supervised classification learning algorithms. *Neural Computation*, 10:1895–1924, 1998.

[145] H. Digabel and C. Lantuejoul. Iterative algorithms. In J. L. Chermant, editor, *Proceedings of the Second European Symposium Quantitative Analysis of Microstructures in Material Science, Biology, and Medicine*, pages 86–99. Riederer Verlag, Stuttgart, 1978.

[146] Digital Globe, Inc., 2008. http://www.digitalglobe.com.

[147] Digital Persona, Inc., 2008. http://www.digitalpersona.com/.

[148] DIII-D National Fusion Facility, 2008. http://fusion.gat.com/global/DIII-D/.

[149] S. Dippel, M. Stahl, R. Wiemker, and T. Blaffert. Multiscale contrast enhancement for radiographies: Laplacian pyramid versus fast wavelet transform. *IEEE Transactions on Medical Imaging*, 21:343–353, 2002.

[150] C. Djeraba et al. Special issue on content-based multimedia indexing and retrieval. *IEEE Multimedia Magazine*, 9:18–60, 2002.

[151] P. M. Djuric et al. Particle filtering. *IEEE Signal Processing Magazine*, 20:19–38, 2003.

[152] A. Dobra and J. Gehrke. SECRET: A scalable linear regression tree algorithm. In *Proceedings of the Eighth ACM SIGKDD International Conference on Knowledge Discovery and Data Mining*, pages 481–487, 2001.

[153] P. Domingos. MetaCost: A general method for making classifiers cost-sensitive. In *Proceedings of the Fifth ACM SIGKDD International Conference on Knowledge Discovery and Data Mining*, pages 155–164, 1999.

[154] G. Dong, J. Li, and L. Wong. On the use of emerging patterns in the analysis of gene expression profiles for the diagnosis and understanding of diseases. In M. M. Kantardzic and J. Zurada, editors, *Next Generation of Data Mining Applications*, pages 331–353. Wiley Interscience, Hoboken, NJ, 2005.

[155] D. L. Donoho and I. M. Johnstone. Ideal spatial adaptation via wavelet shrinkage. *Biometrika*, 81:425–455, 1994.

[156] D. L. Donoho and I. M. Johnstone. Adapting to unknown smoothness via wavelet shrinkage. *Journal of the American Statistical Association*, 90:1200–1224, 1995.

[157] D. Donoho et al., Technical Reports, 2008. Available online at http://www-stat.stanford.edu/~donoho/Reports/index.html.

[158] E. R. Dougherty and R. A. Lotufo. *Hands-on Morphological Image Processing*. SPIE Press, Bellingham, WA, 2003.

[159] A. C. Doyle. The Adventure of the Reigate Squire. In *The Illustrated Sherlock Holmes Treasury*, pages 246–257. Avenel Books, New York, 1984.

[160] A. C. Doyle. A Scandal in Bohemia. In *The Illustrated Sherlock Holmes Treasury*, pages 1–15. Avenel Books, New York, 1984.

[161] C. Drummond and R. C. Holte. C4.5, class imbalance, and cost sensitivity: Why under-sampling beats over-sampling. In *Proceedings of the Workshop on Learning from Imbalanced Datasets*, 2003. http://www.site.uottawa.ca/~nat/Workshop2003/schedule.html.

[162] I. L. Dryden and K. V. Mardia. *Statistical Shape Analysis*. John Wiley, Chichester, England, 1998.

[163] W. Duch. Filter methods. In I. Guyon, S. Gunn, M. Nikravesh, and L. Zadeh, editors, *Feature Extraction: Foundations, and Applications*, pages 89–118. Springer-Verlag, Berlin, 2006.

[164] R. O. Duda, P. E. Hart, and D. G. Stork. *Pattern Classification*. John Wiley, New York, 2001.

[165] G. Durrell. *Two in the Bush*. William Collins Sons & Co. Ltd., Glasgow, 1966.

[166] J. G. Dy and C. E. Brodley. Feature selection for unsupervised learning. *Journal of Machine Learning Research*, 5:845–889, 2004.

[167] Earth Observing System (EOS) Data Products Web page, 2008. http://eospso.gsfc.nasa.gov/index.php.

[168] R. D. Eastman, J. LeMoigne, and N. S. Netanyahu. Research issues in image registration for remote sensing, 2007. *CVPR Workshop on Image Registration and Fusion*, June 23, 2007, Minneapolis, MN.

[169] A. Elgammal, D. Harwood, and L. Davis. Non-parametric model for background subtraction. In *Proceedings of IEEE ICCV '99 Frame-rate Workshop*, 1999.

[170] C. Elkan. The foundations of cost-sensitive learning. In *Proceedings of the XVII International Joint Conference on Artificial Intelligence*, pages 973–978, 2001.

[171] EOS Web page: NASA's Earth Observing System Web page, 2008. http://eospso.gsfc.nasa.gov/.

[172] L. Ertöz, E. Eilertson, A. Lazarevic, P.-N. Tan, V. Kumar, J. Srivastava, and P. Dokas. MINDS—Minnesota Intrusion Detection System. In H. Kargupta, A. Joshi, K. Sivakumar, and Y. Yesha, editors, *Next Generation Data Mining*, pages 199–218. MIT Press, Cambridge, MA, 2004.

[173] European Virtual Observatory Web page, 2008. http://www.euro-vo.org/pub.

[174] Faint Images of the Radio Sky at Twenty Centimeters (FIRST) Web page, 2008. http://sundog.stsci.edu.

[175] A. X. Falcao, P. Miranda, A. Rocha, and F. Bergo. Object detection by k-connected seed competition. In *Proceedings of the XVII Brazilian Symposium on Computer Graphics and Image Processing*, pages 97–104, 2005.

[176] C. Faloutsos. *Searching Multimedia Databases by Content.* Kluwer Academic Publishers, Norwell, MA, 1996.

[177] C. Faloutsos. Multimedia IR: Indexing and searching. In R. Baeza-Yates and B. Ribeiro-Neto, editors, *Modern Information Retrieval*, pages 345–365. Addison-Wesley, Harlow, England, 1999.

[178] C. Faloutsos. Multimedia indexing. In V. Castelli and D. Bergman, editors, *Image Databases: Search and Retrieval of Digital Imagery*, pages 435–464. John Wiley, New York, 2002.

[179] C. Faloutsos and K.-I. Lin. Fastmap: A fast algorithm for indexing, data-mining, and visualization of traditional and multimedia datasets. In *Proceedings of the 1995 ACM SIGMOD International Conference on the Management of Data*, pages 163–174, 1995.

[180] K.-T. Fang, R. Li, and A. Sudjianto. *Design and Modeling for Computer Experiments.* Chapman and Hall/CRC Press, Boca Raton, FL, 2005.

[181] M. Farge, N. Kevlahan, V. Perrier, and E. Goirand. Wavelets and turbulence. *Proceedings of the IEEE*, 84:639–669, 1996.

[182] M. Farge and K. Schneider. Analyzing and compressing turbulent fields with wavelets. Technical Report, Institut Pierre Simon de Laplace, 2002. http://euler.lmd.polytechnique.fr/nai/nai_20.pdf.

[183] O. Faugeras and W.-T. Luong. *The Geometry of Multiple Images.* MIT Press, Cambridge, MA, 2001.

[184] U. Fayyad, G. Djorgovski, and N. Weir. Automating the analysis and cataloging of sky surveys. In *Advances in Knowledge Discovery and Data Mining*, pages 471–493. MIT Press, Cambridge, MA, 1996.

[185] U. Fayyad, G. Piatetsky-Shapiro, and P. Smyth. From data mining to knowledge discovery: An overview. In *Advances in Knowledge Discovery and Data Mining*, pages 1–34. MIT Press, Cambridge, MA, 1996.

[186] U. Fayyad, G. Piatetsky-Shapiro, and P. Smyth. The KDD process for extracting useful knowledge from volumes of data. *Communications of the ACM, Special Issue on Data Mining*, 39:27–34, 1996.

[187] U. Fayyad and P. Smyth. From massive data sets to science catalogs: Applications and challenges. In J. Kettenring and D. Pregibon, editors, *Proceedings of the Workshop on Massive Data Sets*, pages 129–142, National Research Council, Washington, DC, 1995.

[188] J. Ferre-Gine, R. Rallo, A. Arenas, and F. Giralt. Identification of coherent structures in turbulent shear flows with a fuzzy ARTmap neural network. *International Journal of Neural Systems*, 7:559–568, 1996.

[189] R. P. Feynmann. Cargo cult science: Some remarks on science, pseudoscience, and learning how to not fool yourself. In J. Robbins, editor, *The Pleasure of Finding Things Out: The Best Short Works of Richard P. Feynmann*, pages 205–216. Perseus Books, Cambridge, MA, 1999.

[190] R. P. Feynmann. The value of science. In J. Robbins, editor, *The Pleasure of Finding Things Out: The Best Short Works of Richard P. Feynmann*, pages 141–149. Perseus Books, Cambridge, MA, 1999.

[191] FIRST Validation Tool: Information Visualization, Sapphire Web page, 2008. https://computation.llnl.gov/casc/sapphire/infovis/infovis.html.

[192] B. Fishbine. Code validation experiments. *Los Alamos Research Quarterly*, (Fall 2002):6–14, 2002. http://www.lanl.gov/quarterly/q_fall02/.

[193] R. A. Fisher. The use of multiple measurements in taxonomic problems. *Annals of Eugenics*, 7:179–188, 1936.

[194] FITS: Flexible Image Transport System Web page and archive, 2008. http://heasarc.nasa.gov/docs/heasarc/fits.html and http://www.cv.nrao.edu/fits/FITS.html.

[195] J. M. Fitzpatrick, D. L. G. Hill, and C. R. Maurer. Image registration. In M. Sonka and J. M. Fitzpatrick, editors, *Handbook of Medical Imaging, Volume 2: Medical Image Processing and Analysis*, pages 447–513. SPIE Press, Bellingham, WA, 2000.

[196] I. K. Fodor, E. Cantú-Paz, C. Kamath, and N. Tang. Finding bent-double radio galaxies: A case study in data mining. In *Interface: Computer Science and Statistics*, volume 32, 2000.

[197] I. K. Fodor and C. Kamath. A comparison of denoising techniques for FIRST images. In *Proceedings of the Third Workshop on Mining Scientific Data Sets*, pages 13–20, 2001. In conjunction with SIAM Data Mining 2001. See https://computation.llnl.gov/casc/sapphire/pubs.html.

[198] I. K. Fodor and C. Kamath. Denoising through wavelet shrinkage: An empirical study. *SPIE Journal on Electronic Imaging*, 12:151–160, 2003.

[199] I. K. Fodor and C. Kamath. Using independent component analysis to separate signals in climate data. In *Proceedings of the Independent Component Analyses, Wavelets, and Neural Networks, SPIE Proceedings, Volume 5102*, pages 25–36. SPIE Press, Bellingham, WA, 2003.

[200] I. K. Fodor and C. Kamath. Dimension reduction techniques and the classification of bent-double galaxies. *Computational Statistics and Data Analysis*, 41:91–122, 2002.

[201] D. Forsyth, J. Malik, and R. Wilensky. Searching for digital pictures. *Scientific American*, 276:88–93, 1997.

[202] Y. Freund and R. Schapire. Experiments with a new boosting algorithm. In *Machine Learning: Proceedings of the Thirteenth International Conference*, pages 148–156, 1996.

[203] J. H. Friedman. Multivariate adaptive regression splines. *Annals of Statistics*, 19:1–67, 1991.

[204] J. H. Friedman and W. Stuetzle. Projection pursuit regression. *Journal of the American Statistical Association*, 76:817–823, 1981.

[205] J. H. Friedman and J. W. Tukey. A projection pursuit algorithm for exploratory data analysis. *IEEE Transactions on Computers*, C-23:881–890, 1974.

[206] N. Friedman and S. Russell. Image segmentation in video sequences: A probabilistic approach. In *Proceedings of the Thirteenth Annual Conference on Uncertainty in Artificial Intelligence (UAI–97)*, pages 175–181, Morgan Kaufmann, San Francisco, 1997.

[207] N. T. Frink. Assessment of an unstructured-grid method for predicting 3-d turbulent viscous flows. In *Proceedings of the AIAA 34th Aerospace Sciences Meeting*, 1996. http://www.tpub.com/content/nasa1996/NASA-aiaa-96-0292/index.htm.

[208] Fv Web page: The Interactive FITS File Editor, 2008. http://heasarc.gsfc.nasa.gov/lheasoft/ftools/fv/.

[209] V. Gaede and O. Günther. Multidimensional access methods. *ACM Computing Surveys*, 30:170–231, 1998.

[210] J. Gama, R. Fernandes, and R. Rocha. Decision trees for mining data streams. *Intelligent Data Analysis*, 10:23–45, 2006.

[211] X. Gao, T. E. Boult, F. Coetzee, and V. Ramesh. Error analysis of background adaption. In *Proceedings of the IEEE Conference on Computer Vision and Pattern Recognition*, volume 1, pages 503–510, Hilton Head Isand, SC, June 2000.

[212] GeoEye Web page, 2008. http://www.geoeye.com.

[213] A. George and J. W.-H. Liu. *Computer Solutions of Large Sparse Positive Definite Systems*. Prentice Hall, Englewood Cliffs, NJ, 1981.

[214] A. Gersho and R. M. Gray. *Vector Quantization and Signal Compression*. Kluwer Academic Publishers, Boston, MA, 1992.

[215] A. Gezahegne and C. Kamath. Tracking non-rigid structures in computer simulations. In *Proceedings of the 15th IEEE International Conference on Image Processing*, 2008, pp. 1548–1551.

[216] GGobi: Interactive and dynamic graphics, 2008. http://www.ggobi.org/.

[217] J. Ghosh. Scalable clustering. In N. Ye, editor, *The Handbook of Data Mining*, pages 247–278. Lawrence Erlbaum Associates, Mahwah, NJ, 2003.

[218] S. Gibson, R. Harvey, and J. A. Bangham. Multi-dimensional histogram comparison via scale trees. In *Proceedings of the International Conference on Image Processing*, volume 2, pages 709–712, 2001.

[219] M. L. Giger, Z. Huo, M. A. Kupinski, and C. J. Vyborny. Computer-aided diagnosis in mammography. In M. Sonka and J. M. Fitzpatrick, editors, *Handbook of Medical Imaging, Volume 2: Medical Image Processing and Analysis*, pages 915–1004. SPIE Press, Bellingham, WA, 2000.

[220] S. Gilles. *Robust Matching and Description of Images*. PhD thesis, Oxford University, Oxford, UK, 1998.

[221] GIMP Web page: GNU Image Manipulation Program, 2008. http://www.gimp.org/.

[222] B. Gloyer, H. K. Aghajan, K.-Y. Siu, and T. Kailath. Video-based freeway monitoring system using recursive vehicle tracking. In *Proceedings of SPIE*, volume 2421, pages 173–180, 1995.

[223] GMV Web page: The General Mesh Viewer, 2008. http://www-xdiv.lanl.gov/XCM/gmv/GMVHome.html.

[224] Gnuplot Plotting Utility, 2008. http://www.gnuplot.info/.

[225] D. E. Goldberg. *Genetic Algorithms in Search, Optimization, and Machine Learning*. Addison-Wesley, Reading, MA, 1989.

[226] J. Gomes and O. Faugeras. Reconciling distance functions and level sets. *Journal of Visual Communication and Image Representation*, 11:209–223, 2000.

[227] R. C. Gonzalez and R. E. Woods. *Digital Image Processing*. Addison-Wesley, Reading, MA, 1993.

[228] R. Goodwin, R. Miller, E. Tuv, A. Borisov, M. Janakiram, and S. Louchheim. Advancements and applications of statistical learning/data mining in semiconductor manufacturing. *Intel Technology Journal*, 8:1826–1834, 2004.

[229] A. A. Goshtasby. *2-D and 3-D Image Registration*. John Wiley, Hoboken, NJ, 2005.

[230] A. Graps. An introduction to wavelets. *IEEE Computational Science and Engineering*, 2:50–61, 1995.

[231] GRIB: GRIdded Binary Web page, 2008. http://www.wmo.ch/pages/prog/www/WDM/Guides/Guide-binary-2.html.

[232] D. Griffith. Statistical and mathematical sources of regional science theory: Map pattern analysis as an example. *Papers in Regional Science*, 78:21–45, 1999.

[233] J. Gross and J. Yellen. *Graph Theory and Its Applications*. CRC Press, Boca Raton, FL, 1998.

[234] M. Gu and S. C. Eisenstat. A stable and fast algorithm for updating the singular value decomposition. Technical Report YALEU/DCS/RR-966, Yale University, New Haven, CT, 1993.

[235] A. H. Gunatilaka and B. A. Baertlein. Feature-level and decision-level fusion of noncoincidentally sampled sensors for land mine detection. *IEEE Transactions on Pattern Analysis and Machine Intelligence*, 23:577–589, 2001.

[236] I. Guyon and A. Elisseeff. An introduction to variable and feature selection. *Journal of Machine Learning Research*, 3:1157–1182, 2003.

[237] E. Gyftodimos, L. Moss, D. Sleeman, and A. Welch. Analysing PET scans data for predicting response to chemotherapy in breast cancer patients. In *Applications and Innovations in Intelligent Systems XV, Proceedings of AI-2007, the Twenty-seventh SGAI International Conference on Innovative Techniques and Applications of Artificial Intelligence*, pages 59–72, 2007.

[238] J. V. Hajnal, D. L. G. Hill, and D. J. Hawkes, editors. *Medical Image Registration*. CRC Press, London, 2001.

[239] G. Halevy and D. Weinshall. Motion of disturbances: Detection and tracking of multi-body non-rigid motion. *Machine Vision and Applications*, 11:122–137, 1999.

[240] M. Halkidi, Y. Batistakis, and M. Vazirgiannis. On clustering validation techniques. *Journal of Intelligent Information Systems*, 17:107–145, 2001.

[241] D. L. Hall and J. Llinas. An introduction to multisensor data fusion. *Proceedings of the IEEE*, 85:6–23, 1997.

[242] D. L. Hall and J. Llinas. *Handbook of Multisensor Data Fusion*. CRC Press, Boca Raton, FL, 2001.

[243] D. L. Hall and S. A. H. McMullen. *Mathematical Techniques in Multisensor Data Fusion*. Artech House Publishers, Norwood, MA, 2004.

[244] L. Hall, K. Bowyer, W. Kegelmeyer, T. Moore, and C. Chao. Distributed learning on very large data sets. In *Workshop on Distributed and Parallel Knowledge Discovery, in conjunction with KDD2000*, 2000.

[245] M. A. Hall. *Correlation-Based Feature Subset Selection for Machine Learning*. PhD thesis, University of Waikato, New Zealand, 1999.

[246] M. A. Hall. Correlation-based feature selection for discrete and numeric class machine learning. In *Proceedings of the 17th International Conference on Machine Learning*, pages 359–366. Morgan Kaufmann, San Francisco, CA, 2000.

[247] R. W. Hamming. *Numerical Methods for Scientists and Engineers*. McGraw–Hill, New York, 1973.

[248] R. W. Hamming. *The Art of Doing Science and Engineering: Learning to Learn*. Gordon and Breach Science Publishers, Amsterdam, The Netherlands, 1997.

[249] E.-H. Han, G. Karypis, V. Kumar, and B. Mobasher. Hypergraph based clustering in high dimensional data sets: A summary of results. *Bulletin of the IEEE Technical Committee on Data Engineering*, 21:15–22, 1998.

[250] J. Han, M. Kamber, and A. K. H. Tung. Spatial clustering methods in data mining: A survey. In H. Miller and J. Han, editors, *Geographic Data Mining and Knowledge Discovery*, pages 188–217. Taylor and Francis, New York, 2001.

[251] S. Hanash. Disease proteomics. *Nature*, 422:226–232, 2003.

[252] D. Hand, H. Mannila, and P. Smyth. *Principles of Data Mining*. MIT Press, Cambridge, MA, 2001.

[253] H. Hangan, G. A. Kopp, A. Vernet, and R. Martinuzzi. A wavelet pattern recognition technique for identifying flow structures in cylinder generated wakes. *Journal of Wind Engineering and Industrial Aerodynamics*, 89:1001–1015, 2001.

[254] C. D. Hansen and C. R. Johnson, editors. *Visualization Handbook*. Elsevier, Burlington, MA, 2005.

[255] R. M. Haralick, K. Shanmugam, and I. Dinstein. Textural features for image classification. *IEEE Transactions on Systems, Man, and Cybernetics*, 3:610–621, 1973.

[256] R. M. Haralick and L. G. Shapiro. Image segmentation techniques. *Computer Vision, Graphics, and Image Processing*, 29:100–132, 1985.

[257] C. Harris and M. Stephens. A combined corner and edge detector. In *Proceedings of the Fourth Alvey Vision Conference*, pages 147–151, 1988.

[258] R. Hartley and A. Zisserman. *Multiple View Geometry in Computer Vision*. Cambridge University Press, Cambridge, UK, 2003.

[259] T. Hastie. Principal curves and surfaces. Technical Report SLAC-0276, Stanford Linear Accelerator Center, 1984.

[260] T. Hastie and W. Stuetzle. Principal curves. *Journal of the American Statistical Association*, 84:502–516, 1989.

[261] T. Hastie, R. Tibshirani, and J. Friedman. *The Elements of Statistical Learning: Data Mining, Inference, and Prediction*. Springer, New York, 2001.

[262] S. Haykin. *Neural Networks: A Comprehensive Foundation*. Prentice–Hall, Englewood Cliffs, NJ, 1999.

[263] HDF: Hierarchical Data Format Web page, 2008. http://www.hdfgroup.org.

[264] M. D. Heath, S. Sarkar, T. Sanocki, and K. W. Bowyer. A robust visual method for assessing the relative performance of edge-detection algorithms. *IEEE Transactions on Pattern Analysis and Machine Intelligence*, 19:1338–1359, 1997.

[265] M. T. Heath. *Scientific Computing: An Introductory Survey*. McGraw–Hill, New York, 2002.

[266] J. Heikkila and O. Silven. A real-time system for monitoring of cyclists and pedestrians. In *Second IEEE Workshop on Visual Surveillance*, pages 246–252, Fort Collins, CO, 1999.

[267] P. Hein. *Grooks I*. Doubleday Publishing, New York, 1969.

[268] J. H. Heinbockel. *Numerical Methods for Scientific Computing*. Trafford Publishing, Victoria, BC, Canada, 2006.

[269] M. Hemmer and J. Gasteiger. Data mining in chemistry, 2000. Terena Networking Conference, 2000. http://www.terena.nl/tnc2000/proceedings/10B/10b5.html.

[270] F. M. Henderson and Z. G. Xia. SAR applications in human settlement detection, population estimation, and urban land-use pattern analysis: A status report. *IEEE Transactions on Geoscience and Remote Sensing*, 35:79–85, 1997.

[271] B. Hendrickson and R. Leland. An improved spectral graph partitioning algorithm for mapping parallel computations. *SIAM Journal on Scientific Computing*, 16:452–469, 1995.

[272] S. Hettich and S. D. Bay. The UCI KDD archive, 2005. Department of Information and Computer Science, University of California, Irvine. http://kdd.ics.uci.edu.

[273] D. Higdon, M. Kennedy, J. C. Cavendish, J. A. Cafeo, and R. D. Ryne. Combining field data and computer simulations for calibration and prediction. *SIAM Journal on Scientific Computing*, 26:448–466, 2004.

[274] C. J. Hilditch. Linear skeletons from square cupboards. In B. Meltzer and D. Mitchie, editors, *Machine Intelligence*, pages 404–420. University Press, Edinburgh, 1969.

[275] T. K. Ho. Random decision forests. In *Proceedings of the 3rd International Conference on Document Analysis and Recognition*, pages 278–282, 1995.

[276] W. Hobbs. A Relational database management system, 1993. ftp://ftp.rand.org/pub/RDB-hobbs/.

[277] B. K. P. Horn and B. G. Schunk. Determining optical flow. *Artifical Intelligence*, 17:185–203, 1981.

[278] H. Hotelling. Analysis of a complex of statistical variables into principal components. *Journal of Educational Psychology*, 24:417–441, 1933.

[279] M. Hu. Visual pattern recognition by moment invariants. *IRE Transactions on Information Theory*, 2:179–187, 1962.

[280] B. B. Hubbard. *The World According to Wavelets: The Story of a Mathematical Technique in the Making*. A. K. Peters, Wellesley, MA, 1998.

[281] The Hubble space telescope Web page, 2008. http://hubble.nasa.gov/index.php.

[282] P. J. Huber. Projection pursuit. *The Annals of Statistics*, 13:435–475, 1985.

[283] T. J. R. Hughes. *The Finite Element Method: Linear Static and Dynamic Finite Element Analysis*. Dover, Mineola, NY, 2000.

[284] R. Hummel. Image enhancement by histogram transformation. *Computer Vision, Graphics, and Image Processing*, 6:184–195, 1977.

[285] A. Hyvärinen. Survey on independent component analysis. *Neural Computing Surveys*, 2:94–128, 1999.

[286] A. Hyvärinen, J. Karhunen, and E. Oja. *Independent Component Analysis*. John Wiley, New York, 2001.

[287] Image Web page: A public-domain image processing and analysis program, 2008. http://rsb.info.nih.gov/nih-image/.

[288] ImageChecker: CAD for Mammography, R2 Technology, 2008. http://www.r2tech.com/mammography/home/index.php.

[289] ImageMagick Web page, 2008. http://www.imagemagick.org/.

[290] P. Indyk, N. Koudas, and S. Muthukrishnan. Identifying representative trends in massive time series data sets using sketches. In *Proceedings of the Conference on Very Large Data Bases*, pages 363–372, 2000.

[291] Information Technology and Systems Center; University of Alabama at Huntsville, 2008. http://www.itsc.uah.edu/.

[292] A. Inselberg. Visualization and knowledge discovery for high dimensional data. In *Proceedings of the Second International Workshop on User Interfaces to Data Intensive Systems*, pages 5–24, 2001.

[293] A. Inselberg and B. Dimsdale. Parallel coordinates: A tool for visualizing multidimensional geometry. In *Proceedings of the First IEEE Conference on Visualization*, pages 361–378, 1990.

[294] Institute for Mathematics Applied to Geosciences, 2008. http://www.image.ucar.edu/.

[295] V. S. Iyengar, C. Apte, and T. Zhang. Active learning using adaptive resampling. In *Sixth ACM SIGKDD International Conference on Knowledge Discovery and Data Mining*, pages 92–98, 2000.

[296] J. E. Jackson. *A User's Guide to Principal Components.* John Wiley, New York, 1991.

[297] A. K. Jain. *Fundamentals of Digital Image Processing.* Prentice–Hall, Englewood Cliffs, NJ, 1989.

[298] A. K. Jain and R. C. Dubes. *Algorithms for Clustering Data.* Prentice–Hall, Englewood Cliffs, NJ, 1988.

[299] A. K. Jain, M. N. Murty, and P. J. Flynn. Data clustering: A review. *ACM Computing Surveys*, 31:264–323, 1999.

[300] A. K. Jain and A. Ross. Multibiometric systems. *Communications of the ACM*, 47:34–40, 2004.

[301] M. Jansen. *Noise Reduction by Wavelet Thresholding.* Springer, New York, 2001.

[302] N. Japkowicz. Learning from imbalanced data sets: A comparison of various strategies. In *Proceedings of the AAAI Workshop on Learning from Imbalanced Data Sets*, pages 10–15, 2000.

[303] N. Japkowicz and S. Stephen. The class imbalance problem: A systematic study. *Intelligent Data Analysis*, 6:429–449, 2002.

[304] J. Jarvis and J. Tyson. FOCAS: Faint object classification and analysis system. *The Astronomical Journal*, 86:476–495, 1981.

[305] D. J. Jobson, Z. Rahman, and G. A. Woodell. A multi-scale Retinex for bridging the gap between color images and the human observations of scenes. *IEEE Transactions on Image Processing*, 6:965–976, 1997.

[306] D. J. Jobson, Z. Rahman, and G. A. Woodell. Properties and performance of a center/surround Retinex. *IEEE Transactions on Image Processing*, 6:451–462, March 1997.

[307] G. H. John, R. Kohavi, and K. Pfleger. Irrelevant features and the subset selection problem. In *International Conference on Machine Learning*, pages 121–129, 1994.

[308] I. T. Joliffe. Discarding variables in a principal component analysis. I: Artificial data. *Applied Statistics*, 21:160–173, 1972.

[309] I. T. Joliffe. Discarding variables in a principal component analysis. II: Real data. *Applied Statistics*, 22:21–31, 1973.

[310] I. T. Joliffe. *Principal Component Analysis.* Springer, New York, 2002.

[311] T. Kadir and M. Brady. Saliency, scale and image description. *International Journal of Computer Vision*, 45:83–105, 2001.

[312] P. KaewTraKulPong and R. Bowden. An improved adaptive background mixture model for real-time tracking with shadow detection. In *Proceedings of the 2nd European Workshop on Advanced Video-Based Surveillance Systems*, 2001.

[313] C. Kamath and E. Cantú-Paz. Creating ensembles of decision trees through sampling. In *Proceedings of the 33rd Symposium on the Interface: Computing Science and Statistics*, 2001.

[314] C. Kamath, E. Cantú-Paz, S. C. Cheung, I. K. Fodor, and N. Tang. Experiences in mining data from computer simulations. In M. Kantardzic and J. Zurada, editors, *Next Generation of Data Mining Applications*, pages 211–232, John Wiley, New York, 2005.

[315] C. Kamath, E. Cantú-Paz, I. K. Fodor, and N. Tang. Searching for bent-double galaxies in the FIRST survey. In R. Grossman, C. Kamath, W. Kegelmeyer, V. Kumar, and R. Namburu, editors, *Data Mining for Scientific and Engineering Applications*, pages 95–114. Kluwer, Boston, MA, 2001.

[316] C. Kamath, E. Cantu-Paz, I. K. Fodor, and N. Tang. Classification of bent-double galaxies in the FIRST survey. *IEEE Computing in Science and Engineering*, 4:52–60, 2002.

[317] C. Kamath, E. Cantú-Paz, and D. Littau. Approximate splitting for ensembles of trees using histograms. In *Proceedings of the Second SIAM International Conference on Data Mining*, pages 370–383, 2002. Available online at http://www.siam.org/meetings/sdm02/proceedings/sdm02-22.pdf.

[318] C. Kamath, A. Gezahegne, and P. L. Miller. Analysis of Rayleigh-Taylor instability, Part I: Bubble and spike count. Technical Report UCRL-TR-223676, Lawrence Livermore National Laboratory, 2006. http://www.llnl.gov/casc/sapphire/pubs/TR-223676.pdf.

[319] C. Kamath, A. Gezahegne, and P. L. Miller. Analysis of Rayleigh-Taylor instability: Statistics on rising bubbles and falling spikes. Technical Report UCRL-TR-236111-REV-1, Lawrence Livermore National Laboratory, 2007. http://www.llnl.gov/casc/sapphire/pubs/UCRL-TR-236111-REV-1.pdf.

[320] C. Kamath, A. Gezahegne, S. Newsam, and G. M. Roberts. Salient points for tracking moving objects in video. In *Image and Video Communications and Processing, Proceedings of SPIE, Volume 5685*, pages 442–453, 2005.

[321] C. Kamath and P. L. Miller. Image analysis for validation of simulations of a fluid-mix problem. In *IEEE International Conference on Image Processing, Volume III*, pages 525–528, 2007.

[322] C. Kamath and T. Nguyen. Feature extraction from simulations and experiments: preliminary results using a fluid mix problem. Technical Report UCRL-TR-208853, Lawrence Livermore National Laboratory, California, 2005.

[323] C. Kamath, S. K. Sengupta, D. N. Poland, and J. A. H. Futterman. Use of machine vision techniques to detect human settlements in satellite images. In *Image Processing: Algorithms and Systems II, Proceedings of SPIE/IS&T, Volume 5014*, pages 270–280, 2003.

[324] A. Karalic. Linear regression in regression tree leaves. In *Proceedings of the 1992 International School for Synthesis of Expert Knowledge (ISSEK)*, 1992.

[325] K.-P. Karmann and A. von Brandt. Moving object recognition using an adaptive background memory. In V. Cappellini, editor, *Time-Varying Image Processing and Moving Object Recognition*, volume 2, pages 289–307. Elsevier Science Publishers B.V., Amsterdam, The Netherlands, 1990.

[326] G. Karypis and E.-H. Han. Concept indexing: A fast dimensionality reduction algorithm with applications to document retrieval and categorization. Technical Report 00-016, University of Minnesota, Minneapolis, MN, 2000.

[327] G. Karypis, E.-H. Han, and V. Kumar. Chameleon: Hierarchical clustering using dynamic modeling. *Computer*, 32:68–75, 1999.

[328] G. Karypis and V. Kumar. Multilevel *k*-way partitioning scheme for irregular graphs. *Journal of Parallel and Distributed Computing*, 48:96–129, 1998.

[329] G. Karypis and V. Kumar. A fast and high quality multilevel scheme for partitioning irregular graphs. *SIAM Journal on Scientific Computing*, 20:359–392, 1998.

[330] S. Kaski. Dimensionality reduction by random mapping: Fast similarity computation for clustering. In *Proceedings of the IEEE International Joint Conference on Neural Networks*, pages 413–418, 1998.

[331] M. Kass, A. Witkin, and D. Terzopoulos. Snakes: Active contour models. *International Journal of Computer Vision*, 1:321–331, 1987.

[332] L. Kaufman and P. J. Rousseeuw. *Finding Groups in Data: An Introduction to Cluster Analysis*. John Wiley, New York, 1990.

[333] W. P. Kegelmeyer, B. Groshong, M. Allman, and K. Woods. Decision trees and integrated features for computer-aided mammographic screening. Technical Report SAND9708233, Sandia National Laboratory, California, 1997.

[334] W. P. Kegelmeyer, J. M. Pruneda, P. D. Bourland, A. Hillis, M. W. Riggs, and M. Nipper. Computer-aided mammographic screening for spiculated lesions. *Radiology*, 191:331–337, 1994.

[335] E. Keogh, K. Chakrabarti, S. Mehrotra, and M. Pazzani. Locally adaptive dimensionality reduction for indexing large time series databases. In *Proceedings of the ACM SIGMOD Conference on Management of Data*, pages 151–162, 2001.

[336] E. Keogh, K. Chakrabarti, M. Pazzani, and S. Mehrotra. Dimensionality reduction for fast similarity search in large time series databases. *Knowledge and Information Systems*, 3:263–286, 2000.

[337] E. Keogh, Time series tutorials, 2008. http://www.cs.ucr.edu/~eamonn/tutorials.html.

[338] J. R. Kettenring. The practice of cluster analysis. *Journal of Classification*, 23:3–30, 2006.

[339] S. Kichenassamy, A. Kumar, P. Olver, A. Tannenbaum, and A. Yezzi. Gradient flows and geometric active contour models. In *Proceedings of the IEEE Fifth International Conference on Computer Vision*, pages 694–699, 1995.

[340] K. Kira and L. Rendell. A practical approach to feature selection. In *Proceedings of the Ninth International Conference on Machine Learning*, pages 249–256, 1992.

[341] M. Kirby and L. Sirovich. Application of the Karhunen-Loeve procedure for the characterization of human faces. *IEEE Transactions on Pattern Analysis and Machine Intelligence*, 12:103–108, 1990.

[342] A. Kitamoto. Data mining for typhoon image collection. In *Proceedings of the Second International Workshop on Multimedia Data Mining, in conjunction with the ACM SIGKDD Conference*, pages 68–77. ACM, New York, 2001.

[343] R. Klinkenberg. Learning drifting concepts: Example selection vs. example weighting. *Intelligent Data Analysis*, 8:281–300, 2004.

[344] J. R. Koehler and A. B. Owen. Computer experiments. In S. Ghosh and C. R. Rao, editors, *Handbook of Statistics*, pages 261–308. Elsevier Science, New York, 1996.

[345] R. Kohavi and G. John. Wrappers for feature subset selection. *Artificial Intelligence*, 97:273–324, 1997.

[346] T. Kohonen. *Self-Organizing Maps*. Springer, Berlin, 2001.

[347] A. Kokaram. Detection and removal of line scratches in degraded motion picture sequences. In *Proceedings of the Signal Processing VIII*, volume 2, pages 5–8, 1996.

[348] A. Kokaram. *Motion Picture Restoration: Digital Algorithms for Artefact Suppression in Degraded Motion Picture Film and Video*. Springer, New York, 1998.

[349] I. Kononenko. Estimating attributes: Analysis and extensions of RELIEF. In *European Conference on Machine Learning*, pages 171–182, 1994.

[350] G. Korte. *The GIS Book*. OnWord Press, Clifton Park, New York, 2000.

[351] Z. Kou, W. W. Cohen, and R. F. Murphy. Extracting information from text and images for location proteomics. In *Proceedings of the BIOKDD03 Workshop on Data Mining in Bio-Informatics*, pages 2–9. ACM, 2003.

[352] M. Kubat, R. C. Holte, and S. Matwin. Machine learning for the detection of oil spills in satellite radar images. *Machine Learning*, 30:195–215, 1998.

[353] M. Kubat and S. Matwin. Addressing the curse of imbalanced training sets: One-sided selection. In *Proceedings of the XIV International Conference on Machine Learning*, pages 179–186, 1997.

[354] V. Kumar, A. Grama, A. Gupta, and G. Karypis. *Introduction to Parallel Computing: Design and Analysis of Algorithms*. Benjamin/Cummings Publishing Company, Redwood City, CA, 1994.

[355] T. Kurosu, M. Fujita, and K. Chiba. Monitoring of rice crop growth from space using the ERS-1 C-band SAR. *IEEE Transactions on Geoscience and Remote Sensing*, 33:1092–1096, 1995.

[356] A. Laine, J. Fan, and W. Yang. Wavelets for contrast enhancement of digital mammography. *IEEE Engineering in Medicine and Biology Magazine*, 14:536–550, 1995.

[357] E. Land. An alternative technique for the computation of the designator in the retinex theory of color vision. *Proceedings of the National Academy of Science*, 83:3078–3080, 1986.

[358] E. S. Lander et al. Initial sequencing and analysis of the human genome. *Nature*, 409:860–921, 2001.

[359] L. J. Latecki and R. Lakämper. Shape description and search for similar objects in image databases. In R. C. Veltkamp, H. Burkhardt, and H.-P. Kriegel, editors, *State-of-the-Art in Content-Based Image and Video Retrieval*, pages 69–95. Kluwer Academic Publishers, 2001.

[360] M. M. Lazarescu, S. Venkatesh, and H. H. Bui. Using multiple windows to track concept drift. *Intelligent Data Analysis*, 8:29–59, 2004.

[361] A. Lazarevic, R. Kanapady, and C. Kamath. Effective localized regression for damage detection in large complex mechanical structures. In *Proceedings of the ACM International Conference on Knowledge Discovery and Data Mining*, pages 450–459, 2004.

[362] A. Lazarevic, R. Kanapady, C. Kamath, V. Kumar, and K. Tamma. Localized prediction of continuous target variables using hierarchical clustering. In *Proceedings of the Third IEEE International Conference on Data Mining*, pages 139–146, 2003.

[363] J. Le Moigne. Parallel registration of multi-sensor remotely sensed imagery using wavelet coefficients. In *Proceedings of the SPIE OE/Aerospace Sensing, Wavelet Applications Conference*. SPIE Press, Bellingham, WA, 1994.

[364] A. R. Leach and V. J. Gillet. *An Introduction to Chemoinformatics*. Springer, Dordrecht, The Netherlands, 2003.

[365] J. C. Leachtenauer and R. G. Driggers. *Surveillance and Reconnaissance Imaging Systems: Modeling and Performance Prediction*. Artech House, Norwood, MA, 2001.

[366] D.-S. Lee, J. Hull, and B. Erol. A Bayesian framework for Gaussian mixture background modeling. In *Proceedings of IEEE International Conference on Image Processing*, Barcelona, Spain, 2003.

[367] H. Y. Lee, W. K. Park, H. K. Lee, and T. G. Kim. Towards knowledge-based extraction of roads from 1m-resolution satellite images. In *Proceedings of the IEEE Southwest Symposium on Image Analysis and Interpretation*, pages 171–176. IEEE, 2000.

[368] J. Lee, R. C. Weger, S. S. Sengupta, and R. M. Welch. A neural network approach to cloud classification. *IEEE Transactions on Geoscience and Remote Sensing*, 28:846–855, 1990.

[369] L. Leherte, J. Glasgow, K. Baxter, E. Steeg, and S. Fortier. Analysis of three-dimensional protein images. *Journal of Artificial Intelligence Research*, 7:125–159, 1997.

[370] R. B. Lehoucq, D. C. Sorenson, and C. Yang. *ARPACK Users' Guide: Solution of Large-Scale Eigenvalue Problems with Implicitly Restarted Arnoldi Methods*. SIAM, Philadelphia, PA, 1998.

[371] A. M. Lesk. *Introduction to Bio-Informatics*. Oxford University Press, Oxford, UK, 2002.

[372] S. Letourneau, F. Famili, and S. Matwin. Data mining to predict aircraft component replacement. *IEEE Intelligent Systems*, 14:59–66, 1999.

[373] R. J. LeVeque. *Finite Difference Methods for Ordinary and Partial Differential Equations: Steady State and Time-Dependent Problems*. SIAM, Philadelphia, PA, 2007.

[374] D. D. Lewis and W. A. Gale. A sequential algorithm for training text classifiers. In W. B. Croft and C. J. van Rijsbergen, editors, *Proceedings of SIGIR-94, 17th ACM International Conference on Research and Development in Information Retrieval*, pages 3–12, 1994.

[375] J. M. Lewis, S. Lakshmivarahan, and S. Dhall. *Dynamic Data Assimilation: A Least Squares Approach*. Cambridge University Press, Cambridge, UK, 2006.

[376] LHC: The Large Hadron Collider Web page, 2008. http://lhc-new-homepage.web.cern.ch.

[377] H. Li, B. S. Manjunath, and S. K. Mitra. Image fusion using wavelets. *Computer Vision, Graphics & Image Processing: Graphical Models and Image Processing*, 57:36–42, 2007.

[378] X. Li and T. Chen. Nonlinear diffusion with multiple edginess thresholds. *Pattern Recognition*, 27:1029–1037, 1994.

[379] Y. Li, C.-C. Jay Kuo, and X. Wan. Introduction to content-based image retrieval—overview of key techniques. In V. Castelli and D. Bergman, editors, *Image Databases: Search and Retrieval of Digital Imagery*, pages 261–284. John Wiley, New York, 2002.

[380] Z. Li, A. Khananian, R. H. Fraser, and J. Cihlar. Automatic detection of fire smoke using artificial neural networks and threshold approaches applied to AVHRR imagery. *IEEE Transactions on Geoscience and Remote Sensing*, 39:1859–1870, 2001.

[381] T. M. Lillesand and R. W. Kiefer. *Remote Sensing and Image Interpretation*. John Wiley, New York, 2000.

[382] T. Lindeberg. *Scale-Space Theory in Computer Vision*. Kluwer Academic Publishers, Boston, MA, 1994.

[383] H. Ling and K. Okada. Diffusion distance for histogram comparison. In *IEEE Computer Society Conference on Computer Vision and Pattern Recognition (CVPR)*, volume 1, pages 246–253, 2006.

[384] H. Ling and K. Okada. An efficient earth mover's distance algorithm for robust histogram comparison. *IEEE Transactions on Pattern Analysis and Machine Intelligence*, 29:840–853, 2007.

[385] H. Liu and H. Motoda. *Feature Extraction, Construction, and Selection: A Data Mining Perspective*. Kluwer Academic Publishers, Boston, MA, 1998.

[386] H. Liu and H. Motoda. *Feature Selection for Knowledge Discovery and Data Mining*. Kluwer Academic Publishers, Boston, MA, 1998.

[387] B. P. L. Lo and S. A. Velastin. Automatic congestion detection system for underground platforms. In *Proceedings of the 2001 International Symposium on Intelligent Multimedia, Video, and Speech Processing*, pages 158–161, Hong Kong, 2001.

[388] S. L. Lohr. *Sampling: Design and Analysis*. Duxbury Press, Pacific Grove, CA, 1999.

[389] S. Loncaric. A survey of shape analysis techniques. *Pattern Recognition*, 31:983–1001, 1998.

[390] N. S. Love and C. Kamath. An empirical study of block-matching techniques for detection of moving objects. Technical Report UCRL-TR-218038, Lawrence Livermore National Laboratory, 2006.

[391] N. S. Love and C. Kamath. An experimental comparison of block-matching techniques for detection of moving objects. In *Proceedings of the Applications of Digital Image Processing, XXIX, SPIE Conference 6312*, pages 63120C.1–63120C.12, 2006.

[392] N. S. Love and C. Kamath. Image analysis for the identification of coherent structures in plasma. In *Proceedings of the Applications of Digital Image Processing, XXX, SPIE Conference 6696*, paper 66960D, 2007.

[393] D. G. Lowe. Object recognition from local scale-invariant features. In *Proceedings of the International Conference on Computer Vision*, pages 1150–1157, 1999.

[394] D. G. Lowe. Demo software: SIFT keypoint detector, 2008. http://www.cs.ubc.ca/~lowe/keypoints/.

[395] D. G. Lowe. Distinctive image features from scale-invariant keypoints. *International Journal of Computer Vision*, 60:91–110, 2004.

[396] LSST: The large synoptic survey telescope Web page, 2008. http://www.lsst.org.

[397] B. D. Lucas and T. Kanade. An iterative image registration technique with an application to stereo vision. In *Proceedings of the Seventh International Joint Conference on Artificial Intelligence*, pages 674–679, 1981.

[398] J. Lumley and P. Blossey. Control of turbulence. *Annual Review of Fluid Mechanics*, 30:311–327, 1998.

[399] J. L. Lumley. The structure of inhomogeneous turbulent flows. In A.M. Yaglom and V.I. Tatarski, editors, *Atmospheric Turbulence and Radio Propagation*, pages 166–178. Nauka, 1967.

[400] U. Luxborg. A tutorial on spectral clustering. *Statistics and Computing*, 17:395–416, 2007.

[401] F. Lv, J. Kang, R. Nevatia, I. Cohen, and G. Medioni. Automatic tracking and labeling of human activities in a video sequence. In *The Joint IEEE International Workshop on Visual Surveillance and Performance Evaluation of Tracking and Surveillance, in conjunction with ECCV '04*, 2004.

[402] C. Lynnes and R. Mack. KDD services at the Goddard Earth Sciences Distributed Active Archive Center. In R. Grossman, C. Kamath, W. Kegelmeyer, V. Kumar, and R. Namburu, editors, *Data Mining for Scientific and Engineering Applications*, pages 165–181. Kluwer, Boston, MA, 2001.

[403] J. MacCormick. *Stochastic Algorithms for Visual Tracking*. Springer, London, 2002.

[404] R. Machiraju, J. E. Fowler, D. Thompson, B. Soni, and W. Schroeder. EVITA—Efficient Visualization and Interrogation of Tera-Scale Data. In R. L. Grossman, C. Kamath, P. Kegelmeyer, V. Kumar, and R. R. Namburu, editors, *Data Mining for Scientific and Engineering Applications*, pages 257–279. Kluwer Academic Publishers, Norwell, MA, 2001.

[405] MACHO Project Web page, 2008. http://macho.anu.edu.au.

[406] J. Maintz and M. Viergever. A survey of medical image registration. *Medical Image Analysis*, 2:1–36, 1998.

[407] D. Malerba, F. Esposito, M. Ceci, and A. Appice. Top-down induction of model trees with regression and splitting nodes. *IEEE Transactions on Pattern Analysis and Machine Intelligence*, 26:612–625, 2004.

[408] R. Malladi, editor. *Geometric Methods in Bio-Medical Image Processing*. Springer, New York, 2002.

[409] R. Malladi, J. Sethian, and B. Vemuri. Shape modeling with front propagation: A level set approach. *IEEE Transactions on Pattern Analysis and Machine Intelligence*, 17:158–175, 1995.

[410] S. G. Mallat. A theory for multiresolution signal decomposition: The wavelet representation. *IEEE Transactions on Pattern Analysis and Machine Intelligence*, 11:674–693, 1989.

[411] M. A. Maloof. Learning when data sets are imbalanced and when costs are unequal and unknown. In *Proceedings of the Workshop on Learning from Imbalanced Data Sets*, 2003. http://www.site.uottawa.ca/~nat/Workshop2003/schedule.html.

[412] M. A. Maloof, P. Langley, T. O. Binford, R. Nevatia, and S. Sage. Improved rooftop detection in aerial images with machine learning. *Machine Learning*, 53:157–191, 2003.

[413] D. P. Mandal, C. A. Murthy, and S. K. Pal. Analysis of IRS imagery for detecting man-made objects with a multi-valued recognition system. *IEEE Transactions on Pattern Analysis and Machine Intelligence*, 18:837–842, 1996.

[414] V. Mandava, J. Fitzpatrick, and D. Pickens. Adaptive search space scaling in digital image registration. *IEEE Transactions on Medical Imaging*, 8:251–262, 1989.

[415] B. S. Manjunath, S. Chandrasekharan, and Y. F. Wang. An eigenspace update algorithm for image analysis. In *Proceedings of the IEEE International Symposium on Computer Vision*, pages 551–556, 1995.

[416] B. S. Manjunath and W. Y. Ma. Texture features for browsing and retrieval of image data. *IEEE Transactions on Pattern Analysis and Machine Intelligence*, 18:837–842, 1996.

[417] B. S. Manjunath and W. Y. Ma. Texture features image retrieval. In V. Castelli and D. Bergman, editors, *Image Databases: Search and Retrieval of Digital Imagery*, pages 313–372. John Wiley, New York, 2002.

[418] B. S. Manjunath, P. Salembier, and T. Sikora, editors. *Introduction to MPEG-7: Multimedia Content Description Interface*. John Wiley, New York, 2002.

[419] D. J. Marchette. *Computer Intrusion Detection and Network Monitoring: A Statistical Viewpoint*. Springer, New York, 2001.

[420] H. Markram. The blue brain project. *Nature Review Neuroscience*, 7:153–160, 2006.

[421] D. Marr and E. Hildreth. Theory of edge detection. *Proceedings of the Royal Society of London, Series B*, 207:187–217, 1980.

[422] G. Matheron. *Random Sets and Integral Geometry*. John Wiley, New York, 1975.

[423] C. R. Maurer and J. M. Fitzpatrick. A review of medical image registration. In R. J. Maciunas, editor, *Interactive Image-Guided Neurosurgery*, pages 17–44. American Association of Neurological Surgeons, Park Ridge, IL, 1993.

[424] T. McInerney and D. Terzopoulos. Topologically adaptable snakes. In *Proceedings of the Fifth International Conference on Computer Vision*, pages 840–845, Cambridge, MA, 1995.

[425] G. Medioni, M. S. Lee, and C. K. Tang. *A Computational Framework for Segmentation and Grouping*. Elsevier, Amsterdam, The Netherlands, 2000.

[426] G. Medioni, C. K. Tang, and M. S. Lee. Tensor voting: Theory and applications. In *Proc. 12eme Congres Francophone AFRIF-AFIA de Reconnaissance des Formes et Intelligence Artificielle (RFIA)*, 2000.

[427] V. Megalooikomonou and D. Kontos. Medical data fusion for telemedicine. *IEEE Engineering in Medicine and Biology*, 26:823–854, 1998.

[428] METIS: Family of Multilevel Partitioning Algorithms—includes METIS, ParMETIS, and hMETIS, 2007. http://glaros.dtc.umn.edu/gkhome/views/metis.

[429] H. B. Mitchell. *Multi-Sensor Data Fusion: An Introduction*. Springer, Berlin, 2007.

[430] T. M. Mitchell. *Machine Learning*. McGraw–Hill, Boston, MA, 1997.

[431] D. Mladenić, I. Bratko, R. J. Paul, and M. Grobelnik. Using machine learning techniques to interpret results from discrete event simulation. In *Proceedings of the Sixth European Conference on Machine Learning ECML94*. Springer, 1994.

[432] D. Mladenic, M. Grobelnik, I. Bratko, and R. J. Paul. Classification tree chaining in data analysis. In *IJCAI Workshop on Data Engineering for Inductive Learning*, Montreal, 1995.

[433] J. Modersitzki. *Numerical Methods for Image Registration*. Oxford University Press, Oxford, 2004.

[434] F. Mokhtarian and M. Bober. *The Curvature Scale Space Representation: Theory, Applications, and MPEG-7 Standardisation*. Kluwer Academic Publishers, Norwell, MA, 2003.

[435] F. Mokhtarian and A. K. Mackworth. A theory of multiscale, curvature-based shape representation for planar curves. *IEEE Transactions on Pattern Analysis and Machine Intelligence*, 14:789–805, 1992.

[436] F. Mokhtarian and R. Suomela. Robust image corner detection through curvature scale space. *IEEE Transactions on Pattern Analysis and Machine Intelligence*, 20:1376–1381, 1998.

[437] D. C. Montgomery. *Design and Analysis of Experiments*. John Wiley, Hoboken, NJ, 2004.

[438] H. P. Moravec. Towards automatic visual obstacle avoidance. In *Proceedings of the Fifth International Joint Conference on Artificial Intelligence*, page 584, 1977.

[439] P. Moulin. Multiscale image decompositions and wavelets. In A. Bovik, editor, *Handbook of Image and Video Processing*, pages 289–300. Academic Press, San Diego, CA, 2000.

[440] R. Mukundan and K. R. Ramakrishnan. *Moment Functions in Image Analysis: Theory and Applications*. World Scientific, Singapore, 1988.

[441] F. Murtagh. Clustering massive data sets. In J. Abello, P. M. Pardalos, and M. G. C. Reisende, editors, *Handbook of Massive Data Sets*. Kluwer, Norwell, MA, 2000.

[442] K. V. S. Murthy. *On Growing Better Decision Trees from Data*. PhD thesis, Johns Hopkins University, Baltimore, MD, 1997.

[443] R. H. Myers and D. C. Montgomery. *Response Surface Methodology: Process and Product Optimization using Designed Experiments*. Wiley-InterScience, Hoboken, NJ, 2002.

[444] A. Naim. Approaches to automated morphological classification of galaxies. *Vistas in Astronomy*, 38:265–271, 1994.

[445] U. S. Nair, J. F. Rushing, R. P. Ramachandran, K. S. Kuo, R. M. Welch, and S. J. Graves. Detection of cumulus cloud fields in satellite imagery. In *Earth Observing Systems IV, Proceedings of SPIE, Volume 3750*, pages 345–355. SPIE Press, Bellingham, WA, 1999.

[446] G. P. Nason and B. W. Silverman. The stationary wavelet transform and some statistical applications. In A. Antoniadis and G. Oppenheim, editors, *Wavelets and Statistics*, Lecture Notes in Statistics, pages 261–280. Springer, New York, 1995.

[447] National Spherical Torus Experiment, 2008. http://nstx.pppl.gov/.

[448] Nature: Special Issue on Proteomics, March 2003.

[449] NetCDF: Network Common Data Form Web page, 2008. http://www.unidata.ucar.edu/software/netcdf/.

[450] R. Nevatia and K. E. Price. Locating structures in aerial imagery. *IEEE Transactions on Pattern Analysis and Machine Intelligence*, 4:476–484, 1982.

[451] S. Newsam and C. Kamath. Retrieval using texture features in high-resolution, multispectral satellite imagery. In *Data Mining and Knowledge Discovery: Theory, Tools, and Technology, VI, Proceedings of SPIE, Volume 5433*, pages 21–32. SPIE Press, Bellingham, WA, 2004.

[452] S. Newsam and C. Kamath. Comparing shape and texture features for pattern recognition in simulation data. In *Image Processing: Algorithms and Systems IV, Proceedings of SPIE, Volume 5672*, pages 106–117. SPIE Press, Bellingham, WA, 2005.

[453] G. M. Nielson, H. Hagen, and H. Mueller. *Scientific Visualization: Overview, Methodologies, Techniques*. IEEE, Los Alamitos, CA, 1997.

[454] J. Nocedal and S. J. Wright. *Numerical Optimization*. Springer, New York, 1999.

[455] S. C. Odewahn and M. L Nielsen. Star/galaxy separation using neural networks. *Vistas in Astronomy*, 38:281–286, 1994.

[456] S. C. Odewahn, E. B. Stockwell, R. L. Pennington, R. M. Humphreys, and W. A. Zumach. Automated star/galaxy discrimination with neural networks. *The Astronomical Journal*, 103:318–331, 1992.

[457] A. O'Hagan. Bayesian analysis of computer code outputs: A tutorial. *Reliability Engineering and System Safety*, 91:1290–1300, 2006.

[458] A. Okabe, B. Boots, K. Sugihara, and S. N. Chiu. *Spatial Tessellations: Concepts and Applications of Voronoi Diagrams*. John Wiley, New York, 2000.

[459] C. Oliver and S. Quegan. *Understanding Synthetic Aperture Radar Images*. Artech House, Boston, MA, 1998.

[460] OpenCV Web page: Open Source Computer Vision Library, 2008. http://sourceforge.net/projects/opencvlibrary.

[461] OpenGL Web Page: High-Performance Graphics, 2008. http://www.opengl.org/.

[462] D. Opitz and R. Maclin. Popular ensemble methods: An empirical study. *Journal of Artificial Intelligence Research*, 11:168–198, 1999.

[463] R. Osada, T. Funkhouser, B. Chazelle, and D. Dobkin. Shape distributions. *ACM Transactions on Graphics*, 21:807–832, 2002.

[464] S. Osher and R. Fedkiw. *Level Set Methods and Dynamic Implicit Surfaces*. Springer, New York, 2003.

[465] S. Osher and N. Paragios, editors. *Geometric Level Set Methods in Imaging, Vision, and Graphics*. Springer, New York, 2003.

[466] S. Osher and J. A. Sethian. Fronts propagating with curvature dependent speed: Algorithms based on Hamilton-Jacobi formulations. *Journal of Computational Physics*, 79:12–49, 1988.

[467] A. Paccanaro, J. A. Casbon, and M. A. S. Saqi. Spectral clustering of protein sequences. *Nucleic Acids Research*, 34:1571–1580, 2006.

[468] G. Pandey, V. Kumar, and M. Steinbach. Computational approaches for protein function prediction: A survey. Technical Report 06-028, Department of Computer Science and Engineering, University of Minnesota, Twin Cities, 2006.

[469] S. Pankanti, R.M. Bolle, and A. Jain. Biometrics: The future of identification. *IEEE Computer*, 33:46–49, 200. Guest editors' introduction, Special issue on Biometrics.

[470] D. Parikh and R. Polikar. An ensemble-based incremental learning approach to data fusion. *IEEE Transactions on Systems, Man, and Cybernetics—Part B: Cybernetics*, 37:437–450, 2007.

[471] K. Pearson. On lines and planes of closest fit to systems of points in space. *Phil. Mag.*, 2:559–572, 1901.

[472] D. B. Percival and A. T. Walden. *Wavelet Methods for Time Series Analysis*. Cambridge University Press, Cambridge, UK, 2006.

[473] P. Perona and J. Malik. Scale-space and edge detection using anisotropic diffusion. In *Proceedings of the IEEE Computer Society Workshop on Computer Vision*, pages 16–27, 1987.

[474] P. Perona and J. Malik. Scale-space and edge detection using anisotropic diffusion. *IEEE Transactions on Pattern Analysis and Machine Intelligence*, 12:629–639, 1990.

[475] B. Perry et al. *Content-Based Access to Multimedia Information—from Technology Trends to State of the Art*. Kluwer Academic Publishers, Norwell, MA, 1999.

[476] M. Petrou and P. Bosdogianni. *Image Processing: The Fundamentals*. John Wiley, New York, 2000.

[477] R. W. Picard et al. Special issue on digital libraries. *IEEE Transactions on Pattern Analysis and Machine intelligence*, 18, August 1996.

[478] S. M. Pizer et al. Adaptive histogram equalization and its variants. *Computer Vision, Graphics, and Image Processing*, 39:366–368, 1987.

[479] J. Platt. Fast training of support vector machines using sequential minimal optimization. In B. Schölkopf, C. Burges, and A. Smola, editors, *Advances in Kernel Methods—Support Vector Learning*, pages 185–208. MIT Press, Cambridge, MA, 1998.

[480] C. Pohl and J. L. van Genderen. Multisensor image fusion in remote sensing: Concepts, methods, and applications. *International Journal of Remote Sensing*, 19:823–854, 1998.

[481] A. Pothen, H. D. Simon, and K.-P. Liou. Partitioning sparse matrices with eigenvectors of graphs. *SIAM Journal on Matrix Analysis and Applications*, 11:430–452, 1990.

[482] P. W. Power and J. A. Schoonees. Understanding background mixture models for foreground segmentation. In *Proceedings of Image and Vision Computing New Zealand*, pages 267–271, Auckland, New Zealand, 2002.

[483] R. W. Preisendorfer and C. D. Mobley. *Principal Component Analysis in Meteorology and Oceanography*. Elsevier Science, Amsterdam, The Netherlands, 1988.

[484] J. M. S. Prewitt. Object enhancement and extraction. In B. S. Lipkin and A. Rosenfeld, editors, *Picture Processing and Psychopictorics*, pages 75–149. Academic Press, New York, 1970.

[485] Princeton Plasma Physics Laboratory, 2008. http://www.pppl.gov.

[486] F. Provost and T. Fawcett. Robust classification for imprecise environments. *Machine Learning*, 42:203–231, 2001.

[487] F. J. Provost and T. Fawcett. Analysis and visualization of classifier performance: Comparison under imprecise class and cost distributions. In *Proceedings of the Third International Conference on Knowledge Discovery and Data Mining*, pages 43–48, 1997.

[488] Qt Cross-Platform Application Framework, 2008. http://trolltech.com/products/qt.

[489] J. Quinlan. Bagging, boosting, and C4.5. In *Proceedings of the Thirteenth National Conference on Artificial Intelligence*, pages 725–730. AAAI Press, Menlo Park, CA, and MIT Press, Cambridge, MA, 1996.

[490] J. R. Quinlan. Induction of decision trees. *Machine Learning*, 1:81–106, 1986.

[491] J. R. Quinlan. Learning with continuous classes. In *Proceedings of the Fifth Australian Joint Conference on Artificial Intelligence*, pages 343–348, 1992.

[492] J. R. Quinlan and R. L. Rivest. Inferring decision trees using the minimum description length principle. *Information and Computation*, 80:227–248, 1989.

[493] J. R. Quinlan. *C4.5: Programs for Machine Learning*. Morgan Kaufman, San Mateo, CA, 1993.

[494] J. R. Quinlan. Simplifying decision trees. *International Journal of Man-Machine Studies*, 27:221–234, 1987.

[495] The R-Project for Statistical Computing, 2008. http://www.r-project.org/.

[496] A. Rahman, G. A. Woodell, and D. J. Jobson. A comparison of the multiscale retinex with other image enhancement methods. In *Proceedings of the IS&T 50th Anniversary Conference*, pages 426–431, 1997.

[497] Z. Rahman, D. J. Jobson, and G. A . Woodell. Retinex processing for automatic image enhancement. *Journal of Electronic Imaging*, 13:100–110, 2004.

[498] K. Rajan and M. Zaki. Data mining through information association: A knowledge discovery tool for materials science. In *Proceedings of 17th CODATA Conference*, Baveno, Italy, 2000.

[499] K. Rajan, C. Suh, and B. Narasimhan. Informatics methods for combinatorial materials science. In S. K. Mallapragada, B. Narasimhan, and M. D. Porter, editors, *Combinatorial Materials Science*. John Wiley, New York, 2007.

[500] R. Ramachandran, H. Conover, S. Graves, and K. Keiser. Challenges and solutions to mining earth science data. In *Data Mining and Knowledge Discovery, SPIE Proceedings*, volume 4057, pages 259–264, 2000.

[501] K. Rangarajan and M. Shah. Establishing motion correspondence. *CVGIP: Image Understanding*, 54:56–73, 1991.

[502] R. M. Rangayyan. *Biomedical Image Analysis*. CRC Press, Boca Raton, FL, 2001.

[503] C. R. Rao. The use and interpretation of principal component analysis in applied research. *Sankhya*, 26:329–358, 1964.

[504] C. E. Rasmussen and C. K. I. Williams. *Gaussian Processes for Machine Learning*. MIT Press, Cambridge, MA, 2006.

[505] K. V. RaviKanth, D. Agrawal, and A. D. Singh. Dimensionality reduction for similarity searching in dynamic databases. In *Proceedings of the 1998 ACM SIGMOD International Conference on the Management of Data*, pages 166–176, 1998.

[506] A. Refregier. Shapelets: I. A method for image analysis. *Monthly Notices of the Royal Astronomical Society*, 338:35, 2003.

[507] J. M. Reinhardt, R. Uppaluri, W. E. Higgins, and E. A. Hoffman. Pulmonary imaging and analysis. In M. Sonka and J. M. Fitzpatrick, editors, *Handbook of Medical Imaging, Volume 1: Medical Image Processing and Analysis*, pages 1005–1060. SPIE Press, Bellingham, WA, 2000.

[508] A. M. Reza. Realization of the contrast limited adaptive histogram equalization (CLAHE) for real-time image enhancement. *Journal of VLSI Signal Processing*, 38:35–44, 2004.

[509] J. A. Richards and X. Jia. *Remote Sensing Digital Image Analysis: An Introduction.* Springer, Berlin, 1999.

[510] W. Rider et al. Using Richtmyer-Meshkov driven mixing experiments to impact the development of numerical methods for compressible hydrodynamics. In *Proceedings of the Ninth International Conference on Hyperbolic Problems: Theory, Numerics, Applications*, pages 84–88, 2002.

[511] L. G. Roberts. Machine perception of three-dimensional solids. In J. T. Tippett et al., editors, *Optical and Electro-Optical information Processing*, pages 159–197. MIT Press, Cambridge, MA, 1965.

[512] M. Robnik-Šikonja and I. Kononenko. An adaptation of Relief for attribute estimation in regression. In *Proceedings of the 14th International Conference on Machine Learning*, pages 296–304, 1997.

[513] J. B. T. M. Roerdink and A. Meijster. The watershed transform: Definitions, algorithms, and parallelization strategies. *Fundamenta Informaticae*, 41:187–228, 2000.

[514] R. Rojas. *Neural Networks: A systematic introduction.* Springer, Berlin, 1996.

[515] ROOT Web page: An Object-Oriented Data Analysis Framework, 2008. http://root.cern.ch.

[516] RoSuDa: Interactive statistical software, 2008. http://rosuda.org/software/.

[517] S. Roweis and L. Saul. Nonlinear dimensionality reduction by locally linear embedding. *Science*, 290:2323–2326, 2000.

[518] Y. Rubner, C. Tomasi, and L. J. Guibas. The earth mover's distance as a metric for image retrieval. *International Journal of Computer Vision*, 40:99–121, 2000.

[519] J. P. Sacha, L. S. Goodenday, and K. J. Cios. Bayesian learning for cardiac SPECT image interpretation. *Artificial Intelligence in Medicine*, 26:109–143, 2002.

[520] J. Sacks, W. J. Welch, T. J. Mitchell, and H. P. Wynn. Design and analysis of computer experiments. *Statistical Science*, 4:409–435, 1989.

[521] H. Samet. *Foundations of Multidimensional and Metric Data Structures.* Morgan Kaufmann, San Francisco, CA, 2006.

[522] T. J. Santner, B. J. Williams, and W. Notz. *The Design and Analysis of Computer Experiments.* Springer, New York, 2003.

[523] G. Sapiro. *Geometric Partial Differential Equations and Image analysis.* Cambridge University Press, Cambridge, UK, 2001.

[524] Sapphire Scientific Data Mining Project Web page, 2008. https://computation.llnl.gov/casc/sapphire/.

[525] B. Schölkopf, A. Smola, and K.-R. Müller. Nonlinear component analysis as a kernel eignevalue problem. *Neural Computation*, 10:1299–1319, 1998.

[526] M. Scholz and R. Klinkenberg. Boosting classifiers for drifting concepts. *Intelligent Data Analysis*, 11:3–28, 2007.

[527] R. A. Schowengerdt. *Remote Sensing: Models and Methods for Image Processing*. Academic Press, San Diego, CA, 1997.

[528] SDSS: The Sloan Digital Sky Survey Web page, 2008. http://www.sdss.org.

[529] N. Sebe and M. S. Lew. Texture features for content-based retrieval. In M. S. Lew, editor, *Principles of Visual Information Retrieval*, pages 51–85. Springer, London, 2001.

[530] S. Sengupta and P. Edwards. *Data Structures in ANSI C*. Academic Press, San Diego, CA, 1991.

[531] S. Sengupta and C. P. Korobkin. *C++ Object-Oriented Data Structures*. Springer, New York, 1994.

[532] S. K. Sengupta and J. S. Boyle. Nonlinear principal component analysis of climate data. Technical Report PCMDI No. 29, Program for Climate Model Diagnosis and Intercomparison (PCMDI), 1995.

[533] K. Sentz and S. Ferson. Combination of evidence in Dempster-Shafer theory. Technical Report, Sandia National Laboratory, SAND 2002-0835, 2002. http://www.sandia.gov/epistemic/Reports/SAND2002-0835.pdf.

[534] J. Serra. *Image Analysis and Mathematical Morphology*. Academic Press, London, 1982.

[535] J. A. Sethian. Tracking interfaces with level sets. *American Scientist*, 85:254–263, 1997.

[536] J. A. Sethian. *Level Set Methods and Fast Marching Methods: Evolving Interfaces in Computational Geometry, Fluid Mechanics, Computer Vision and Materials Science*. Cambridge University Press, Cambridge, UK, 1999.

[537] A. Sha'ashua and S. Ullman. Structural saliency: The detection of globally salient structures using a locally connected network. In *Proceedings of the International Conference on Computer Vision*, pages 321–327, 1988.

[538] G. Shafer. The Dempster-Shafer theory. In S. C. Shapiro, editor, *Encyclopedia of Artificial Intelligence, Second edition*, pages 330–331. John Wiley, Hoboken, NJ, 1992.

[539] L. G. Shapiro and G. C. Stockman. *Computer Vision*. Prentice–Hall, Upper Saddle River, NJ, 2001.

[540] M. J. Shensa. The discrete wavelet transform: Wedding the à Trous and Mallat algorithms. *IEEE Transactions on Signal Processing*, 40:2464–2482, 1992.

[541] J. Shi and J. Malik. Normalized cuts and image segmentation. *IEEE Transactions on Pattern Analysis and Machine Intelligence*, 22:888–905, 2000.

[542] A. B. Shiflet and G. W. Shiflet. *Introduction to Computational Science: Modeling and Simulation for the Sciences*. Princeton University Press, Princeton, NJ, 2006.

[543] M. C. Shin, D. B. Goldgof, and K. W. Bowyer. Comparison of edge detector performance through use in an object recognition task. *Computer Vision and Image Understanding*, 84:160–178, 2001.

[544] N. Short. RST: The Remote Sensing Tutorial, 2008. http://rst.gsfc.nasa.gov/Homepage/Homepage.html.

[545] K. Siddiqui, Y. Lauzière, A. Tannenbaum, and S. Zucker. Area and length minimizing flows for shape segmentation. *IEEE Transactions on Image Processing*, 7:158–175, 1998.

[546] W. Siedlecki and J. Sklansky. On automatic feature selection. *International Journal of Pattern Recognition and Artificial Intelligence*, 2:197–220, 1988.

[547] A Siegel and J. B. Weiss. A wavelet-packet census algorithm for calculating vortex statistics. *Physics of Fluids*, 9:1988–1999, 1997.

[548] E. Simoudis. Reality check for data mining. *IEEE Expert*, 11:26–33, 1996.

[549] A. Singh, D. Goldgof, and D. Terzopoulos, editors. *Deformable Models in Medical Image Analysis*. IEEE Computer Society Press, Los Alamitos, CA, 1998.

[550] D. Smith and S. Singh. Approaches to multisensor data fusion in target tracking: A survey. *IEEE Transactions on Knowledge and Data Engineering*, 18:1696–1710, 2006.

[551] S. M. Smith. Reviews of optic flow, motion segmentation, edge finding and corner finding. Technical Report TR97SMS1, Oxford University, 1997. http://users.fmrib.ox.ac.uk/~steve/review/.

[552] S. M. Smith and J. M. Brady. ASSETT-2—real-time motion segmentation and shape tracking. *IEEE Transactions on Pattern Analysis and Machine Intelligence*, 17:814–820, 1995.

[553] S. M. Smith and J. M. Brady. SUSAN—a new approach to low-level image processing. *International Journal of Computer Vision*, 23:45–78, 1997.

[554] A. J. Smola and B. Schölkopf. A tutorial on support vector regression. Technical Report NC-TR-98-030, NeuroCOLT2 Technical Report series, 1998. http://www.neurocolt.com/abs/1998/abs98030.html.

[555] P. Smyth. Data mining: Data analysis on a grand scale? *Statistical Methods in Medical Research*, 9:309–327, 2000.

[556] L. Sodré and H. Cuevas. Spectral classification of galaxies. *Vistas in Astronomy*, 38:287–291, 1994.

[557] P. Soille, editor. *Morphological Image Analysis: Principles and Applications*. Springer, New York, 1999.

[558] M. Sonka and J. M. Fitzpatrick, editors. *Handbook of Medical Imaging, Volume 1: Medical Image Processing and Analysis*. SPIE Press, Bellingham, WA, 2000.

[559] M. Sonka, V. Hlavac, and R. Boyle. *Image Processing, Analysis, and Machine Vision*. International Thompson Publishing, Pacific Grove, CA, 1999.

[560] M. E. Spear. *Charting Statistics*. McGraw–Hill, New York, 1952.

[561] Stanford National Accelerator Laboratory Web page, 2008. http://www.slac.stanford.edu.

[562] R. J. Stanley, P. D. Gader, and K. C. Ho. Feature and decision level sensor fusion of electromagnetic induction and ground penetrating radar sensors for landmine detection with hand-held units. *Information Fusion*, 3:215–223, 2002.

[563] J.-L. Starck, F. Murtagh, and A. Bijaoui. *Image Processing and Data Analysis. The Multiscale Approach*. Cambridge University Press, Cambridge, UK, 1998.

[564] J.-L. Starck, E. J. Candès, and D. J. Donoho. The curvelet transform for image denoising. *IEEE Transactions on Image Processing*, 11:670–684, 2002.

[565] C. Stauffer and W. E. L. Grimson. Learning patterns of activity using real-time tracking. *IEEE Transactions on Pattern Analysis and Machine Intelligence*, 22:747–57, 2000.

[566] D. Stekel. *Micro-Array Bio-informatics*. Cambridge University Press, Cambridge, UK, 2003.

[567] J. Stoeckel and G. Fung. SVM feature selection for classification of SPECT images of Alzheimer's disease using spatial information. In *Proceedings of the IEEE International Conference on Data Mining*, pages 410–417, 2005.

[568] E. J. Stollnitz, T. D. DeRose, and D. H. Salesin. *Wavelets for Computer Graphics: Theory and Applications*. Morgan Kaufmann Publishers, San Francisco, CA, 1996.

[569] P. Stolorz and C. Dean. Quakefinder: A scalable data mining system for detecting earthquakes from space. In *Proceedings of the Second International Conference on Knowledge Discovery and Data Mining*, pages 208–213. AAAI Press, 1996.

[570] M. C. Storrie-Lombardi, M. J. Irwin, T. von Hippel, and L. J. Storrie-Lombardi. Spectral classification with principal component analysis and artificial neural networks. *Vistas in Astronomy*, 38:331–340, 1994.

[571] M. C. Storrie-Lombardi, O. Lahav, L. Sodré, and L. J.Storrie-Lombardi. Morphological classification of galaxies by artificial neural networks. *Monthly Notices of the Royal Astronomical Society*, 259:8–12, 1992.

[572] J. Suri, L. Liu, S. Singh, S. Laxminarayan, X. Zeng, and L. Reden. Shape recovery algorithms using levels sets in 2-D/3-D medical imagery: A state-of-the-art review. *IEEE Transactions on Information Technology in Biomedicine*, 6:8–28, 2002.

[573] A. Szalay. On the Virtual Observatory Project, 2003. In article, "Telescopes of the World, Unite! A Cosmic Database Emerges," by Bruce Schechter, New York Times, May 20, 2003.

[574] P.-N. Tan, M. Steinbach, and V. Kumar. *Introduction to Data Mining*. Addison-Wesley, Boston, MA, 2006.

[575] Tcl (tool command language) Developer Xchange Web site, 2008. http://www.tcl/tk/.

[576] A. M. Tekalp. Image and video restoration. In V. K. Madisetti and D. B. Williams, editors, *The Digital Signal Processing Handbook*, pages 53-1–53-20. CRC Press, Boca Raton, FL, 1998.

[577] A. M. Tekelp. *Digital Video Processing*. Prentice–Hall, Upper Saddle River, NJ, 1995.

[578] J. B. Tenenbaum, V. de Silva, and J. C. Langford. A global geometric framework for nonlinear dimensionality reduction. *Science*, 290:2319–2323, 2000.

[579] Terascale Supernova Initiative Web page, 2008. http://www.phy.ornl.gov/tsi/.

[580] M. C. Testik and G. C. Runger. Mining manufacturing quality data. In N. Ye, editor, *Handbook of Data Mining*, pages 657–668. Lawrence Erlbaum Associates, Mahwah, NJ, 2003.

[581] S. Theodoridis and K. Koutroumbas. *Pattern Recognition*. Academic Press, San Diego, CA, 2003.

[582] P. M. Thompson, M. S. Mega, K. L. Narr, E. R. Sowell, R. E. Blanton, and A. W. Toga. Brain image analysis and atlas construction. In M. Fitzpatrick and M. Sonka, editors, *SPIE Handbook of Medical Image Processing and Analysis*, pages 1061–1130. SPIE Press, Bellingham, WA, 2000.

[583] S. K. Thompson. *Sampling*. John Wiley, Hoboken, NJ, 2002.

[584] S. K. Thompson and G. A. F. Seber. *Adaptive Sampling*. Wiley-Interscience, Hoboken, NJ, 1996.

[585] C. Tomasi and R. Manduchi. Bilateral filtering for gray and color images. In *Proceedings of the IEEE International Conference on Computer Vision*, pages 839–846, 1998.

[586] L. Torgo. Functional models for regression tree leaves. In *Proceedings of the 14th International Machine Learning Conference*, pages 385–393, 1997.

[587] K. Toyama, J. Krumm, B. Brumitt, and B. Meyers. Wallflower: Principles and practice of background maintenance. In *ICCV (1)*, pages 255–261, 1999.

[588] T. G. Trucano, L. P. Swiler, T. Igusa, W. L. Oberkampf, and M. Pilch. Calibration, validation, and sensitivity analysis: What's what. *Reliability Engineering and System Safety*, 91:1331–1357, 2006.

[589] Y.-H. Tseng, P.-H. Hsu, and H. Chen. Automatic detecting rice fields by using multitemporal satellite images, land-parcel data and domain knowledge. In *Asian Conference on Remote Sensing*, 1998.

[590] M. Tuceryan and A. K. Jain. Texture analysis. In C. H. Chen, L. F. Pau, and P. S. P. Wang, editors, *Handbook of Pattern Recognition and Computer Vision*, pages 235–276. World Scientific, Singapore, 1993.

[591] E. R. Tufte. *Envisioning Information*. Graphics Press, Cheshire, CT, 1990.

[592] E. R. Tufte. *Visual Explanations: Images and Quantities, Evidence and Narrative*. Graphics Press, Cheshire, CT, 1997.

[593] E. R. Tufte. *The Visual Display of Quantitative Information*. Graphics Press, Cheshire, CT, 2001.

[594] J. W. Tukey. *Exploratory Data Analysis*. Addison-Wesley, Reading, MA, 1977.

[595] M. Turk and A. Pentland. Eigenfaces for recognition. *Journal of Cognitive Neurosience*, 3:71–86, 1991.

[596] S. Umbaugh. *Computer Vision and Image Processing: A Practical Approach Using CVIPtools*. Prentice–Hall, Upper Saddle River, NJ, 1998.

[597] United States Virtual Observatory Web page, 2008. http://www.us-vo.org.

[598] A. Unwin, M. Theus, and H. Hofmann. *Graphics of Large Datasets: Visualizing a Million*. Springer, Singapore, 2006.

[599] P. P. Vaidyanathan. *Multirate Systems and Filter Banks*. Prentice–Hall, Englewood Cliffs, NJ, 1993.

[600] N. Vaswani, A. Roy Chowdhury, and R. Chellappa. Activity recognition using the dynamics of the configuration of interacting objects. In *Proceedings of the Computer Vision and Pattern Recognition Conference*, volume 2, pages 633–640. IEEE Press, 2003.

[601] R. C. Veltkamp, H. Burkhardt, and H.-P. Kriegel, editors. *State-of-the-Art in Content-Based Image and Video Retrieval*. Kluwer Academic publishers, New York, 2001.

[602] R. C. Veltkamp and M. Hagedoorn. State-of-the-art in shape matching. In M. S. Lew, editor, *Principles of Visual Information Retrieval*, pages 87–119. Springer, Berlin, 2001.

[603] J. C. Venter et al. The Sequence of the Human Genome. *Science*, 291:1304–1351, 2001.

[604] L. Vincent and P. Soille. Watersheds in digital spaces: An efficient algorithm based on immersion simulations. *IEEE Transactions on Pattern Analysis and Machine Intelligence*, 13:583–598, 1991.

[605] VisIt Web page: Interactive Parallel Visualization, 2008. http://www.llnl.gov/visit.

[606] VLA: The Very Large Array Telescope Web page, 2008. http://www.vla.nrao.edu.

[607] N. Wale and G. Karypis. Comparison of descriptor spaces for chemical compound retrieval and classification. In *Proceedings of the IEEE International Conference on Data Mining*, pages 678–689, 2006.

[608] D. C. C. Wang, A. H. Vagnucci, and C. C. Li. Gradient inverse weighted smoothing scheme and the evaluation of its performance. *Computer Graphics and Image Processing*, 15:167–181, 1981.

[609] J. Wang, M. J. Zaki, H. T. T. Toivonen, and D. Shasha, editors. *Data Mining in Bioinformatics*. Springer, London, 2004.

[610] Xmdv Tool home page: Parallel coordinates. http://davis.wpi.edu/~xmdv/vis_parcoord.html.

[611] C. Ware. *Information Visualization: Perception for Design*. Morgan Kaufmann, San Francisco, CA, 2004.

[612] J. B. Weaver, X. Yansun, Jr., D. M. Healy, and L. D. Cromwell. Filtering noise from images with wavelet transforms. *Magnetic Resonance in Medicine*, 24:288–195, 1991.

[613] A. Webb. *Statistical Pattern Recognition*. John Wiley, Chichester, UK, 2005.

[614] S. K. Weeratunga and C. Kamath. PDE-based non-linear diffusion techniques for denoising scientific and industrial images: An empirical study. In *Image Processing: Algorithms and Systems, SPIE Proceedings, Volume 4667*, pages 279–290. SPIE Press, Bellingham, WA, 2002.

[615] S. K. Weeratunga and C. Kamath. A comparison of PDE-based non-linear anisotropic diffusion techniques for image denoising. In *Image Processing: Algorithms and Systems, SPIE Proceedings, Volume 5014*, pages 201–212. SPIE Press, Bellingham, WA, 2003.

[616] S. K. Weeratunga and C. Kamath. Investigation of implicit active contours for scientific image segmentation. In *Visual Communications and Image Processing, SPIE Proceedings, Volume 5308*, pages 210–221. SPIE Press, Bellingham, WA, 2004.

[617] J. Weickert. *Anisotropic Diffusion in Image Processing*. B. G. Teubner, Stuttgart, 1998.

[618] J. Weickert and G. Kühne. Fast methods for implicit active contour models. Technical Report, preprint no. 61, Universität des Saarlandes, 2002.

[619] J. Weickert and H. Scharr. A scheme for coherence-enhancing diffusion filtering with optimized rotation invariance. *Journal of Visual Communication and Image Representation*, 13:103–118, 2002.

[620] J. Weickert, B. ter Haar Romeny, and M. Viergever. Efficient and reliable schemes for nonlinear diffusion filtering. *IEEE Transactions on Image Processing*, 7:398–410, 1998.

[621] WEKA Web page: The Waikato Environment for Knowledge Analysis, 2008. http://weka.sourceforge.net/wekadoc/index.php/Main_Page.

[622] R. Whitaker and S. M. Pizer. A multi-scale approach to non-uniform diffusion. *CVGIP: Image Understanding*, 57:99–110, 1993.

[623] R. L. White, R. H. Becker, D. J. Helfand, and M. D. Gregg. A catalog of 1.4 GHz radio sources from the FIRST survey. *Astrophysical Journal*, 475:479, 1997.

[624] R. L. White and J. W. Percival. Compression and progressive transmission of astronomical images. *Bulletin of the American Astronomical Society*, 26:1513, 1994.

[625] M. V. Wickerhauser. *Adapted Wavelet Analysis from Theory to Software*. IEEE Press, Piscataway, NJ, 1994.

[626] G. Widmer and M. Kubat. Learning in the presence of concept drift and hidden contexts. *Machine Learning*, 23:69–101, 1996.

[627] C. L. Wilson and R. M. McCabe. Simple test procedure for image-based biometric verification systems. NIST Interagency Report 6336, National Institute of Standards and Technology, 1999. http://www.itl.nist.gov/iad/pubs2.html.

[628] J. A. Withers and K. A. Robbins. Tracking cell splits and merges. In *Proceedings of the IEEE Southwest Symposium on Image Analysis and Interpretation*, pages 117–122, 1996.

[629] A. P. Witkin. Scale-space filtering. In *Proceedings of the Eighth International Joint Conference on Artificial Intelligence*, pages 1019–1022, 1983.

[630] I. H. Witten and E. Frank. *Data Mining: Practical Machine Learning Tools and Techniques*. Morgan Kaufmann, San Francisco, CA, 2005.

[631] C. Wren, A. Azabayejani, T. Darrel, and A. Pentland. Pfinder: Real-time tracking of the human body. *IEEE Transactions on Pattern Analysis and Machine Intelligence*, 19:780–785, 1997.

[632] E. P. Xing, M. I. Jordan, and R. M. Karp. Feature selection for high-dimensional genomic microarray data. In *Proceedings of the 18th International Conference on Machine Learning*, pages 601–608. Morgan Kaufmann, San Francisco, CA, 2001.

[633] C. Xu, D. L. Pham, and J. L. Prince. Image segmentation using deformable models. In M. Fitzpatrick and M. Sonka, editors, *SPIE Handbook of Medical Image Processing and Analysis*, pages 129–174. SPIE Press, Bellingham, WA, 2000.

[634] C. Xu and J. L. Prince. Snakes, shapes, and gradient vector flow. *IEEE Transactions on Image Processing*, 7:359–369, March 1998.

[635] R. Xu and D. Wunsch. Survey of clustering algorithms. *IEEE Transactions on Neural Networks*, 16:645–678, 2005.

[636] M. Yeung et al. Special section on storage, processing, and retrieval of digital media. *Journal of Electronic Imaging*, 10:833–834, 2001.

[637] A. Yezzi, S. Kichenassamy, A. Kumar, P. Olver, and A. Tannenbaum. A geometric snake model for segmentation of medical imagery. *IEEE Transactions on Medical Imaging*, 16:199–209, 1997.

[638] K. M.-K. Yip. *KAM: A System for Intelligently Guiding Numerical Experimentation by Computer.* MIT Press, Cambridge, MA, 1991.

[639] A. Yu and C. Bajaj. A fast and adaptive method for image contrast enhancement. In *Proceedings of the IEEE International Conference on Image Processing, Volume 2*, pages 1001–1004, 2004.

[640] Z. Yu and C. Bajaj. Normalized gradient vector diffusion and image segmentation. In *Proceedings of the 7th European Conference on Computer Vision*, volume 3, pages 517–530, 2002.

[641] L. Zagorchev and A. Goshtasby. A comparative study of transformation functions for nonrigid image registration. *IEEE Transactions on Image Processing*, 15:529–538, 2006.

[642] C. T. Zahn. Graph-theoretical methods for detecting and describing gestalt clusters. *IEEE Transactions on Computers*, C-20:68–86, 1971.

[643] M. J. Zaki. Mining data in bioinformatics. In N. Ye, editor, *The Handbook of Data Mining*, pages 573–596. Lawrence Erlbaum Associates, Mahwah, NJ, 2003.

[644] H. Zha and H. D. Simon. On updating problems in latent semantic indexing. *SIAM Journal on Scientific Computing*, 21:782–791, 1999.

[645] L. Zhang, D. Samaras, D. Tomasi, N. Volkow, and R. Goldstein. Machine learning for clinical diagnosis from functional magnetic resonance imaging. In *Proceedings of the IEEE Computer Society Conference on Computer Vision and Pattern Recognition*, volume 1, pages 1211–1217, 2005.

[646] H. Zhao. A fast sweeping method for Eikonal equations. *Mathematics of Computation*, 74:603–627, 2005.

[647] H. Zhao, J. K. Aggarwal, C. Mandal, and B. C. Vemuri. 3-D shape reconstruction from multiple views. In A. Bovik, editor, *Handbook of Image and Video Processing*, pages 243–257. Academic Press, San Diego, CA, 2000.

[648] L. Zhou, C. Kambhamettu, D. B. Goldgof, K. Palaniappan, and A. F. Hasler. Tracking non-rigid motion and structure from 2-D satellite cloud images without correspondences. *IEEE Transactions on Pattern Analysis and Machine Intelligence*, 23:1330–1336, 2001.

[649] Q. Zhou and J. K. Aggarwal. Tracking and classifying moving objects from videos. In *Proceedings of IEEE Workshop on Performance Evaluation of Tracking and Surveillance*, 2001.

[650] X. Zhu. Semi-supervised learning literature survey. Technical Report number 1530, Department of Computer Science, University of Wisconsin, Madison, 2005. http://pages.cs.wisc.edu/~jerryzhu/research/ssl/semireview.html.

[651] D. Ziou and S. Tabbone. Edge detection techniques: An overview. *International Journal on Pattern Recognition and Image Analysis*, 8:537–559, 1998.

[652] B. Zitová and J. Flusser. Image registration methods: A survey. *Image and Vision Computing*, 21:977–1000, 2003.

[653] C. A. Zoldi. *A Numerical and Experimental Study of a Shock-Accelerated Heavy Gas Cylinder*. PhD thesis, State University of New York at Stony Brook, 2002.

[654] X. Zong, A. F. Laine, E. A. Geiser, and D. C. Wilson. De-noising and contrast enhancement via wavelet shrinkage and nonlinear adaptive gain. In *Wavelet Applications III, SPIE Proceedings, Volume 2762*, pages 566–574. SPIE Press, Bellingham, WA, 1996.

Index